# The New Taxonomy

Today molecular data is part of many biological studies, including taxonomic works. Such data is embraced by taxonomists for good reasons. When combined with comparative morphology, palaeontology, and embryology, it creates a rich, integrated overview of the history of life. This book is intended as a clear articulation of the mission, goals, and needs of fundamental taxonomists and a planetary-scale inventory of species by revisiting the idea of taxonomy as a fusion of the traditional questions asked by taxonomists and the latest technologies. It is a clear roadmap to a taxonomic renaissance and world species inventory.

**Key Features:**

- Establishes the role and responsibilities of natural history museums to baseline taxonomic studies
- Emphasizes the potential of 'descriptive' taxonomy
- Proposes a cyberinfrastructure specifically designed to meet the needs of taxonomists to do taxonomy
- Provides a clear statement of taxonomy's mission, goals, and prospects
- Reviews taxonomic philosophies and codes of nomenclature from an historical perspective

**David M. Williams** is a diatom systematist–taxonomist. His research is divided between empirical studies on the systematics and biogeography of diatoms and theoretical studies related to advances in systematic theory, especially as it relates to cladistics. In addition to his work on diatom phylogeny, systematics, and biogeography, he has focused on the role fossils have in determining evolutionary relationships in diatoms.

**Quentin D. Wheeler** is an insect taxonomist, author, columnist, and podcaster. He was professor of entomology in Cornell University, Dean of the College of Liberal Arts and Sciences in Arizona State University, Keeper and Head of Entomology at the Natural History Museum in London, Director of the Division of Environmental Biology of the U.S. National Science Foundation, and President of the State University of New York College of Environmental Science and Forestry.

# The New Taxonomy
## A Science Reimagined

Edited by
David M. Williams and Quentin D. Wheeler

CRC Press
Taylor & Francis Group
Boca Raton London New York

CRC Press is an imprint of the
Taylor & Francis Group, an **informa** business

Cover image: Shutterstock.

First edition published 2025
by CRC Press
2385 NW Executive Center Drive, Suite 320, Boca Raton FL 33431

and by CRC Press
4 Park Square, Milton Park, Abingdon, Oxon, OX14 4RN

*CRC Press is an imprint of Taylor & Francis Group, LLC*



ISBN: 9781032249209 (hbk)
ISBN: 9781032249216 (pbk)
ISBN: 9781003280743 (ebk)

DOI: 10.1201/b22822

Typeset in Times
by KnowledgeWorks Global Ltd.

# Contents

# Preface

The editors may appear to have made what Aristotle would call a category mistake when they invited me to write a preface for a collection of articles by biologists, for I have no credentials in any science. My profession is history, a discipline securely classed with the humanities. Yet David and Quentin have long known my work, so their mistake was not about my field of competence. Perhaps they were misled by the casual term *the judgement of history*, as if an historian of science could adjudicate the struggles that systematists today are facing. But that kind of judgement requires perspective, which will become available to some future historian, after the dust has settled.

Mistaken or not, their invitation has nudged me to offer a few remarks, based not only on my reading of these papers but on my attendance at the Systematics Association meeting where they were presented. My first comment may be surprising: I was glad to see scant references in these papers to the history of taxonomy. Reading about the past of one's discipline can benefit a scientist, as can the reading of other kinds of history. Yet I find that scientists' references to history, especially when they are not describing their research but pleading a case, are often rhetorical, in the negative sense of that fine old word. In evolutionary biology, name dropping is terribly common because Charles Darwin's name is golden. Textbook writers and teachers find him irresistible, even though he developed his theory in London rather than on the Galapagos Islands (Sulloway, 1982). His private letters and notes show him to have been a person of admirable character, and the way Darwin arrived at his ideas can be a good way to explain evolution to a beginner. His life story can inspire as well as inform, and we are all better off for having available the rich details about his life that historians have uncovered. Yet science teachers should beware of the great-names approach to history, for Darwin could easily have fallen off the *Beagle* and drowned, and what then? If your answer is that without his genius, we would not now have an understanding of evolution pretty close to the one we do have, you do not appreciate the way the scientific community values nature's input above the voice of any single person. Referring to the names of founding fathers in science, although usually considered a harmless practice, should always be viewed with suspicion, in my opinion.

Within the science of the diversity of life, the situation about name dropping is quite different from the situation in evolutionary biology. Taxonomy, the core of systematics, is an information science that serves a variety of users, within and outside of science. The founder of this discipline was Carl von Linné, whose memory has always been honored by taxonomists, including Darwin. No, wait: not always. Early in the 19th century, some British botanists were struggling to replace the artificial classification of flowering plants invented by Linnaeus with a more natural classification, so they used the adjective 'Linnaean' to mean 'not natural' (Mursinna, 2023). This was not fair, since Linnaeus himself, and several other 18th century taxonomists, had begun to look for nature's own groups, but that connotation nevertheless became common in English. The eminent German botanist Julius von Sachs viewed the growth of natural history museums as the wrong direction for biology's future, which lay in experiment; his authoritative history of botany portrayed Linnaeus as scarcely worthy to be called a scientist. In the mid-20th century, evolutionary biologists who understood the value of taxonomy expressed their appreciation for one of his innovations by calling our very long stack of taxonomic ranks 'the Linnaean hierarchy,' which is misleading, because his list of ranks was purposely short. Some people are now using the term 'Linnaean name' to mean the scientific name of a species as distinct from the vernacular name, but this introduces new confusion, because most currently valid names cannot be found in his volumes. I wish that taxonomists would not use the name of their field's founder as an adjective.

The mantra that nothing in biology makes sense without evolution was chiseled in stone by 20th-century biologists, sometimes quite literally. Linnaeus's location a hundred years before Darwin must always leave him forever stuck, I am sorry to say, on the wrong side of history, so that with respect to public relations, his name is a liability. Scholars can find in Linnaeus's later works the idea that all the species in an order could be descended from one original species, but this cannot clear his reputation as a believer in divine creation. Biographers struggle to show his complex personality in a favorable light (Broberg, 2023). Most often his claim to fame is said to be the invention of binominal nomenclature, but good luck explaining to a student what's so wonderful about that. I have noticed that people who call for radical changes in taxonomy make effective use of the negative connotations of Linnaeus's name in their polemic. If an adjective is needed to differentiate modern practices that trace their origin to Linnaeus from competing schemes of nomenclature, the term *classical taxonomy* might do, but in most contexts, the word *taxonomy* alone should serve.

My second comment is that controversy among scientists is neither new nor unusual, indeed quite the contrary. Fresh from the *Beagle* voyage, Darwin confided in a letter his disgust with zoologists, after he attended a scientific meeting where 'the speakers were snarling at each other, in a manner anything but like that of gentlemen.' Any historian who reads the private notes or letters of a scientist will uncover disagreements hidden from public view, and the background to Darwin's unhappy experience is an interesting story (McOuat, 1996). Much later, on the eve of World War II, the Systematics Association was founded by biologists concerned about dissension among biologists in different disciplines (Winsor, 1995, 2000). Of course, there will always be competing views within science, for the way forward into the unknown is never obvious, and the people who comprise the scientific endeavor are not chosen, as astronauts are, for their cooperative temperaments. Scientists are self-selected, driven by human motives like curiosity or ambition, and because nature gives up her secrets unwillingly, different views are inevitable. In the stories I've researched, the moral is that courtesy, the virtue Darwin identified as gentlemanly behaviour, benefits the scientific enterprise. When people disagree, refraining from uncivil words is more productive of progress than descending, metaphorically, into warfare. There is wisdom in such ancient qualities as tact, empathy, and tolerance.

This leads me to my final comment. To me, what is most important about this volume are things you cannot see here: the personal interactions that took place at the University of Reading where these papers, along with many others, were delivered.

Systematists of widely different views gathered there for several days in June 2024. In all the sessions, coffee breaks, lunches, and dinners, there was not a snarl to be heard. Occasions like that, where experts listen respectfully and thoughtfully to a variety of opinions, are essential to the health of any scientific discipline. The founders of the Systematics Association understood this, and their heirs are to be congratulated for carrying their spirit forward.

**Mary P. Winsor**
*Institute for the History and Philosophy of Science and Technology, University of Toronto*

## REFERENCES

Broberg, G. (2023) *The Man who Organized Nature: The Life of Linnaeus.* Princeton University Press, Princeton.

McOuat, G.R. (1996) Species, rules and meaning: The politics of language and the ends of definitions in 19th century natural history. *Studies in History and Philosophy of Science*, 27(4): 473–519.

Mursinna, J.T. (2023) A "Mean Quarrelsome Spirit:" Controversy in British systematics, 1822–1836. *Journal of the History of Biology*, 56(4): 673–714. https://doi.org/10.1007/s10739-023-09743-7

Sulloway, F.M. (1982) Darwin's conversion: The *Beagle* voyage and its aftermath. *Journal of the History of Biology*, 15: 325–396.

Winsor, M.P. (1995) The English debate on taxonomy and phylogeny, 1937–1940. *History and Philosophy of the Life Sciences*, 17: 227–252. http://www.jstor.org/stable/23331855

Winsor, M.P. (2000) Species, demes, and the omega taxonomy: Gilmour and *The New Systematics*. *Biology and Philosophy*, 15: 349–388. https://doi.org/10.1023/A:1006774217770

# List of Contributors

**Donat Agosti**
Plazi
Bern, Switzerland

**Gavin Broad**
The Natural History Museum
London, United Kingdom

**Les Christidis**
Southern Cross University
Coffs Harbour, Australia

**Stijn Conix**
Institut supérieur de philosophie
Université catholique de Louvain
Louvain-La-Neuve, Belgium
Centre for Logic and Philosophy of Science
KU Leuven
Leuven, Belgium

**Manuel B. Crespo**
CIBIO (Instituto de la Biodiversidad)
Universidad de Alicante
Alicante, Spain

**Alain Dubois**
Institut de Systématique, Evolution Biodiversité (ISYEB)
Muséum national d'Histoire naturell
CNRS
Sorbonne Université
EPHE
Université des Antilles
Paris, France

**Stephen Garnett**
Research Institute for the Environment and Livelihoods
Charles Darwin University
Casuarina, NT, Australia

**Anthony Gill**
Macleay Collections
Chau Chak Wing Museum
University Avenue
The University of Sydney
Sydney, New South Wales, Australia
School of Life and Environmental Sciences
The University of Sydney
Sydney, New South Wales, Australia
Ichthyology
Australian Museum Research Institute
Sydney, New South Wales, Australia

**Anaïs Grand**
Colombes, France

**Michael D. Guiry**
AlgaeBase, Ryan Institute
University of Galway
Galway, Ireland

**Frank-Thorsten Krell**
Gillette Museum of Arthropod Diversity
Department of Agricultural Biology
Colorado State University
Fort Collins, Colorado

**A.M. Lien**
Research Institute for the Environment and Livelihoods
Charles Darwin University Casuarina
NT, Australia
School of Natural Resources and the Environment
University of Arizona
Tucson, Arizona

**Evgeny Mavrodiev**
Florida Museum of Natural History
University of Florida
Gainesville, Florida

**Pablo Muñoz-Rodríguez**
Faculty of Biological Sciences
Department of Biodiversity
Universidad Complutense de Madrid
Madrid, Spain

**Michelle J. Price**
Conservatory and Botanical Garden of Geneva
Geneva, Switzerland

**Peter Uetz**
Center for Biological Data Science
Virginia Commonwealth University
Richmond, Virginia

**Visotheary Ung**
Institut de Systématique Évolution Biodiversité
Centre National de la Recherche Scientifique
MNHN
Sorbonne Université
EPHE-PSL
Université des Antilles
Paris, France

**Frank E. Zachos**
Natural History Museum Vienna
Vienna, Austria
Department of Evolutionary Biology
University of Vienna
Vienna, Austria
Department of Genetics
University of the Free State
Bloemfontein, South Africa
Research Institute for the Environment and Livelihoods
Charles Darwin University
Casuarina, NT, Australia

# Acknowledgments

We thank the Systematics Association for the invitation to produce a second edition of *The New Taxonomy*. Most chapters are based on contributions to a symposium at the Systematics Association biennial meeting on the campus of the University of Reading (19th–21st June 2024); all presented thoughtful perspectives on taxonomy's status, needs and opportunities. Clear-eyed about the unprecedented challenges facing biodiversity and the traditional goals of taxonomists to explore, make known, and phylogenetically classify all species—a mission more relevant and urgent today than ever, due to a mass extinction already underway—authors express concern and inspire hope for the future. This book is a call to reimagine taxonomy's purpose and possibilities; to create a *new* taxonomy, not in the image of other sciences or shaped only to meet immediate needs to identify species, but as a fusion of the best ideas, accomplishments and traditions from its past with the latest theories and technologies. Every indicator suggests that the century ahead will be the last best opportunity to explore, document, and learn about earth species, their attributes and history, before millions have gone extinct. The price of failing to reimagine taxonomy as an independent, curiosity-driven, fundamental science would be enormous and long-enduring, with unthinkable opportunity costs for biodiversity conservation, biomimetic problem-solving, and science. We have, for the imaginable future, only one chance to thoroughly explore the life forms, biosphere and evolutionary history of a hyper-biodiverse planet. Only taxonomic research and exploration can minimize biodiversity losses, maximize knowledge of the world's diverse life forms, and memorialize evolutionary history. The decimation of biodiversity over the next two centuries now appears inevitable, but it is not yet too late to reimagine taxonomy and engage in an urgent campaign to explore, discover, document, describe, classify and understand the fantastic abundance of life forms about to be lost.

We are grateful to Charles Crumley and Daina Habdankaite, of Taylor and Francis, for guiding this project to completion, to the audience in Reading for constructive comments and debate, and to the following for reviews of chapters: Marcelo de Carvalho, Malte Ebach, Kelly B. Miller, Alex Munro, Alessandro Minelli, Thomas Pape, Edward Theriot, Barbara Thiers, Jon Todd, Bart Van de Vijver, and Francisco Welter-Schultes.

The meeting and this book are dedicated to the memory of Norman Ira Platnick (1951–2020).

Norman I. Platnick in his natural habitat at the American Museum of Natural History. Photographer: Steve Thurston (American Museum of Natural History, New York; image supplied by Lorenzo Prendini, American Museum of Natural History, New York).

# 1 Introduction
## *Toward a New Taxonomy: Means, Motives, and Opportunities*

*Quentin D. Wheeler and David M. Williams*

> 'In other words, a study of the diversity of life, when viewed through the lens of a theory of evolution, becomes equivalent to a study of the history of life'
>
> *Nelson and Platnick, 1981, p. 5*

In the years since the first edition of *The New Taxonomy* (Wheeler, 2008), a great deal has transpired, very little of it good for taxonomy as a science. Advances in molecular approaches have been impressive. However, instead of complementing traditional sources of evidence, they have replaced them. The result, encouraging for users of taxonomic information who merely want to identify species, has been devastating for taxonomy's own mission. DNA barcodes tell species apart while telling us almost nothing about species themselves and the so-called minimalist taxonomy is, true to its name, only minimally better (Sharkey et al., 2021a, 2021b; Ahrens et al., 2021; see also Broad, this volume, Chapter 2).

An objective assessment of the state of taxonomy as a fundamental, independent science is nothing short of tragic (Wheeler, 2024). Natural history museums have failed to grow, develop, and use collections in a manner proportionate to the rate of extinction (Paknia et al., 2015); taxonomic curricula and graduate programs have been gutted in many leading universities (Britz et al., 2020); in spite of inadequate data, it is clear that the average number of species described each year has been more or less steady as the rate of extinctions has accelerated; DNA data is favored over other, more informative, evidence of evolutionary history; grants to do so-called descriptive taxonomy, that is, comprehensive revisions and monographs, are few and dwindling in number; taxon experts are rarely replaced in kind as they retire (Engel et al., 2021; Löbl et al., 2023); and taxonomy is increasingly treated as though it exists as a service to other biologists—how else do you justify a 'taxonomy' consisting of little more than a DNA barcode and a photograph (Sharkey et al., 2021a, 2021b)? Unabated, these trends will lead to a death spiral for taxonomy (see Krell, this volume, Chapter 3).

No one could question the urgency of addressing the biodiversity crisis, from accelerated extinction rates to the diminution, disruption, or destruction of habitats around the world (Dubois, 2003; and this volume, Chapter 6). Learning more about

DOI: 10.1201/b22822-1

ecology and ecosystems, and developing effective conservation strategies, are urgently needed to address immediate environmental problems (Thomson et al., 2018). But such short-term needs, no matter how grave or significant, should not be permitted to blind us to other, equally important, things that are at stake. In the long term, science and humankind's ultimate knowledge and understanding of the diversity of life and its remarkable history—including the origins of our own species' attributes—depend on a taxonomy that takes a broader and longer view of the challenge at hand.

Taxonomists will, as they always have, deliver knowledge with which other biologists may identify species and see them in the context of their evolutionary relationships. But these profound benefits are nonetheless by-products of taxonomic research, not its primary aim. Taxonomy is, when practiced to its highest level of excellence, a science quite distinct from the experimental biology disciplines that dominate today. It is instead observational, comparative, and historical in perspective. And while it is nonexperimental, it is nonetheless based on rigorous, explicitly testable hypotheses when practiced at its highest levels.

After two millennia of more or less steady progress toward exploring species and their characters, relationships, distributions, and histories, it is indefensible to abandon taxonomy now. Many criticize taxonomy because it has made known only about two of an estimated ten or more million species. But these critics represent the same disciplines that have denigrated and failed to respect, support, or defend taxonomy for many decades. For taxonomy's part, through most of its history, it was assumed that we had as long as it would take to discover and describe species. With knowledge of mass extinction now upon us, that assumption is clearly wrong. Taxonomy was always critically important but suddenly its agenda is urgent, too.

The development of an increasingly molecular-based taxonomy is widely welcomed by other life sciences, because it appears to provide minimally sufficient information without the need to share funds for the traditional goals of taxonomy, or for the growth and use of collections. Molecular taxonomy is seen as modern and technologically sophisticated. Its prestige comes from the amount of funding it attracts and the association of DNA sequencing with more popular fields, including biomedicine. To many, molecular taxonomy looks more like genetics than taxonomy and, to them at least, that is a good thing (Ebach and Williams, 2006).

This decline of taxonomy, as we understand the science, is not new. It has suffered a long, slow decay over many decades. More than a little blame rests at the feet of taxonomists themselves. We, as a community, chose to make finding the shortest trees a higher priority than careful interpretation of individual characters, thus shifting priority away from the latter (Nelson, 2004; Platnick, 2016, 2017; see also Williams and Gill, this volume, Chapter 2). In this context, we emphasized quantity of data over quality, thus paving the way for sequence data dominance. And we went along with a divorce of systematics (associated with evolutionary considerations) from taxonomy (relegated to dry rules for naming and classifying), raising the profile of the former while diminishing that of descriptions, classifications, revisions, and monographs. Taxonomy (a.k.a. systematics), as we know it, is an integration of all of the above. Descriptive work provides critically important evidence for understanding evolutionary history, and classifications reflect phylogenetic knowledge; neither is complete without the other.

Such compromises seemed to make sense at the time and be appropriate for winning this or that clash of ideas. In retrospect, elevating some aspects of taxonomy at the expense of others helped win battles while losing the war. All of this was played out in a biological *zeitgeist* that held up experimentalism as the gold standard for science itself. Add to these developments a growing concern about ecosystems around the globe, and the urgent need to identify species, and strong incentives exist to cut corners.

Lost in this rush to meet the environmental challenges of our day is the mission of taxonomy itself. While it has a proud and accomplished tradition of making species identifiable, this is not its primary motive. The core aim of taxonomy is easily said, but difficult to achieve:

> *To discover what kinds of living things exist, describe and interpret the characters that make each unique or provide evidence of common ancestry, map their geographic and ecological distributions, provide unique identifiers in the form of Linnaean binomials, and summarize all that is learned in an informative, predictive phylogenetic classification.*

This entails detailed descriptive and comparative studies, explicitly testable hypotheses, synthesis of all available and relevant evidence, and scientifically rigorous analyses of individual characters as well as of historic branching patterns.

> *The new taxonomy should be a fusion of these traditional aims with the latest theories, technologies, and methods.*

And, it should include a renewed commitment to describe, compare, interpret, and understand characters and suites of homologous characters:

> 'The classification is not based on a historical narrative; rather, any historical narratives that we may choose to present are based, as is the classification, on the characters of organisms.'
>
> *(Platnick, 1978, p. 365)*

Both an emphasis on molecular data and the computational ability to analyze large datasets have helped create a preference for quantity of data over quality of evidence. While there is much to say about access to as much data as possible, it remains that quality of evidence matters very much. First, only by knowing novel and complex characters of species and clades can we engage with the most fascinating and unexpected results of evolutionary history. Second, it remains that even one properly interpreted complex character can provide compelling evidence of relationships, such as presence of a backbone among vertebrates or flowers among angiosperms. And third, it is in understanding complex characters themselves that we bridge evidence from the fossil record, sequencing, ontogeny, and comparative morphology.

We have moved substantially toward a molecular-based taxonomy. Before we progress further, it is time to ask: 'Is this the kind of taxonomy that we want or need?' With species going extinct faster than at any time in the past sixty-five million years, our goal should be to learn as much as possible about life on earth in the time remaining. This means thoroughly exploring the biosphere, and documenting as many species, and as much about each, as possible. It means preserving as much evidence of species and evolutionary history as possible; and that includes whole specimens in collections as well as tissues, observations, and other supporting evidence. Our mission should be to assemble as much taxonomic information as possible for science and posterity. These goals are only possible if taxonomy is unleashed, pursued for its own sake, and to its own high standards. After all, this is, for the imaginable future, the only opportunity to explore, study, and understand the details, origins, and history of a species-rich planet.

Before we permit taxonomy to become further reliant on a single data source and shaped by the applications of its information and knowledge, we should carefully examine the means, motives, and opportunities for a different kind of taxonomy—a taxonomy that is a fusion of a centuries-old mission with the latest theories, methods, and possibilities. A taxonomy that values and synthesizes all relevant evidence; that refuses to compromise excellence; that carries on intellectual and scientific traditions begun by Aristotle, Linnaeus, Owen, Cuvier, Darwin, and Hennig, among others; that rises to meet the challenge of documenting, describing, and understanding the full diversity of life on a planet undergoing rapid changes; a taxonomy laser-focused on its own aims while faithfully providing information required by biology in general, from species identifications to a general reference system in the form of a phylogenetic classification.

Authors in this volume provide creative and realistic suggestions for creating a bright future for taxonomy, preserving the best of its long, distinguished history while embracing the latest tools to accelerate its mission. In order to have a new taxonomy fit to the extraordinary challenges of our time, and avoid being distracted by others' priorities, we must have clarity of vision and courage of purpose. After so much progress toward deep knowledge and highly informative classifications, it would be tragic were taxonomy dumbed down as a mere service or arbitrarily limited to one source of evidence. Taxonomy's traditional goals are more relevant, impactful, and urgently needed now than ever before. What could be dismissed as an impossibly large challenge—to *discover, describe, name, and phylogenetically classify* the millions of species on earth—we prefer to see as one of humankind's greatest opportunities to understand itself, its world, and their origins. No subsequent generation will have the chance to fully explore or make sense of the diversity and history of life on earth, and we may never reach another planet on which a comparable diversity of life exists.

Rather than diminishing taxonomy to meet short-term needs more quickly and easily, we should resolve to pursue taxonomy in its full form to assure that we create as much knowledge, preserve as much evidence, and gain as much understanding as possible. It is our hope that this volume plays a small role in awakening the scientific community and society to what we stand to lose if we continue down the road toward a molecular-based taxonomy. Molecular data has an important role to play in the future of taxonomy—but it is a supporting, not leading, one. We must rise above fads, fashions, and finances to see the big picture regarding what will be lost if we fail to support taxonomy on its own terms, modernize its research infrastructure, and return priority to natural history collection growth and development during the greatest extinction event in human history.

Let's briefly examine the means, motives, and opportunities that exist for revitalizing taxonomy.

## MEANS

Taxonomy's ambitions were always ahead of their time, until now. For centuries, taxonomy's vision of discovering, knowing, describing, analyzing relationships among, and classifying every kind of living thing seemed impossibly audacious. As time passed, estimates of the total number of species increased by orders of magnitude, making the goal seem, to many, even more unattainable. But everything has changed. Everything, that is, except perceptions about what taxonomy is and what it is capable of achieving.

Other big science projects, from the Webb telescope to Martian rovers and particle accelerators, demonstrate what is possible given vision, will, and appropriate funding. Taxonomy suffers from familiarity and its long history. Many see only what has been in the past, not what it can be going forward. Theoretical advances, by and since Hennig (1966), have made taxonomy as rigorous as any other science. And advances in digital imaging, information science and cyberinfrastructure, communications among taxonomists and natural history museums, numerous examples of collaborative, community-scale projects, and other evidence suggest that Linnaeus' impossible vision is now very much within reach (e.g., Wheeler et al., 2012). Compared to the engineering complexities of a mission to Mars, ramping up taxonomy for an all-out inventory of earth life over a period of decades is clearly possible.

We now have the means to scale taxonomy to meet an urgent, planetary-level need for knowledge. With off the shelf technologies and a little imagination, we can radically change how specimens in natural history collections are stored and accessed; how detailed, illustrated, morphological descriptions, prohibitively expensive to publish in the past, can become commonplace; how taxon experts and collections-holding institutions communicate and coordinate among themselves; how both diverse and large datasets are assembled and analyzed; and how species are both readily identified *and* backed by deep and reliable knowledge regarding their attributes, relationships, and distributions.

Efforts to reconstruct the tree of life have so far focused on major branches. A renewed taxonomy means complete knowledge of the history of life, down to twigs and leaves. It means a wealth of information about each species and clade. And it means scientific names and a phylogenetic classification for all living things, with the information retrieval, communication, and predictive capabilities that come with them.

No previous generation has faced a deadline for exploring, documenting, and classifying the diversity and history of life. Neither has any prior generation had the theories, methods, and tools to successfully do so. Because of a convergence of advances in thought, technology, and recognition of the dimensions and implications of the biodiversity crisis, we are in a perfect position to create a legacy of specimens and knowledge that will enrich science for millennia to come, long after, and without regard to the ultimate number of species and clades lost. This is a fantastic, never to be repeated opportunity. It would be foolish to settle for molecular diagnoses of species when we can have exhaustive descriptions of their characters; to accurately estimate the number of species when we could have intimate knowledge of each; to arbitrarily limit knowledge to one data source rather than synthesizing multiple lines of evidence; and to passively witness the decimation of biodiversity when we have the means to thoroughly explore and document it.

The planetary scale of taxonomy's mission and the lack of appropriate theories to make its findings rigorously scientific led to its dismissal as an attainable goal. Advances in the past half dozen decades have put the idea of a comprehensive, worldwide inventory within reach and made taxonomy scientifically rigorous. The means to efficiently and rapidly explore, collect, describe, analyze, and classify millions of species now exist. What has disappeared are credible excuses to continue marginalizing taxonomy's agenda.

## MOTIVES

We cannot imagine stronger motives to reimagine and support taxonomy than the challenges and opportunities we now face.

An accelerated rate of extinction means that the time remaining to preserve evidence of the diversity and history of life grows short. In practical terms, we can vastly improve our chances of conserving as much, and as diverse, life as possible if we invest in learning what species exist to begin with (Thomson et al., 2018). Also, knowledge of the attributes of species will reveal millions of potential biomimetic models for solving countless problems. In scientific terms, we can continue to study and learn about the amazing evolution and diversification of life on earth if we make natural history collections full and accurate reflections of species and clades. And in human terms, we can continue our journey of self-awareness by piecing together the evolutionary history of which we are one among millions of outcomes. Opportunities, knowledge, and our humanity are all impoverished to the extent that we permit evidence of life to disappear undocumented.

The current focus on molecular data and simply identifying species and estimating their relationships is dangerously short-sighted. Both immediate needs, such as developing a global plan for conservation, and long-term aspirations to understand the origins of biodiversity, including humans, are compromised by expedient shortcuts. We can, if we so choose, both address the urgent needs of the present and ensure that a legacy of evidence and knowledge exists long after the current extinction event has passed.

Various communities will no doubt find particular motives more compelling than others. For example, ecologists should wish to have evidence of the inhabitants of ecosystems and their interactions to better understand the systems themselves, support future restoration efforts, and maximize the success of preserving natural areas and managing human-shaped ones. Conservation biologists should want to be able to monitor the status of biodiversity at the granularity of species, rather than relying on a few indicator species. And they should want to develop strategies that take into account the diversity of species conserved, not just their numbers (e.g., Vane-Wright, Humphries and Williams, 1991).

A taxonomic inventory would create baseline knowledge of a biosphere undergoing rapid change. It would enhance our ability to recognize invasive species and the emergence of diseases and their vectors. It would open millions of species as a source of biomimetic ideas. And it would make the full diversity of life forms accessible to anyone wishing to experience, observe, enjoy, or study them.

Like learning what kinds of rocks exist in the lithosphere, the elements in the periodic table, or the objects in our solar system, there is no more basic knowledge of the biosphere than answering the question 'What species exist?' Being a curious species, we next want to know what makes each unique, how they are related, and the step-by-step history of transformations responsible for the universe of characters. With respect to ourselves, it is impossible to fully understand our humanness until we trace the origins of our own characters through the ancestral species that preceded us. These are not life or death questions; it is entirely possible to function and live while remaining ignorant of most species and most details of the story of evolution. We did so prior to Linnaeus, Darwin, and Hennig. But given how much it enriches our lives to know and appreciate the living world, and to discover our own evolutionary heritage, why would we choose ignorance over enlightenment?

With knowledge of no more than twenty percent of living species, and with thousands of species going extinct each year, this is no time to accept ignorance or minimal information. It is less than two hundred years since the recognition of the fact of evolution; we have only begun to explore the details of its results and history. It is easy to list the practical benefits of knowledge as motivation to revive taxonomy, but we should not underestimate the value of knowledge for its own sake. Our earliest ancestors wondered about the complex pattern of similarities among the many organisms they encountered. That curiosity eventually led Aristotle to group species, Linnaeus to enumerate them, and Darwin to propose an explanation for their attributes. It is a fundamental part of human nature to want to understand what makes us human and what accounts for the biodiversity surrounding us. It is central to our intellectual destiny to explore and understand. This should be motive enough to refuse to settle for molecular data alone or to fail to explore life in the biosphere while time remains to do so.

## OPPORTUNITIES

The mass extinction of species comes with unprecedented challenges and opportunities. While we must face and adapt to the former, we should not lose sight of the latter. Among opportunities are the chance to:

- Preserve specimens, observations, and information as a permanent record of species, characters, and clades;
- Explore the organization of the biosphere, creating baseline knowledge of the inhabitants of ecosystems so that change may be monitored, losses known, and arising problems mitigated;

- Gather evidence of the attributes of species as potential biomimetic models;
- Understand the diversity and history of life on the only species-rich planet so far known to exist.

Given what we know of our corner of the Milky Way, we may never set foot on another world with so many varied life forms. But if we do, our first impulse will be to ask how its species and evolutionary history compare to those of earth. Unless we seize the opportunity to explore, describe, document, and classify our planet's species, we unnecessarily limit our understanding of biodiversity, self, and history.

Taxonomy is among the great fundamental sciences, so broad in its vision and implications that it rivals physics or cosmology as essential for human understanding. To selfishly constrain taxonomy in service to immediate environmental needs is shortsighted and risks a level of ignorance that will forever haunt us. We in no way diminish the significance of meeting the needs of ecology, agriculture, conservation, and many other fields stressed by global events. But meeting such challenges need not come at the expense of taxonomy's agenda. On the contrary, by supporting taxonomists in their own work, by expecting museums to rise to the challenge of documenting species in their collections, and by pressing every relevant technology and source of evidence into the service of taxonomy's goals we can deliver the support so urgently needed by others with no sacrifice in the quality or depth of taxonomic knowledge itself.

Species and their history represent the greatest unexplored frontier on earth. What matters now is how we respond to what may be our last chance to gain knowledge of species, clades, characters, and phylogenetic history both to inform efforts to mitigate losses and to avoid an unnecessary depth of permanent ignorance.

## RESISTANCE, RESOLVE, AND RECIPROCITY

It is important to science and humankind that our commitment to the traditional aims of taxonomy be renewed and accelerated. There should be no more apologies for a science as rigorous, information-rich, intellectually exciting, and practically impactful as taxonomy. We must resist pressures to recast taxonomy in the image of, or in service to, genetics, ecology, or any other field; to conform to fads and fashions in order to secure funding; to compromise or dilute our efforts in order to align with popular science; or, to allow the infrastructure and workforce of our science to further erode.

We must resolve to explain, advocate for, and practice taxonomy at its highest levels of excellence without equivocation, apology or compromise. Our generation is called by history to be visionary leaders, not opportunistic conformers. Our mission is second to none in its scientific importance and in its impact on human understanding of themselves and the world. We must reject the claim that taxonomy has importance only in the services it provides, as incredibly valuable as they are.

Every science should be evaluated on its own merits. Taxonomy, however, is expected to bend its priorities to align with those of other sciences, but where is reciprocity? Ecologists are not asked to amend their agenda to facilitate taxonomy, nor should they be. A major problem for taxonomy is that the knowledge it creates is required by the rest of biology, leading non-taxonomists to the erroneous conclusion that taxonomy exists to serve them. Compounding taxonomy's problem, it is observational, comparative, historical, and non-experimental, making it appear foreign to much of biology that colors within the narrow lines of the experimental method. Taxonomists recognize and respect differences between their science and experimental sciences and deserve the same understanding and respect in return.

## WHAT NEXT?

Time grows short for an inventory of life on earth. By inventory we do not mean a simple list of names or database of DNA sequences, but a thorough accounting of species and clades *and* their characters, relationships, and distributions. Unless taxonomy is supported to reinforce its ranks, modernize its infrastructure, solidify its plan of action, and function as an independent, curiosity-driven, fundamental science, then we may lose far more knowledge of life and evolution than we ever gain.

Because so little time remains for action, minds, perceptions and priorities must be changed post-haste. While many details are involved in a taxonomic revival and a successful inventory, a few general steps deserve mention:

- Taxonomists must find the clarity of vision and courage of purpose to resist conforming to popular fads and forcefully advocate for their own agenda;
- We must educate colleagues, administrators, and decision-makers who do not understand taxonomy or its mission;
- We must share the exciting and impactful discoveries of taxonomy with the public;
- Natural history museums must recommit themselves to representing the full diversity of species in their collections;
- We must accelerate the fundamental agenda of taxonomy while, at the same time, translating knowledge, making species identifiable and knowledge about them accessible;
- And, we must educate and inspire the next generation of taxon experts to engage in a fusion of the best of taxonomy's past with the latest theories, methods and technologies.

## DARING GREATLY

There are many to blame for the current state of taxonomy: experimental biologists so narrowly trained that they don't recognize other, even more strenuous, tests of hypotheses; environmental biologists so consumed by immediate problems that they cannot see the bigger picture; university administrators who sacrifice balanced curricula to pursue the popular; natural history museum leaders who shamefully follow fads and forego the opportunities and unique contributions of growing, developing, and using collections to their highest purposes; funding agencies that prioritize new technologies over fundamentally important and urgently needed knowledge.

And, there are taxonomists themselves who have gone along with fads, allowed the lines between taxonomy and other, more popular, sciences to be blurred, permitted a distinction to be made between 'systematics' and 'taxonomy,' rather than defending the terms as synonymous for one and same science in all its activities, and allowed themselves to be bullied by those with too little understanding, integrity, curiosity, imagination, or generosity to see and respect taxonomy for the vitally important, fundamental, independent science that it is.

The time for compliance is over. No more going along with fads and fashions. No more allowing taxonomy to be treated as servant rather than science. No more watching museums waste limited resources pretending to be backwater universities while millions of species go extinct uncollected, undescribed, and unknown. No more accepting any one source of evidence over equally or more informative ones. No more charades, trying to pass taxonomy off as something it is not. No more.

It is time to draw a sharp line between *real* taxonomy and imposters, clearly and unapologetically restating the aims, theories, and methods of the field. Time to be a reliable partner to other sciences and society, generously making species identifiable and knowledge about them openly accessible, while, at the same time, focusing, without compromise, on the unique mission of taxonomy. Time to put molecular data into proper perspective as one among several important sources of evidence. Time to reengage with Hennig's conceptual revolution. Time to plan, organize and implement a rapid, worldwide inventory, description, and classification of all remaining species. And time to cross a bio-political Rubicon from which there may be no return: a wholehearted commitment to realizing the vision of deep knowledge of earth's species, their characters, and their history; a final battle for the soul of taxonomy which, if lost, will result in irreversible ignorance of species, clades, characters, the organization of the biosphere, and the sequence of character transformations that is life's history.

For the timid who may fear what might follow such a declaration, we ask how much worse things could become. Are we really prepared to forego in-depth knowledge of species, clades, characters and history for the ease and temporary popularity of DNA sequence data? Is humankind willing to forego ever truly understanding its evolutionary heritage and position among earth's biodiversity? Are we prepared to see taxonomy further denigrated and marginalized, collections mothballed, and knowledge of species limited to DNA sequences? Before listening to those who say that taxonomy is out of fashion, that describing all species an impossibly big task, that telling species apart with molecular data is knowledge enough; before we give in to the nay-sayers, before we surrender taxonomy's centuries-old vision to the whims of fashion, before we sacrifice understanding on the altar of technology, perhaps to a point of no return, we must ask if taxonomy is worth fighting for and, if yes, consider the words of U.S. President Theodore Roosevelt (1910) in an address at the Sorbonne in Paris:

> It is not the critic who counts; not the man who points out how the strong man stumbles, or where the doer of deeds could have done them better. The credit belongs to the man who is actually in the arena … who at the best knows in the end the triumph of high achievement, and who at the worst, if he fails, at least fails while daring greatly, so that his place shall never be with those cold and timid souls who neither know victory nor defeat.

Do we dare defend taxonomy and its synthesis of evidence as diverse as morphology, fossils, ontogeny, and DNA sequences? Do we dare commit to an immediate, full-scale mission to inventory, describe and classify every kind of living thing on planet

Earth? Can museums find the clarity of purpose to become leaders of such an inventory? Are there universities bold enough to reject conformity and rebuild taxonomic curricula and graduate programs? And will taxonomists dare to rise to the defense of their own science, refusing further compliance, compromise, and subservience?

We hope that the ideas in this book will inspire commitment and action against withering headwinds. We live during an unprecedented time. Failure to revitalize taxonomy now means certain and irreversible ignorance of earth life and its history, to the great loss of science and humankind. Best that we make our stand now and go for broke.

Given current circumstances and trends, we have rather little left to lose. But if we win, we have a world to save, evidence of evolution to preserve, conservation goals to enable, models for biomimicry to exploit, a science to revitalize, and knowledge to gain with which to enrich our intellectual lives. There are, of course, some causes for optimism. There are taxonomists who expertly integrate molecular data with traditional evidence who are already in the fight, and some number of biologists with a broad appreciation for science and what is at stake who would be sympathizers cheering from the sidelines. This particular battle, however, must be fought by taxonomists themselves, and institutions of special vision and courage. Taxonomy, pursued on its own terms, is a science worth fighting for: *Alea jacta est!*

Taxonomy is grounded in an idea as simple as it is profound: to know and understand the attributes, relationships among, and evolutionary origins of the millions of species living on, under, and above the surface of our planet. It grew from innate human curiosity to understand ourselves and our world, to make sense of the incredible diversity of life forms around us; the same curiosity that drove Linnaeus to attempt describing all species, Darwin to explain their origins, and Hennig to make classifications phylogenetic and rigorously testable. Meeting the utilitarian goal of identifying species is a noble and necessary task, but it need not come at the expense of the grand scientific vision and rich intellectual traditions of taxonomy. We owe it to science, future generations and ourselves to dream and achieve much more. The chapters in this volume represent tentative steps toward reawakening the idea behind taxonomy, reigniting our curiosity and inspiring humankind to seize the ephemeral opportunity to document life before much of its diversity has been decimated and its history erased. Advances in theory and technology make taxonomy's ancient mission efficient and its bold vision within reach. Reengaging this idea, we can create a new taxonomy that *is* the old taxonomy, only better.

## REFERENCES


Ahrens, D., Ahyong, S.T., Ballerio, A., Barclay, M.V.L., Eberle, J., Espeland, M., Huber, B.A., Mengual, X., Pacheco, T.L., Peters, R.S., Rulik, B., Vaz-de-Mello, F., Wesener, T. and Krell, F.-T. (2021) Is it time to describe new species without diagnoses?—A comment on Sharkey *et al.* (2021) *Zootaxa*, 5027: 151–159, https://doi.org/10.11646/zootaxa.5027.2.1

Britz, R., Hundsdörfer, A.K. and Fritz, U. (2020) Funding, training, permits—The three big challenges of taxonomy. *Megataxa*, 1: 49–52, https://doi.org/10.11646/megataxa.1.1.10

Dubois, A. (2003) The relationships between taxonomy and conservation biology in the century of extinctions. *Comptes Rendus Biologies*, 326: s9–s21, https://doi.org/10.1016/S1631-0691(03)00022-2

Ebach, M.C. and Williams, D.M. (2006) Molecular systematics is not genetics. *Rivista di Biologia*, 98: 373–376.

Engel, M.S., Ceríaco, L.M.P., Daniel, G.M., Dellapé, P.M., Löbl, I., Marinov, M., Reis, R.E., Young, M.T., Dubois, A., Agarwal, I., Lehmann, A.P., Alvarado, M., Alvarez, N., Andreone, F., Araujo-Vieira, K., Ascher, J.S., Baêta, D., Baldo, D., Bandeira, S.A., Barden, P., Barrasso, D.A., Bendifallah, L., Bockmann, F.A., Böhme, W., Borkent, A., Brandão, C.R.F., Busack, S.D., Bybee, S.M., Channing, A., Chatzimanolis, S., Christenhusz, M.J.M., Crisci, J.V., D'Elía, G., Costa, L.M. da, Davis, S.R., Lucena, C.A.S. de, Deuve, T., Fernandes Elizalde, S., Faivovich, J., Farooq, H., Ferguson, A.W., Gippoliti, S., Gonçalves, F.M.P., Gonzalez, V.H., Greenbaum, E., Hinojosa-Díaz, I.A., Ineich, I., Jiang, J., Kahono, S., Kury, A.B., Lucinda, P.H.F., Lynch, J.D., Malécot, V., Marques, M.P., Marris, J.W.M., Mckellar, R.C., Mendes, L.F., Nihei, S.S., Nishikawa, K., Ohler, A., Orrico, V.G.D., Ota, H., Paiva, J., Diogo Parrinha, D., Pauwels, O.S.G., Pereyra, M.O., Pestana, L.B., Pinheiro, P.D.P., Prendini, L., Prokop, J., Rasmussen, C., Rödel, M.-O., Rodrigues, M.T., Rodríguez, S.M., Salatnaya, H., Sampaio, Í., Alba Sánchez-García, A., Shebl, M.A., Bruna S. Santos, B.S., Solórzano-Kraemer, M.M., Sousa, A.C.A., Stoev, P., Teta, P., Trape, J.-F., Van-Dúnem Dos Santos, C., Vasudevan, K., Vink, C.J., Vogel, G., Wagner, P., Wappler, T., Ware, J.L., Wedmann, S. and Kusamba, C.Z. (2021) The taxonomic impediment: a shortage of taxonomists, not the lack of technical approaches. *Zoological Journal of the Linnean Society*, 22: 381–387, https://doi.org/10.1093/zoolinnean/zlab072

Hennig, W. (1966) *Phylogenetic Systematics*. University of Illinois Press, Urbana.

Löbl, I., Klausnitzer, B., Hartmann, M. and Krell, F.-T. (2023) The silent extinction of species and taxonomists—an appeal to science policymakers and legislators. *Diversity*, 15: 1053 (17 pp.), https://doi.org/10.3390/d15101053

Nelson, G. J. (2004) Cladistics: Its arrested development, pp. 127–147. In *Milestones in Systematics* (eds D. M. Williams and P. L. Forey), Systematics Association Special Volume 67. CRC Press, Boca Raton.

Nelson, G.J. and Platnick, N.I. (1981) *Systematics and Biogeography: Cladistics and Vicariance*. Columbia University Press, New York.

Paknia, O., Rajaei Sh., H. and Koch, A. (2015) Lack of well-maintained natural history collections and taxonomists in megadiverse developing countries hampers global biodiversity exploration. *Organisms Diversity Evolution*, 15: 619–629, https://doi.org/10.1007/s13127-015-0202-1

Platnick, N.I. (1978) Classifications, historical narratives, and hypotheses. *Systematic Zoology*, 27: 365–369, https://doi.org/10.2307/2412891

Platnick, N. I. (2016) Foreword: Willi Hennig and systematics: A personal view, pp. xi–xvi. In *The Future of Phylogenetic Systematics: The Legacy of Willi Hennig* (eds D. M. Williams, M. Schmitt and Q. Wheeler), Systematics Association Special Volume Series, Cambridge University Press, Cambridge.

Platnick, N. I. (2017) Foreword: Donn Rosen and inhibiting assumptions, pp. vii–ix. In *Assumptions Inhibiting Progress in Comparative Biology* (eds B. I. Crother and L. R. Parenti), CRC Press, Boca Raton.

Roosevelt, T. (1910) "Citizenship in a Republic," address delivered at the Sorbonne, Paris, 23 April. The American Presidency Project, University of California, Santa Barbara. https://www.presidency.ucsb.edu/documents/address-the-sorbonne-paris-france-citizenship-republic

Sharkey, M.J., Baker, A., McCluskey, K., Smith, A., Naik, S., Ratnasingham, S., Manjunath, R., Perez, K., Sones, J., D'Souza, M., St. Jacques, B., Hebert, P., Hallwachs, W., and Janzen, D. (2021a) Addendum to a minimalist revision of Costa Rican Braconidae: 28 new species and 23 host records. *ZooKeys*, 1075: 77–136, https://doi.org/10.3897/zookeys.1075.72197

Sharkey, M.J., Janzen, D.H., Hallwachs, W., Chapman, E.G., Smith, M.A., Dapkey, T., Brown, A., Ratnasingham, S., Naik, S., Manjunath, R., Perez, K., Milton, M., Hebert, P., Shaw, S.R., Kittel, R.N., Solis, M.A., Metz, M.A., Goldstein, P.Z., Brown, J.W., Quicke, D.L.J., Achterberg, C.V., Brown, B.V. and Burns, J.M. (2021b) Minimalist revision and description of 403 new species in 11 subfamilies of Costa Rican braconid parasitoid wasps, including host records for 219 species. *ZooKeys*, 1013: 1–665, https://doi.org/10.3897/zookeys.1013.55600

Thomson, S.A., Pyle, R.L., Ahyong, S., Alonso-Zarazaga, M., Ammirati, J., Araya, J.F., Ascher, J.S., Audisio, T.L., Azevedo-Santos, V.M., Bailly, N., Baker, W.J., Balke, M., Barclay, M.V.L., Barrett, R.L., Benine, R.C., Bickerstaff, J.R.M., Bouchard, P., Bour, R., Bourgoin, T., Boyko, C.B., Breure, A.S.H., Brothers, D.J., Byng, J.W., Campbell, D., Ceríaco, L.M.P., Cernák, I.,Cerretti, P., Chang, C.-H., Cho, S., Copus, J.M., Costello, M.J., Cseh, A., Csuzdi C., Culham, A., D'Elía, G., d'Udekem d'Acoz, C., Daneliya, M.E., Dekker, R., Dickinson, E.C., Dickinson, T.A., van Dijk, P.P., Dijkstra, K.-D.B., Dima, B., Dmitriev, D.A., Duistermaat, L., Dumbacher, J.P., Eiserhardt, W.L., Ekrem, T., Evenhuis, N.L., Faille, A., Fernández-Triana, J.L., Fiesler, E., Fishbein, M., Fordham, B.G., Freitas, A.V.L., Friol, N.R., Fritz, U., Frøslev, T., Funk, V.A., Gaimari, S.D., Garbino, G.S.T., Garraffoni, A.R.S., Geml, J., Gill, A.C., Gray, A., Grazziotin, F.G., Greenslade, P., Gutiérrez, E.E., Harvey, M.S., Hazevoet, C.J., He, K., He, X., Helfer, S., Helgen, K.M., van Heteren, A.H., Hita Garcia, F., Holstein, N., Horváth, M.K., Hovenkamp, P.H., Hwang, W.S., Hyvönen, J., Islam, M.B., Iverson, J.B., Ivie. M.A., Jaafar, J., Jackson, M.D., Jayat, J.P., Johnson, N.F., Kaiser, K., Klitgård, B.B., Knapp, D.G., Kojima, J.-I., Kõljalg U., Kontschán, J., Krell, F.-T., Krisai-Greilhuber, I., Kullander, S., Latella, L., Lattke, J.E., Lencioni, V., Lewis, G.P., Lhano, M.G., Lujan, N.K., Luksenburg, J.A., Mariaux, J., Marinho-Filho, J., Marshall, C.J., Mate, J.F., McDonough, M.M., Michel, E., Miranda, V.F.O., Mitroiu, M.-D., Molinari, J., Monks, S., Moore, A.J., Moratelli, R., Murányi, D., Nakano, T., Nikolaeva, S., Noyes, J., Ohl, M., Oleas, N.H., Orrell, T., Páll-Gergely, B., Pape, T., Papp, V., Parenti, L.R., Patterson, D., Pavlinov, I.Ya., Pine, R.H., Poczai, P., Prado, J., Prathapan, D., Rabeler, R.K., Randall, J.E., Rheindt, F.E., Rhodin, A.G.J., Rodríguez, S.M., Rogers, D.C., Roque, F. de O., Rowe, K.C., Ruedas, L.A., Salazar-Bravo, J., Salvador, R.B., Sangster, G., Sarmiento, C.E., Schigel, D.S., Schmidt, S., Schueler, F.W., Segers, H., Snow, N., Souza-Dias, P.G.B., Stals, R., Stenroos, S., Stone, R.D., Sturm, C.F., Štys, P., Teta, P., Thomas, D.C., Timm, R.M., Tindall, B.J., Todd, J.A., Triebel, D., Valdecasas, A.G., Vizzini, A., Vorontsova, M.S., Vos, J.M. de, Wagner, P., Watling, L., Weakley, A., Welter-Schultes, F., Whitmore, D., Wilding, N., Will, K., Williams, J., Wilson, K., Winston, J.E., Wüster, W., Yanega, D., Yeates, D.K., Zaher, H., Zhang, G., Zhang, Z.-Q. and Zhou, H.-Z. (2018) Taxonomy based on science is necessary for global conservation. *PLoS Biology*, 16: e2005075 (12 pp.), https://doi.org/10.1371/journal.pbio.2005075

Vane-Wright, R.I., Humphries, C.J. and Williams, P.H. (1991) What to protect? – Systematics and the agony of choice. *Biological Conservation*, 55: 235–254, https://doi.org/10.1016/0006-3207(91)90030-D

Wheeler, Q.D., S. Knapp, D.W. Stevenson, J. Stevenson, S.D. Blum, B.M. Boom, G.G. Borisy, et al. (2012) Mapping the biosphere: Exploring species to understand the origin, organization and sustainability of biodiversity. *Systematics and Biodiversity*, 10: 1–20, doi: 10.1080/14772000.2012.665095.

Wheeler, Q.D. ed (2008) *The New Taxonomy.* The Systematics Association Special Volume Series, 76. CRC Press, Boca Raton.

Wheeler, Q. (2024) *Species, Science and Society: The Role of Systematic Biology.* Routledge, Oxon.

# 2 Norman Platnick, the Development of Cladistics, 'Integrative' Taxonomy, and Modern Monography

*David M. Williams and Anthony Gill*

'Men fight and lose the battle, and the thing that they fought for comes about in spite of their defeat, and when it comes turns out not to be what they meant, and other men have to fight for what they meant under another name'

*William Morris, London Review of Books, 8th Feb. 2024.*
*[From William Morris (1888), A Dream of John Ball]*

'The point is made that diffusion (the process by which new technology is adopted) can occur not just for good ideas but misconceptions. Incorrect notions can be subject to diffusion. "Bad ideas" can be "catchy" when the conditions are right for them to be received into the "channels and repositories" of the knowledge shared by people'

*Ruth Williams, 2009, p. 53*

'I am sceptical about integration and synthesis, perhaps because of the "new synthesis" regarding systematics'

*Gareth Nelson in Mattox, 2011, p. 54*

It would be a gross understatement to describe Norman Platnick (1951–2020) as simply a spider taxonomist, even though he documented nearly 2200 arachnid taxa (Prendini, 2020d). During his career, tragically cut short (Prendini, 2020a–e), Platnick published over 300 papers, eight books, and two edited volumes dealing with a range of scientific matters. These stark figures hide the fact that 42 of those 300+ papers were detailed spider monographs and one of the eight books was *Systematics and Biogeography: Cladistics and Vicariance* (1981), co-authored with Gareth Nelson, an ichthyologist (both then at the American Museum of Natural History) – this book was a challenging and, to some, the most significant book written on systematics in the 20th century (see Ung and Grand, this volume; Janvier, 2003; Ebach and Williams, 2011). Spiders to one side, Platnick's intellectual journey included all the ingredients of an interactive approach to the subjects that captured


DOI: 10.1201/b22822-2

his interest, with, on occasion, some necessary contradictions that always emerge with active scientific research.

This chapter details progress in systematics during the half-century of Platnick's working life.

## DEVELOPMENTS IN CLADISTICS: PLATNICK'S JOURNEY

Late in his career, in a short preface to a book commemorating what would have been Willi Hennig's 100th birthday (Williams et al., 2010), Platnick offered an account of his introduction to cladistics via his American Museum of Natural History (AMNH) colleague, the entomologist Pedro Wygodzinsky (1916–1987, for biographical details on Wygodzinsky, see Schuh and Herman, 1988[1]): 'Hennig's writings were nothing short of a revelation. Once one worked through Hennig's somewhat arcane terminology (which was perhaps even more intrusive in the English translations than in the original German) and grasped his arguments, it became easy to see "how to do systematics"'(Platnick, 2016, p. xii). He continued: 'Every synapomorphic resemblance among organisms fits at some node on the cladogram connecting them all, and it fits at only one such node. Any resemblances that do not fit in this way are just not synapomorphic' (Platnick, 2016, p. xii). That is, for every character discovered, for every node on any tree, the sum total will provide a hierarchical classification – a classification supported, node by node, with available (and appropriate) evidence (see below). Such a view is related to the somewhat cumbersome term *heterobathmy* (meaning 'different steps on a stair', Nelson, 2004, p. 133), 'borrowed from the Armenian botanist Armen Takhtajan, from a book translated from the Russian, published in Germany in 1959' (Nelson, 2004, p. 133; Takhtajan, 1959; Takhtajan, 1991, p. 228). And, like synapomorphy, the concept, as opposed to the name, has a long and varied history (Takhtajan, 1991, p. 228; Nelson, 2004; Williams and Ebach, 2020: Rieppel, 2023).

The phrase 'heterobathmy of characters' (Takhtajan, 1959, p. 13, Rieppel, 2023, p. [04]) was used by Hennig like so: 'heterobathmy of characters is therefore a precondition for the establishment of the phylogenetic relationship of species and hence a phylogenetic system' (Hennig, 1965, p. 107, fig. 3, derived from Hennig, 1957, p. 66, abb. 9, the latter reproduced here as Figure 2.1). With these two concepts in mind – *synapomorphy* and *heterobathmy* – one simply needs to study specimens in a comparative sense and come to some understanding as to what 'step on the stair' any particular character occupies. And to repeat: *synapomorphy* and *heterobathmy*, as concepts, have had a long history, pre-dating what might be referred to as the cladistic episode in systematics (Nelson and Platnick, 1981; Williams and Ebach, 2020, and some references given in Goloboff, 2022, p. 12). But this 2016 account reflects Platnick's mature position, after some 40+ years of experience.

Let us examine Platnick's beginnings. The first 'phylogeny' he produced was published in his fourth monograph – which was only his seventh publication (Platnick and Shadab, 1974, p. 13, figure 31, reproduced here as Figure 2.2). Of this 'phylogeny' the authors wrote, 'A phylogenetic tree reflecting the probable evolutionary history of the family is shown in figure 31 [see Figure 2.2]. It must be remembered that the known distribution of these species indicates that additional species, and possibly genera,

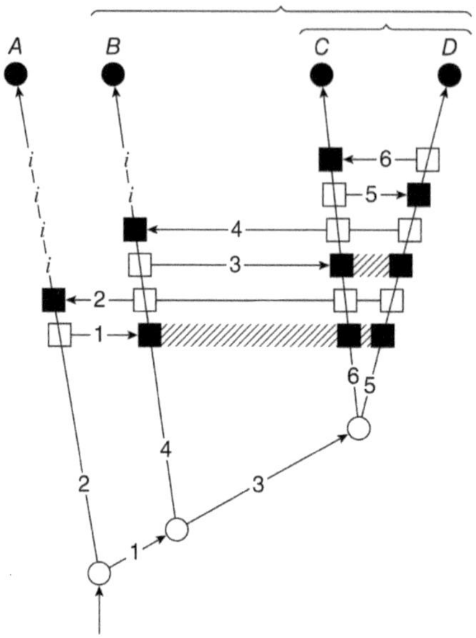


**FIGURE 2.1** A simple cladogram with characters having derived (black boxes) and 'primitive' states (white boxes). (After Hennig, 1957, p. 66, abb. 9, also in Hennig, 1965, p. 107, fig. 3).

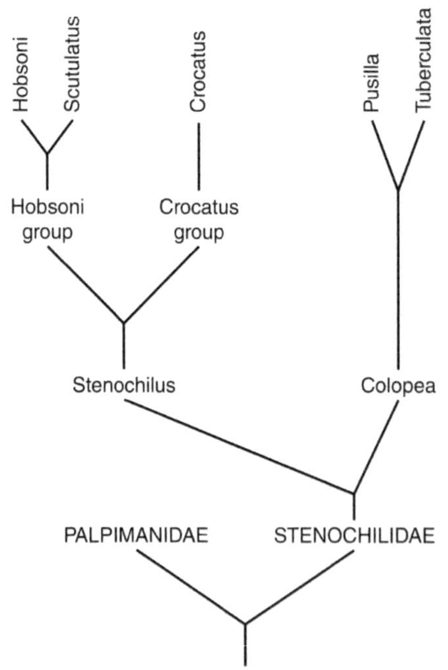


**FIGURE 2.2** Platnick's first 'phylogeny.' (After Platnick and Shadab, 1974, p. 13, figure 31: 'A phylogenetic tree reflecting the probable evolutionary history of the family is shown in figure 31').

may eventually be found on other Pacific islands' (Platnick and Shadab, 1974, p. 6). In another revisionary monograph, published a year later, Platnick included a diagram of relationships (Platnick, 1975, p. 5). The diagram is labelled a 'Probable phylogeny of *Callilepis* species', where the 'Dark squares denote derived character states, light squares primitive ones; N and P indicate Nearctic and Palaearctic, respectively' (Platnick, 1975, p. 5, figure 7, reproduced here as our Figure 2.3, compare with Figure 2.1).

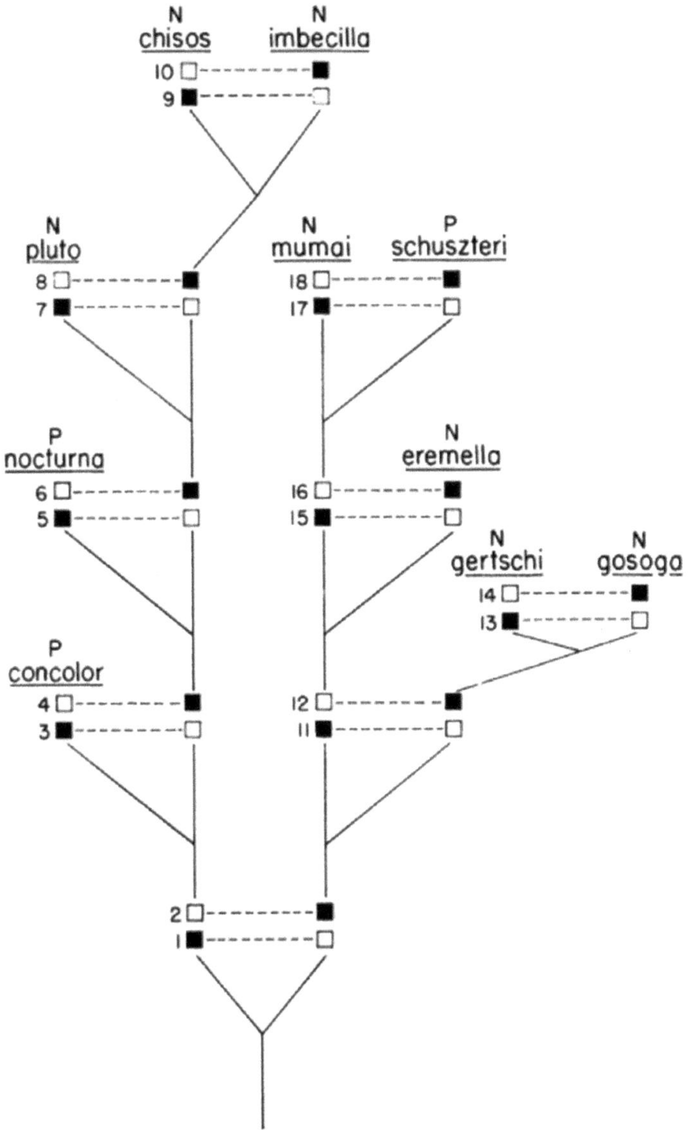


**FIGURE 2.3** Platnick's first cladogram: a 'Probable phylogeny of *Callilepis* species', where the 'Dark squares denote derived character states, light squares primitive ones; N and P indicate Nearctic and Palaearctic, respectively.' (After Platnick, 1975, fig. 7, compare with Figure 2.1).

Although no direct reference was made to any of Willi Hennig's publications, Platnick wrote that 'Figure 7 shows the probable phylogeny of the known *Callilepis* as reconstructed by a *cladistic analysis* of genitalic characters, based on the premise that only shared, derived ("*synapomorphic*") characters, and not shared, primitive ones, can indicate monophyly' (Platnick, 1975, p. 2, our italics; Platnick, like most of us, was not entirely free from the prison of his training: 'In the absence of fossil *Callilepis* it is of course not possible to make absolute statements about what character states are derived'). That diagram of relationships was later re-drawn and used for his first discussion of biogeography: 'Drifting spiders or continents?: Vicariance biogeography of the spider subfamily Laroniinae (Araneae: Gnaphosidae)' (Platnick, 1976, p. 105, fig. 2, reproduced as Figure 2.4). Between then, the mid-1970s, and his final studies, Platnick mostly wrote about spiders, biogeography and Karl Popper (Prendini, 2020c). Popper to one side (Platnick and Gaffney, 1977, 1978; Platnick and Rosen, 1987; Rieppel, 2003, 2004, 2008; Santis, 2020, and many more), biogeography would loom large as a discipline in need of attention (Platnick, 1991, 2017).

For systematics, at least, certain issues had been solved, or at least appeared to have been solved, many years prior to the advent of cladistics: 'From this, one might suspect that work done during the ensuing decades [from the 1970s onwards] would demonstrate this clarity [the role of synapomorphy and heterobathmy], but that does not appear always to have been the case. Workers like Steve Farris showed that *one could construct a matrix of similarities (characters x taxa), of the sort that had*

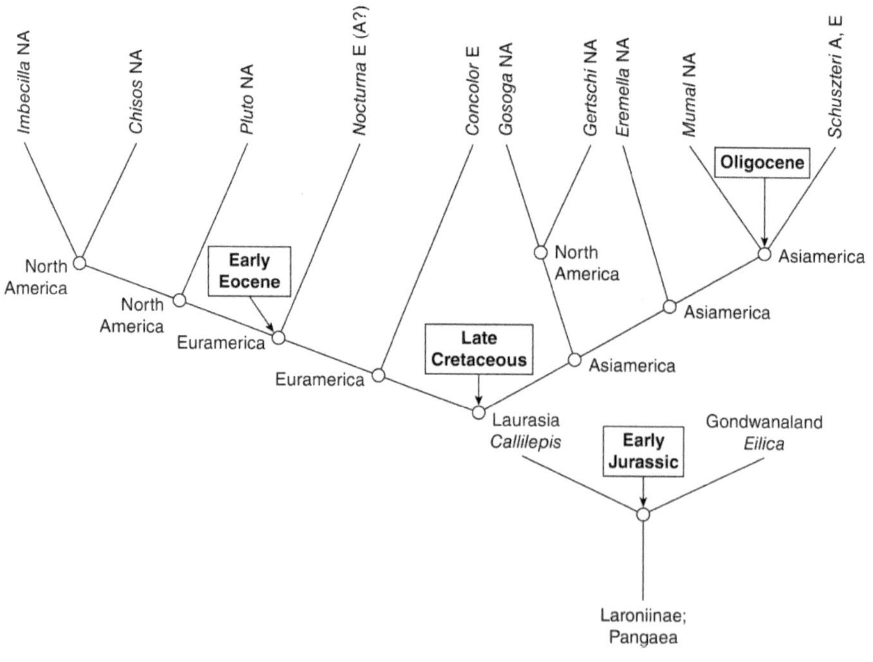


**FIGURE 2.4** A re-drawn phylogeny from Platnick (1975, fig. 7) in 'Drifting spiders or continents?: Vicariance biogeography of the spider subfamily Laroniinae (Araneae: Gnaphosidae).' (Platnick, 1976, fig. 2).

*been favored by pheneticists*, and analyse it parsimoniously to produce the equivalent of Hennig's character schemes (cladograms) [...] (Platnick, 2016, p. xiv, our italics) – but are those trees so produced really the 'equivalent of Hennig's character schemes'? And does it matter?

## CLADISTICS AND COMPUTERS

If we fast forward to 2017 and examine Platnick's last published comments on systematics in another preface, this time for the book *Assumptions Inhibiting Progress in Comparative Biology*, a series of essays that focused on Donn Rosen's teaching notes (Crother and Parenti, 2017; Rosen was an ichthyologist at the AMNH; for biographical details see Nelson et al., 1987; Nelson, 2017), Platnick made a comment concerning the view, popular still, that 'cladistics = parsimony'; Platnick meant 'parsimony analysis' (sensu Goloboff, 2022, p. 72; Farris, 1970[2]), the algorithm, rather than the general concept of parsimony. He wrote: 'for those of us who [...] became cladists *only after we first became practicing systematists*, cladistics = clustering by synapomorphy, not parsimony' (Platnick, 2017, p. viii, our italics). The difference might not have been obvious to all readers of this preface. Platnick was, in fact, indirectly referring to a contribution made earlier by Kevin Nixon and James Carpenter who suggested that 'Parsimony analysis minimizes homoplasy, whether it occurs in apomorphic or plesiomorphic states, and indeed, it is irrelevant which is which during branch-swapping and selection of most parsimonious trees' (Nixon and Carpenter, 2012b, p. 225) and 'This was the beginning of post-Hennigian, modern cladistics. Since that revelation, we no longer "group by synapomorphy" nor construct trees by hand; instead, we read synapomorphies from trees that minimize homoplasy over all character states, be they zeros or ones, A, C, G or T' (Nixon and Carpenter, 2012b, p. 225; in an earlier piece, one of their sub-headings is 'Plesiomorphy is homology', Nixon and Carpenter, 2012a, p. 162, see Platnick, 2013) – 'we no longer "group by synapomorphy"': in spite of their protestations, this does seem to be a version of phenetics, an approach to systematics that makes no evident statements of synapomorphy and assumes 'Plesiomorphy is homology'. One might suggest that 'grouping by synapomorphy' is *data-driven*, 'reading synapomorphies from trees' is *algorithm-driven* – and thus arguments over method dissolve into arguments concerning how good (or bad) a method, any method, might be.

Goloboff recently argued that '[...] some presentations of cladistics talk about synapomorphies as if they could be identified without reference to a tree, synapomorphies cannot be identified in the abstract' (Goloboff, 2022, p. 139) – a comment that might achieve greater focus by inspection of the achievements of taxonomists prior to the numerical onslaught (see, for one example, Hilton and Forey, 2005, p. 644: 'Ridewood ["Walter George Ridewood (1867–1921), an English comparative anatomist"], clearly understood the distinction between shared primitive (= symplesiomorphies) and shared derived characters (= synapomorphies), and that symplesiomorphies were incapable of defining taxonomic groups'; 'Mitchell (1901 [Carolin, 1984]) and Hennig (1950) may be considered among the many independent inventors of the concepts of synapomorphy and of classification as a nested set of synapomorphies, at least in the phyletic context', Nelson and Platnick, 1981, p. 326).

## 'MODERN CLADISTICS': SOME HISTORICAL COMMENT

### ROBERT SOKAL AND WILLI HENNIG: AN EARLY CLADISTIC-PHENETIC CONNECTION

Nixon and Carpenter (and Goloboff, 2022) may very well 'no longer "group by synapomorphy"', but is their version really that 'modern'? This is worth exploring.

During the 1950s, the University of Kansas held regular 'evolutionist club' meetings usually centered on a dinner followed by a presentation from an invited guest. The club was apparently founded by Charles Michener (1918–2015), a world expert on bees. In December 1960 (or maybe 1961, see below), Robert Sokal, a leading figure in the creation of numerical taxonomy, gave a talk on the works of Willi Hennig – Sokal spoke fluent German and would have understood Hennig's major contribution *Grundzüge einer Theorie der phylogenetischen Systematik*, his first general book on systematics, written and published in German (Hennig 1950 – it was discussed in Sokal and Sneath, 1963). Sokal joined the University of Kansas in 1951, having graduated from the University of Chicago (Hull, 1988; Bell and Gurevitch, 2012; Michener wrote an obituary for Sokal, see Michener, 2012). In his Kansas presentation on Hennig, Sokal spoke of Hennig's approach to systematics, differing with him only in that *Hennig wished to distinguish between apomorphic and plesiomorphic characters, while Sokal did not.* Hans-Peter Schultze (Cloutier, 2002), who later joined the University of Kansas, noted the 1960 (1961) meeting in the acknowledgements to a 1981 paper of his (Schultze, 1981, p. 72): "Dr. C. D. MICHENER was kind enough to ask Dr. R. R. SOKAL when he first knew of HENNIG'S book (Dr. C. D. MICHENER and Dr. R. JOHNSTON, both of the University of Kansas, Lawrence, recall a lecture given by Dr. SOKAL on HENNIG'S principle in 1960/61 at the 'evolutionist club' in Lawrence. SOKAL no longer remembers it)" (Schultze, 1981, p. 83, translated).

Michener had already played an influential role in the early exploration of numerical methods in taxonomy collaborating with Sokal (Michener and Sokal, 1957): 'The Sokal-Michener approach came to be known as "numerical taxonomy", and it led to much controversy and debate among taxonomists. But the phenetic (*an adjective referring to observable similarities and differences between organisms without regard to assumed genealogy*) system employed was a major conceptual and methodological leap for biological systematics' (West-Eberhard and Engel, 2017, also West-Eberhard, 2009, pp. 732–733).

It is frequently said that *phenetics* is defined by its approach to analysis, by an algorithm, a method, so that '*phenetics* = overall similarity': 'The term *phenetics* is used here in its strict connotation, that of grouping taxa on the basis of overall similarity [...]' (Goloboff, 2022, p. 23) – and yet, in its more general form, concerning principles, it is about *apomorphic and plesiomorphic characters* and not distinguishing them from one another. All data are relevant.

### ARNOLD KLUGE AND GEORGE GAYLORD SIMPSON: THE PHENETIC-CLADISTIC CONNECTION

In 1971, Arnold Kluge, an early enthusiast and avid promoter of 'parsimony analysis' (the 'Quantitative Phyletics' of Kluge and Farris, 1969), published a review chapter in the book *Chordate Structure and Function* dealing with 'Concepts and principles of morphologic and functional studies' (Kluge, 1971, 1977). In his review, Kluge discusses, among other things, morphoclines, chronoclines and other such evolutionary transformational notions, noting that esteemed palaeontologist George Gaylord Simpson was 'one of the most outstanding students of evolutionary morphology' (Kluge, 1971, p. 7).

In 1976, Kluge published a monograph on the lizard family Pygopodidae: 'Phylogenetic Relationships in the Lizard Family Pygopodidae: an Evaluation of Theory, Methods and Data' (Kluge, 1976; see Figure 2.5). He contrasted 'Phylogenetic Systematics' (sensu Hennig, 1966) and 'Quantitative Phyletics' (sensu Kluge and Farris, 1969), referring to the former as 'cladistics':

> Phylogenetic Systematics, often referred to as cladistics (Farris, Kluge and Eckardt, 1970), and Quantitative Phyletics (Kluge and Farris, 1969) are prominent among the many sets of methods which purport to provide the best phylogenetic inference

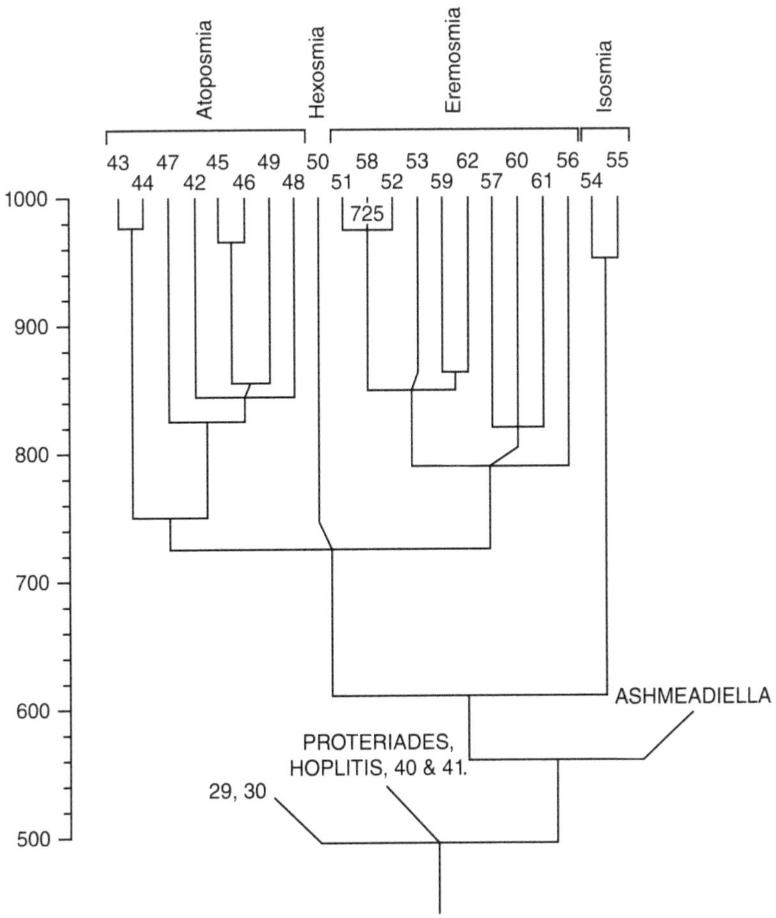


**FIGURE 2.5** Phylogeny from 'Phylogenetic relationships in the lizard family pygopodidae: An evaluation of theory, methods and data.' (Kluge, 1976).

> possible [...] The details of the Phylogenetic Systematic and Quantitative Phyletic methods are described elsewhere (Kluge and Farris, 1969; Farris, Kluge and Eckardt, 1970) and their assumptions and limitations have received considerable review and debate (Estabrook, 1972; Sneath and Sokal, 1973).

Kluge continued:

> The most significant difference between the two approaches concerns the fact that Phylogenetic Systematics estimates only the *cladistic parameter* and in doing so it uses only derived states. Quantitative Phyletics estimates both cladistics and patristics and it *uses all states, derived and primitive*, in those estimations [emphasis ours].
>
> The proponents of Phylogenetic Systematics assert that the primitive state cannot be used to infer relative recency of common ancestry because no evolutionary change is implied by the primitive state. The logic of their philosophy is unassailable; however, I do not believe that it necessarily has to be applied in practice. Moreover, as I will argue below, to *ignore similarity based on shared primitive states can lead to significantly different phylogenetic hypotheses.*
>
> (Kluge, 1976, p. 43, emphasis ours)

One might puzzle over Kluge's comment that 'The logic of their philosophy is unassailable; however, I do not believe that it necessarily has to be applied in practice'. That is, one might suppose that if there is a computer program, such as one for 'parsimony analysis' ('applied in practice'), one might adjust the logic to fit the algorithm (see Farris, 1979).

George Gaylord Simpson contributed the opening chapter to the book *Phylogeny of the Primates* entitled 'Recent Advances in Methods of Phylogenetic Inference' (Simpson, 1975). He referred to Kluge and Farris's method as 'phenetic', which may simply have been a mistaken assumption of his – or, instead, an appropriate comparison: '[...] Kluge and Farris (1969) did select and weigh characters on phylogenetic criteria in a relatively early application of phenetics to [the] construction of a phylogram of a real group (anurans)' (Simpson, 1975, p. 13).

Nevertheless, Sokal's problem with Hennig appeared to be the same as Kluge's: what should be considered relevant data?

## Peter Ashlock, Ernst Mayr, and James Carpenter

Peter Ashlock (1929–1989) was also at the University of Kansas, from 1968 until his death in 1989 (Byers, 1989; Slater and Polhemus, 1990). Ashlock was clear on his approach to systematics: 'I am an evolutionary systematist, a member of the Simpson-Mayr school of systematics, which has profound objections, principally in the area of classification, to the cladistic or so-called phylogenetic school of Hennig' (Ashlock, 1974, p. 81). Ashlock later collaborated with Ernst Mayr to produce the 2nd edition of *Principles of Systematic Zoology* (Mayr and Ashlock, 1991). Some details concerning this 2nd edition are relevant. In the acknowledgements Mayr and Ashlock offer the following: 'Special thanks are due to J. Carpenter, J. Felsenstein, and David Maddison [...] Chapter 11 was completely revised by David Maddison

[...]' (Mayr and Ashlock, 1991, p. xx) – 'It is rumoured that James Carpenter, after commenting extensively on the near-complete manuscript, was asked to replace Ashlock as junior author but he resisted that "kiss of death"' (Nelson, 1991, p. 305); '[...] he [Mayr] avoids mention of cladistics altogether, leaving it entirely to Ashlock' (Groves, 1993, p. 251); and 'Because the latter [Ashlock] fell ill, Mayr did most of the work' (Haffer, 2007, p. 455).

### Total Evidence and Simultaneous Analysis

Broadly speaking, the idea of 'Total Evidence' was to suggest that all data should be subject to the same kind of examination so that the final result would be a product of 'character congruence' (Kluge, 1989, 2004). This was contrasted with the notion that different data sets, such as morphology and molecules, should be analyzed separately and the optimal solution would be achieved through congruence between the separate results. As Nixon and Carpenter later wrote, with respect to simultaneous analysis, 'The position that datasets should be analysed separately is clearly based on a rejection of the principle of parsimony in cladistics' (Nixon and Carpenter, 1996a, p. 237). A view neatly contrasted with Nelson: 'With total evidence, need no such asymptote be expected? If it were we might witness the reemergence, even the vindication, of phenetics as the overall similarity of synapomorphy' (Nelson, 2004, p. 139).

What can be concluded from this? Is 'The past a foreign country', did 'they do things differently there'? (Hartley, 1953, *The Go-Between*)? That Hennig may possibly have, even if unwittingly, influenced the beginnings of Sokal's numerical taxonomy programme – their fundamental disagreement being over what data were to be considered relevant; Kluge's eventual introduction of the 'total evidence' idea was part of the assault on cladistics when understood as 'grouping by synapomorphy' with the promotion of one specific method of analysis ('parsimony analysis', also Nixon and Carpenter, 1996a), which deftly turned a body of thought (cladistics) into a series of arguments about methodology and implementation. As noted above, the viewpoint that 'modern cladists' 'no longer "group by synapomorphy" [but] read[s] synapomorphies from trees' (Nixon and Carpenter, 2012b, p. 225) – the program (the assumptions of the algorithm) is now the decisive factor. One might see that if '*phenetics* = overall similarity' and 'cladistics = parsimony', and with the subsequent introduction of numerous statistical model-based programs, the discussion was now about algorithms not principles: 'The conclusion to be drawn is not that a particular clustering procedure necessarily leads to the truth, as if truth depends upon mere conformity to procedure. The conclusion is that a particular clustering procedure yields a result that might be true or false, and that a judgment as to its truth or falsity may be considered an independent problem that can be investigated in other ways' (Nelson and Platnick, 1981, p. 193).

## AN END TO MORPHOLOGICAL DATA ANALYSIS

While the quest for a method (or methods) that will be appropriate enough and accurate enough, etc., has created a veritable mountain of literature, an alternative viewpoint was developing almost simultaneously which might be summarized as so: maybe, after all, when considering morphology and molecules one form of data really is less informative than the other, with morphology being the less informative (we pass over what informative might mean at this stage). This viewpoint was proposed by Scotland et al. who concluded: '[...] in contrast to the view that "much more [morphological] variation could be included in phylogenetic analyses than is used presently" (Poe and Wiens, 2000, pp. 33–34), we take the view that rigorous and critical anatomical studies of fewer morphological characters *in the context of a molecular phylogeny* is the way that *integrated studies* will and should develop' (Scotland et al., 2003, p. 545, italics ours; see also Wortley and Scotland, 2006).

A justification for the view that '[...] morphology is being superseded by DNA data for phylogenetic studies is because much of the useful morphological diversity has already been scrutinised' (Scotland et al., 2003, p. 545) is perhaps a highly doubtful assumption when considering organisms other than, say, vascular plants and vertebrate animals (Platnick, 2017, p. ix). Nevertheless, Wiens responded to Scotland et al. by stating, somewhat baldly, that '[Scotland et al.] basically advocate giving up on using morphological data to reconstruct phylogenies' (Wiens, 2004, p. 659). The views of Scotland et al. (2003) and Wortley et al. (2002) are more nuanced than a simple rejection of a particular dataset – although Wiens may be correct in assuming the doubts in Scotland et al. concerned the use of morphology for 'reconstructing phylogenies'. One might be tempted to re-word Wiens' comment as follows: Scotland et al. advocated giving up using morphological data as *evidence for reconstructing phylogenies but not for classification*. But this might not be entirely accurate either, as Scotland et al. wanted to plot their established defining (diagnostic) characters – found or established from 'rigorous and critical anatomical studies of fewer morphological characters' – onto the molecular 'phylogeny' and, where they fit the nodes ('in the context of a molecular phylogeny'), claim them to be synapomorphies of each group so identified. Nevertheless, '[...] just because there are enough morphological synapomorphies for careful observers to recognize many monophyletic groups over the years in traditional taxonomic studies, further dissection of morphology by present or future scientists may still not be able to resolve the full branch structure of the tree of life' (Scotland et al., 2003, p. 543).

Here we have to take a step back as, in our view, regardless of the dataset, the outcome, the solution, and the cladogram (the trees) if you will, are all the same: they are *classifications* and only interpretation can 'change' them into a phylogeny (Williams and Ebach, 2020). With classification in mind, one might inspect Scotland et al. a little more closely:

> 'It is our view that the recognition of taxa on the basis of morphological data has occurred largely in those parts of phylogeny in which there are morphological characters that *fully diagnose taxa*, i.e., taxic homologues sensu Patterson (1982), which are equivalent to synapomorphies sensu Hennig (1966). [...] there remain several obstacles to simply equating readily identifiable diagnostic characters with synapomorphy. First, readily identified homologues may diagnose nonmonophyletic groups (as with plesiomorphies). Second, even when the number of putative taxic homologues is reduced to very few by stringent character selection, they may conflict in the groups they define'.
>
> *Scotland et al., 2003, p. 548*

One might first consider some words of Wortley and Scotland, presumably chosen very carefully: 'Nonetheless, there remains considerable debate about the precise role of morphology in one particular area of biology – *computer-based phylogenetic inference*' (Wortley and Scotland, 2006, p. 677, our italics, and see Wortley et al., 2002; Stevens, 2002).

The first objection in Scotland et al. 2004 is that '[...] readily identifiable homologues may diagnose nonmonophyletic groups (as with plesiomorphies)'. This recall's Nixon and Carpenter's view that 'Plesiomorphy is homology'. The second objection of Scotland et al. that characters 'may conflict in the groups they define', is most likely a problem of almost every dataset, only almost all DNA studies never mention it. Remarkably it has been argued that in some DNA datasets homoplasy (noise) increases resolution (Källersjö et al., 1999, '[...] although rapidly evolving and highly homoplastic, third positions contain most of the phylogenetic structure in the data'; see also Goloboff and Wilkinson, 2018) and recently for morphology Santis writes that he 'will discuss the idea that homoplasy is more complex than just nonhomology, sometimes even constituting evidence of common ancestry' (de Santis, 2024).

Again, remarkably we now learn that 'Plesiomorphy is homology', that 'homologues may diagnose nonmonophyletic groups (as with plesiomorphies)' and 'homoplasy [...] sometimes [...] constitutes evidence of common ancestry'. For 'symplesiomorphy as evidence' Goloboff offers the somewhat surprising justification: 'Plesiomorphies *are* informative: they inform us that the bearer of the plesiomorphic condition does *not* belong to the group defined by the alternative state' (Goloboff, 2022, p. 12 – or, it is simply a synapomorphy at a higher level?). In the end, one might question the method applied or question the parameters adopted, or question any aspect of the analytical process, but in the end, this is only possible if data are viewed *phenetically*, 'a matrix of similarities (characters x taxa), of the sort that had been favored by pheneticists [...]' (Platnick, 2016, p. xiv) – that is: a matrix of similarities rather than a matrix of synapomorphies.

Platnick again: '[...] the fascination with matrices soon became bound up with the illusory notion of "total evidence"'(Platnick, 2017, p. xiv).There is little point in re-hashing the tortured history of 'total evidence' and its relationship to the apparent binary choice of using either consensus methods or applying one or another method to combined data sets (Williams and Ebach, 2008, 2020, and our brief comments above). Suffice to say that this, too, mutated into making choices of this or that algorithm and with the introduction of 'simultaneous analysis' and the promotion of 'parsimony analysis' as the algorithm of choice (Nixon and Carpenter, 1996a, see Goloboff et al., 2017) set against the many alternative statistical approaches of maximum likelihood algorithms (Lewis, 2001; Spade, 2020; see Goloboff et al., 2019) or Bayesian Inference methods (Khakurel et al., 2024), one might conclude that of methods there is no end.

For Nixon and Carpenter (2012a, 2012b, and Goloboff, 2022), the justification for promoting a particular algorithm that shuffles and sorts these similarities in the quest for the shortest tree was that '[...] the guiding principle of cladistics is parsimony and maximization of explanatory power, *not grouping by synapomorphy*' (Goloboff, 2022, p. 13, our italics). This array of methods has been recently summarized, accurately in our view, by Wheeler as 'a pseudoscience based on quantitative manipulations of questionable phenetic data [that] has focused on cladograms at the exclusion of character analysis. Reversing this trend will require a return to the refinement of Hennig's theories and methods and the integration of diverse available data' (Wheeler, 2008, p. 2).

## 'INTEGRATIVE TAXONOMY': ANOTHER EXCURSION IN FUTILITY?

Before discussing *Integrative Taxonomy*, let us recall a little more about phenetics. The first comprehensive summary of *numerical taxonomy* was published in 1963 (Sokal and Sneath, 1963, but see Vinarski, 2022). Herbert H. Ross, an entomologist (Unzicker and Wallace, 1979), when reviewing Sokal and Sneath's book was moved enough to write: 'After reading this book carefully, it is my considered opinion that numerical taxonomy is an excursion in futility' (Ross, 1964, p. 108; Nelson and Ladiges, 2009, p. 40). Ross's review is still worth reading today: 'This means that numerical taxonomy as outlined in this book is simply a highly sophisticated method of producing groupings of similar taxa arranged in a tentative fashion, usually agreeing fairly closely (as stated) with prior groupings and arrangements made on other grounds' (Ross, 1964, p. 108). Much later, Richard Jensen, when reviewing the progress of phenetics from a 40-year perspective, wrote: 'This anniversary meeting allowed reflection on the impact that numerical taxonomy has had on systematics and comparative biology. Although few would agree with Herbert Ross's opinion that "numerical taxonomy is an excursion in futility", it is clear that its role in systematics has not evolved as proponents projected: The methods are rarely used as the foundations for classifications' (Jensen, 1993, p. 599). And yet, in 2010, nearly 20 years after Jensen's obituary, Williams et al. had to address the subject yet again in 'The Steady Resurrection of Phenetics' (Williams et al., 2010).

*Integrative Taxonomy* has now become popular, regardless of how it is conceived (or misconceived) or how it is implemented (Padial et al., 2010; Vinarski, 2020; Yeates et al., 2021, as a few conflicting examples). Originally, Benoît Dayrat 'defined [Integrative Taxonomy] as the science that aims to delimit the units of life's diversity from multiple and complementary perspectives (phylogeography, comparative morphology, population genetics, ecology, development, behaviour, etc.)' (Dayrat, 2005, p. 407). Another formulation, appearing at almost the same time, was in Wills et al., who, while arguing the pros and cons of the then new DNA bar-coding initiative, wrote 'Note that we are not defending "traditional taxonomy" here, but instead we argue that the real cutting-edge future for systematics and biodiversity research is *integrative taxonomy*, which uses a large number of characters including DNA and many other types of data, to delimit, discover, and identify meaningful, natural species and taxa at all levels' (Wills et al., 2005, p. 845; they included a note appended to the end of their paper: 'Concurrent with the submission of our final manuscript of this article, but unknown to us at that time, B. Dayrat published a paper, coining and defining the term "Integrative taxonomy" [...] Our use of the term is consistent with his, although we would differ in details of its implementation', Wills et al., 2005, p. 851).

If, as now seems obvious, the common thrust behind both versions concerns large amounts of data, then the issue that still remains is: Are all data equal, are all data relevant?

First, one must distinguish between *intrinsic* and *extrinsic* data. The former are considered to be parts of organisms (their homologues, if you will, Nelson and Platnick, 1981; Williams and Ebach, 2020; Goloboff, 2022, p. 1), the latter describe where and how the organism lives (Goloboff, 2022, p. 1). Dayrat's list, then, has a number of items that are *extrinsic* and not actual properties of the organisms themselves. A few other items listed are neither extrinsic nor intrinsic but disciplines (e.g., population genetics, ecology) or ideologies (phylogeography) and, once removed leaves, as usual, comparative biology with their data sources primarily derived from morphology and molecules – and the issue of 'What are relevant data?'

## 'WHAT ARE RELEVANT DATA?'

In the early 1990s, Nelson and Ladiges were working on the complex biogeography of the plant genus *Eucalyptus* and, as a consequence, had to address some general problems in biogeographic analysis related to area-cladogram combination. Among the many issues, one that demanded attention was the general question of 'What are relevant data?' (Nelson and Ladiges, 1992, p. 470) – and related to this concern was how can area (or taxon) relationships be expressed directly? In short, one might represent area data as either binary variables, as in the classic character x taxon matrix ('a matrix of similarities (characters x taxa), of the sort [...] favored by pheneticists'), or as direct statements of relationship, as in branching diagrams, the minimal type of diagram was the three-item statement, A(BC). That is, systematic data could be represented in more than one way. Combination of the latter are normally associated with consensus techniques. This was recognized by Nixon and Carpenter (1996b). In their table 1, 'Types of Trees', they include under 'Indirect (summary trees, derived from trees)', 'Other component approaches (for example, the three-taxon statement trees of Nelson and Platnick, 1991, are implicitly tree-derived)' (Nixon and Carpenter, 1996b, p. 307).

For the few of us who came to understand the problem differently, '[...] it is not consensus versus "combined data sets" but rather the nature of data and how they are combined' (Nelson and Ladiges, 1991 [1992], p. 481). That is, the problem did not rest with the methods, whatever they might be, however sophisticated, but with the nature of the data themselves: 'To see evidence of geographic differentiation - relationship - where none is known transforms symplesiomorphy into "evidence". We are familiar with the transformation, *which is the heart of phenetics* and its enthusiasm to embrace all of the data *whether they be relevant or not*' (Nelson and Ladiges, 1991, p. 56, our italics). Phenetics raises its head once again with 'its enthusiasm to embrace all of the data' (maybe it never really went away: Nelson and Platnick, 1978, 'The perils of plesiomorphy [...] and phenetic biogeography'; see Lewens 2012: 'This article defends a modest pheneticism, understood as part of a pluralist view of taxonomy').

## HOMOLOGUES AND HOMOLOGY

There is conflict (or possibly disagreement) in the understanding of the terms 'homologue' and 'homology'. *Homologue* refers to the parts of organisms, the 'visible' parts, Owen's 'namesakes', those parts of organism we study: 'Here it is helpful to remember that evidence is "visible" – stuff you can see, feel, hear, taste or smell' (Nelson, 2011, p. 138). The general nature of comparative biology is that for any organism, an understanding of their parts is required: what they are, how might they be defined, how the different parts relate one to another, and how widely distributed each part is among different taxa. Investigation of the parts, the homologues, is the primary focus of taxonomic specialists, be they ornithologists, ichthyologists, entomologists, botanists or whatever – and 'proper homologues are synapomorphies, not symplesiomorphies' (Nelson, 2011, p. 139). All taxonomists/systematists engage in this part of specimen investigation regardless of what they do with these data once acquired. Many, by default, or habit, will code these observations (one way or another) and add them to their ever-growing matrix. Thus, one might understand 'a matrix of similarities (characters x taxa), of the sort [...] favored by pheneticists', to be a matrix of homologues. What is important is that the homologues, by themselves, are not the same thing as homology; those two words do not mean the same. *Homology* is a statement of relationship, which may or may not be true, derived from the homologues, the observations. By 'true', we mean corroborated by other such statements, not some absolute truth.

Goloboff recently expressed concerns over this part of taxonomic studies, which he referred to as 'homology-by-revelation': 'The problem with this approach is that neither they, nor any of several other advocates of homology-by-revelation [...], provided any clues as to what kind of reasoning or method one could use to correctly identify homologies, other than carefully studying the organisms to classify and engaging in meditation until the truth is revealed' (Goloboff, 2022, p. 74, ellipses replace references). Goloboff's statement might gain a little more precision if the word 'homologies' is replaced with 'homologues': 'what kind of reasoning or method one could use to correctly identify homologues, other than *carefully studying the organisms to classify*' (the latter part of his sentence seems best ignored). Taxonomists do this all the time, with no recourse to either matrix or cladogram. For example, Ríos-Tamayo et al., in their revision of the spider genus *Hermacha* (Mygalomorphae: Entypesidae), without matrix or cladogram, just with well-documented morphology, write: 'Regarding the morphological characters previously defined as diagnostic for the genus [...], several were corroborated: the straight copulatory bulb, an apical megaspine on a low mound on tibia I of males, and a small, pale intercheliceral tumescence' (Ríos-Tamayo et al., 2021, p. 75, under the assumption that diagnostic characters and synapomorphies are one and the same thing, something that might be debatable).

The many books and papers on morphology (e.g., Tamborini, 2022, an otherwise excellent account of the history of developments in morphological studies) and homology (e.g., Love, 2007; Rieppel, 2013; Mabee et al., 2020; Peterson, 2020) more or less concentrate on homologues, the parts, rather than homology, the implied relationship. Here, a diagram might help – Figure 2.6 is taken from an early publication of Willi Hennig (Hennig, 1957, abb. 8), but many others from many different authors could suffice. In his Figure, B and C share the homologue to the exclusion of all other organisms. (The black boxes represent the homologue.) The implied relationship is A(BC).

If one codes for just the BC part, the A remains, but also receives a code, usually a zero. If other taxa are added (D–F, Figure 2.6), all lacking the derived state of B

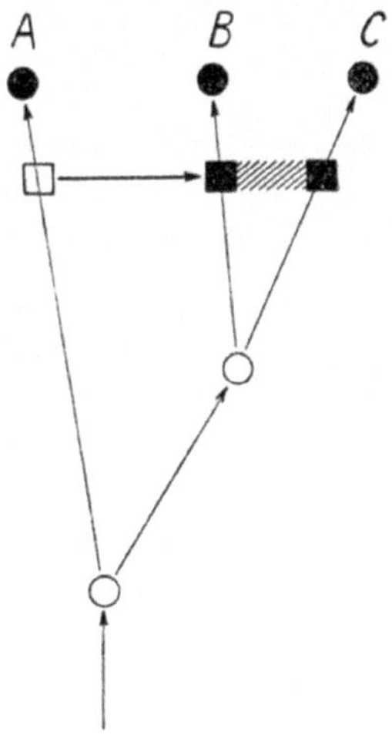


| | |
|---|---|
| A | 0 |
| B | 1 |
| C | 1 |
| | |
| D | 0 |
| E | 0 |
| F | 0 |
| etc. | |

**FIGURE 2.6** A simple three-taxon cladogram from Hennig (1957, abb. 8). In this figure, B and C share the homologue to the exclusion of all other organisms. The table is of six taxa (A–F) with one character scored with 0 as primitive, 1 as derived.

and C, they also receive the code zero. And that becomes evidence in the phenetic matrix, as if AD–F might be a 'group' of some kind. Any applied algorithm, then, simply shuffles around the homologues until they fit some desired parameter (the shortest tree, for example). The solution is not a property of the data, but a property of whatever algorithm or criterion is applied. And so, once again, of methods, there can be no end.

If one codes the relationship A(BC) directly, then to find an optimal tree for all the data, one need only find the tree that includes (accommodates) the most. The solution, then, is a property of the data.

Understanding data as relationships when applied to biogeography, the results were of some interest (Nelson and Ladiges, 1991 [1992]; Morrone and Carpenter, 1994). But could the idea be applied to character data? 'We believe that the relational approach to component analysis merits further study and application to character data' (Nelson and Ladiges, 1992, p. 481). Nelson had discussed the matter with Platnick who went on to create the vast majority of the matrices that were published in the first exploration of systematic data as relationships – as three-item statements – rather than binary variables (Nelson and Platnick, 1991; when asked later, Platnick replied '[I] am sorry to say that my memory is far too spotty for me to be able to corroborate that story. It is certainly possible, as I found the three-taxon approach eminently sensible [...]', Platnick to DMW 2018, pers. comm.).

For taxonomy, 'what in general, then, is to count as evidence of this relationship (of homology and taxon)? The answer is obvious, and we have known it ever since Aristotle. Parts of organisms are the things that resemble: the things that Aristotle and Geoffroy called analogues, the things that Owen called homologues, the things that Hennig called synapomorphies. The parts are the evidence: *parts of organisms* as *evidence* of *relationship*' (Nelson, 2011, p. 140).

## MONOGRAPHS, 'INTEGRATIVE TAXONOMY', AND CLASSIFICATION

There are now a number of revised approaches to monographic studies, many of which have incorporated some of the ideals of 'Integrative Taxonomy', or at least a nod towards that ideal (e.g., 'A taxonomic monograph of *Ipomoea* integrated across phylogenetic scales', Muñoz-Rodríguez et al., 2019). Monographs are indeed seen as comprehensive: they '[…] include species names, descriptions, data on distribution and habitat, taxonomic nomenclature, specimens consulted, maps, photographs, and illustrations. Various other data types may be included, such as a Tree of Life (phylogeny) and data pertaining to the genome, cytology, micromorphology, conservation, extinction risk, georeferenced localities, and adaptive traits of the species' (Olwen et al., 2021, p. 434). For any modern monograph, 'A complete phylogeny prepared in advance expedites the subsequent taxonomic revision and data synthesis needed to produce an integrated monograph that builds on multiple data sources, as well as providing the evolutionary framework for research beyond the monograph' (Olwen et al., 2021, p. 435) and 'Phylogenetic trees estimated from DNA sequences are the framework for integrated monography' (Olwen et al., 2021, p. 435, also Muñoz-Rodríguez et al., 2019, and in a different context, Moonlight et al., 2024).

Here, once again, we may be in the arena of interpretation, even that of semantics. DNA data, when analysed, will produce a tree (or trees) or, if you will, a cladogram (or cladograms) – whatever they are called, they are simply diagrams of relationships. One might interpret this diagram of relationships as a phylogeny, which is fine – but, strictly speaking, it is a classification, albeit a classification without names: it is a *hierarchical classification*. It is a phylogeny by interpretation. Classification, the ordering of nature, a subject Wallace tangentially addressed a long time ago, 'though often despised as unscientific, happened to be *the only one* which would lead us towards a solution of the problem of species' (Wallace, 1908, emphasis in the original).

Leaving this aside for the moment, monographic studies, modern or otherwise, almost exclusively focus on species and species-level diversity, with modern monographs linking species diversity to the phylogenetic framework as if one (species) is found in a different way, or at a different time, to the other (phylogeny). But consider, again, the role of classification and its hierarchy of species, genera, families, etc., rather than just the species: 'Controversy springs from a reductionist view of biodiversity as a species-, or population-, level phenomenon, as if species, or populations, were the only units of life and evolution. Why not consider taxa in general as units, and embrace the full measure of biodiversity?' (Nelson and Ladiges, 1990, p. 559). To consider 'taxa in general' is to consider classification as a prior requirement. Consider again the notion of 'heterobathmy of characters' discussed above. Inspection of any particular specimen for its characters will reveal homologues of the genus, of the family, of the order, and so on. And recall Platnick's words: 'Every synapomorphic resemblance among organisms fits at some node on the cladogram connecting them all, and it fits at only one such node. Any resemblances that do not fit in this way are just not synapomorphic' (Platnick, 2016, p. xii). The result is a classification, however primitive or unresolved it may appear to be.

Thus, we suggest the opposite of what is now becoming popular: 'Phylogenetic trees estimated from DNA sequences are the framework for integrated monography' (Olwen et al., 2021, p. 435). This can be re-stated: 'Classifications from DNA sequences can be the framework for integrated monography'. Of course, it seems obvious that monographs should indeed be based upon a classification of some sort. But we see it as preferably to first derive the classification from morphological data, however unresolved that classification may be. The advantage of this approach is that when new specimens are encountered, from the field, the herbarium, or any collection of whatever kind, knowledge of what characters fit at what taxonomic level (rank) can guide what we already know and place the unknown specimen in the classification at whatever level is appropriate; species diversity need not be the endpoint required. DNA data, as it lacks any specific indication of what characters support each node on the tree, is a secondary, but immensely useful, consideration, one that offers an additional level of congruence and, possibly, an extra dimension on the nodes resolved.

## DISCUSSION

Systematics is not a series of techniques that, given copious amounts of data, when applied will eventually yield the truth, whatever that truth might be. Simply put, systematics is a scientific endeavor that deals with available *evidence* (homologues) and *conclusions* derived from that evidence (homology, taxa, at whatever rank). Systematics has but two parameters, if you will: *synapomorphy* and *heterobathmy*, both known long before the 'cladistic episode' in systematics, even if the meaning of the concept has developed. The results are classifications. Classifications represent schemes of relationship.

In a review of *Animal Evolution: Genomes, Fossils, and Trees* (Telford and Littlewood, 2009), another account of the potential for data integration ('Modern approaches to the study of animal evolution *integrate* palaeontology, evo-devo, phylogenetics, and data from genomes each in the pursuit of a greater understanding of homology as a means of revealing patterns and processes through time and across multiple levels of biological organization', our italics), Joel Cracraft made the following remarks: 'With the rise of Tree of Life research, phylogenetic trees have become ubiquitous, which has resulted in a cottage industry of investigators taking their favorite tree and optimizing their data on it'. He continued: 'This certainly has led to new insights about the history of structural/functional systems, but it is more *using, than integrating*, something akin to a systematist optimizing morphology on a tree built from gene sequences and then making inferences about morphological evolution [e.g., Jeiter and Smets, 2023]. There is nothing wrong with this, but as a healthy literature on this attests, it is not the same as combining ("integrating"?) the data and seeing how they synergistically can be brought to bear on a problem' (Cracraft, 2011, p. 543, italics ours). As far as we understand, that problem is a new version of an old one: is natural classification possible? – and if so, what is its relevance to biology in general?

Over the years, from Linnaeus onward (Larson, 1967, 1971), natural classification has been variously known: comparative biology, taxonomy, systematics, phylogenetics, phylogenomics, and now 'Integrated Taxonomy' – perhaps there is yet more to come (Krell, Chapter 4). One might, with some justification, add 'cladistics' to this list ('[…] the next fashionable movement in systematics, which I suppose, alas, is cladistics', Nelson, 1978a, p. 112). Although some have proposed retaining a separation between taxonomy and 'phylogeny reconstruction' ('Two distinct research agendas in systematics', Wortley et al., 2002) using data 'kinds' to separate the two subjects: 'A key difference between taxonomy and phylogeny reconstruction lies in the data used. DNA sequences and gross morphology currently predominate in phylogeny reconstruction and taxonomy respectively' (Wortley et al., 2002, pp. 341–342, most of their arguments are political rather than scientific) – maybe those two research agendas have now been bought together via 'integrative taxonomy' (Muñoz-Rodríguez et al., 2019). Still, the notion of integration (and pluralism) applied to many facets of biology is rife (Donoghue and Moore, 2003; Mitchell, 2002; Mitchell and Dietrich, 2006; Rozanov et al., 2010; Pavlinova, 2020; Favret, 2024).

The desire for an integrative approach, one that can integrates data of all sorts is an understandable one, its quest, like many others in biology, has a long tradition, perhaps going back to Agassiz' threefold parallelism (Agassiz, 1857, 1859), given an evolutionary perspective by Haeckel ('threefold parallelism between the embryological, the systematic, and the paleontological development of organisms', Haeckel, 1864, p. 29, Haeckel 1866; for recent commentary see Olsson et al., 2017, the source of the translated text above, and Werneburg et al., 2022). Darwin was impressed: "Professor Häckel in his '*Generelle Morphologie*' and in other works, has recently brought his great knowledge and abilities to bear on what he calls phylogeny, or the lines of descent of all organic beings. […] He has thus boldly made a great beginning, and shows us how classification will in the future be treated" (Darwin, 1876, p. 381) – neglecting the fact that Agassiz had done so earlier. Maybe the three-fold parallelism was in need of a re-fresh within a systematics framework (cf., Nelson, 1978b; Williams and Ebach, 2004).

It might seem too much to some to suggest that biodiversity (biological diversity, National Academy of Sciences, 1988) – which effortlessly mutated into *phylodiversity* (phylogenetic diversity, Srivastava et al., 2012; Kline et al., 2019) is yet another parameter. And yet, '[…] what beyond biogeography is "biodiversity" about? (Nelson and Ladiges, 1990, p. 559; Mattox, 2011, p. 54; Ray, 1996) – 'Although Agassiz' concept of parallelism is usually cited as "threefold parallelism" […], it is instead a "fourfold parallelism" […]. The various authors citing the concept as "threefold" omit Agassiz' biogeographic component [He] clearly includes "geographical distribution" as a coequal to "structure", "embryonic growth", and "geologic succession"'(Grande and Bemis, 1998, p. 647, ellipses replace citations). And biogeography is yet another study in classification (Ebach, 2015).

For systematics relevant data are homologues (evidence) and there are '[…] some homologs (organisms and their parts) are ontogenetic, and other homologs (monophyletic taxa and their parts) are phylogenetic […] Molecular homologs constitute a third class of homologs […] and spatial (geographic) homologs […] a fourth' (Nelson, 1989, p. 279).

Finally, and perhaps the main point of this chapter – and something strenuously promoted by Platnick throughout his career (e.g., Platnick and Nelson, 1982; Platnick and Rosen, 1987; Platnick 1989) – rather than situate cladistics within the post-1950s 'numerical taxonomy' paradigm, or the post-1960s 'phylogenetic reconstruction' paradigm, or even within the merger of those two paradigms in the post-models (post-modern?) of the 1980s–90s, it is best understood as a continuation of, or better still, a *reform* of, the 18th-century quest for a natural classification, one that might reflect accurately the data (the characters, whatever their source, whatever their quantity) upon which relationships (classifications) are discovered. Appropriate comparisons, instead, should be made with the efforts of Carl Linnaeus (1707–1778), Michel Adanson (1727–1806), Antoine Laurent de Jussieu (1748–1836), and Augustin Pyramus de Candolle (1778–1841), and other seekers of natural classification (Nelson, 1979) – and then consider what these classifications tell us about the world.

## NOTES

1 'Wygodzinsky was the first scientist at the American Museum (and one of the first in North America) to introduce, employ, and argue for the efficacy of Hennig's methods of phylogenetics systematics' (Schuh and Herman, 1988, p. 229).
2 Goloboff has written that previously we referred to 'parsimony analysis' 'incorrectly [as] "Wagner parsimony"' (Goloboff, 2022, p. 72). The point is well taken (Goloboff, 2015) but it might still be confusing to refer to 'parsimony analysis' when there have been very different versions of parsimony; that is different algorithms based on, or claimed to be, parsimonious – it might even be argued that *all* methods implement one or another aspect of parsimony. To assist, one might refer to the 2nd edition of *Cladistics: Theory and Practice of Parsimony Analysis* (Kitching et al., 1998) and the comments on that book's evolution from its 1st edition in Williams and Ebach (2020).

## REFERENCES


Agassiz, L. (1857) *Contributions to the Natural History of the United States of America. First Monograph. In Three Parts. I. Essay One. II. North American Testudinata. III. Embryology of the Turtle; with Thirty-four Plates.* Volume I. Little, Brown and Company, Boston.

Agassiz, L. (1859) *An Essay on Classification.* Longman, Brown, Green, Longmans, and Roberts, London.

Ashlock, P.D. (1974) The uses of cladistics. *Annual Review of Ecology and Systematics*, 5: 81–99, https://doi.org/10.1146/annurev.es.05.110174.000501

Bell, M.A. and Gurevitch, J. (2012) Robert Rueven Sokal 1926–2012. *The Bulletin of the Ecological Society of America*, 93: 187–190, https://doi.org/10.1890/0012-9623-93.3.187

Byers, G.W. (1989) Peter D. Ashlock 1929–1989. *Bulletin of the Entomological Society of America*, 35: 70, https://doi.org/10.1093/besa/35.2.70

Carolin, R. (1984) Mitchell's theorem. *Australian Systematic Botany Society Newsletter*, 40: 1–9.

Cloutier, R. (2002) Hans-Peter Schultze, a great paleoichthyologist for whom work is synonymous to enjoyment. *Mitteilungen Museum für Naturkunde Berlin, Geowiss. Reihe, Band*, 5: 5–17, https://doi.org/10.1002/mmng.20020050102

Cracraft, J. (2011) Animal evolution all over the place, or what is integrative biology? *Evolution and Education Outreach*, 4, 542–543, https://doi.org/10.1007/s12052-011-0352-3

Crother, B.I. and Parenti, L.R. (eds) (2017) *Assumptions Inhibiting Progress in Comparative Biology.* CRC Press, Taylor and Francis Group, Boca Raton.

Darwin, C. (1872) *The Origin of Species by Means of Natural Selection, or the Preservation of Favoured Races in the Struggle for Life.* 6th edition. John Murray, London.

Dayrat, B. (2005) Towards integrative taxonomy. *Biological Journal of the Linnean Society*, 85: 407–415, https://doi.org/10.1111/j.1095-8312.2005.00503.x
de Santis, M.D. (2020) Scientific explanation and systematics. *Systematics and Biodiversity*, 19(3): 312–321, https://doi.org/10.1080/14772000.2020.1844339
de Santis, M.D. (2024) Homoplasy as an evolutionary process: An optimistic view on the recurrence of similarity in evolution. *Biological Theory*, 19: 267–278, https://doi.org/10.1007/s13752-024-00470-8
Donoghue, M.J. and Moore, B.R. (2003) Toward an integrative historical biogeography. *Integrative and Comparative Biology*, 43: 261–270, https://doi.org/10.1093/icb/43.2.261
Ebach, M.C. (2015) *Origins of Biogeography: The Role of Biological Classification in Early Plant and Animal Geography*. Springer, Dordrecht. 173 pp.
Ebach, M.C. and Williams, D.M. (2011) Book Review [Systematics and Biogeography: Cladistics and Vicariance. – Gareth Nelson and Norman I. Platnick. Columbia University Press, 1981. Xi+567 pp. ISBN 0-231-04574-3 (hardback). Electronic version 2009: http://www.ucpress.edu/books/series/spsy.php]. *Systematic Biology*, 59: 612–614.
Estabrook, G.F. (1972) Cladistic methodology: A discussion of the theoretical basis for the induction of evolutionary history. *Annual Review of Ecology and Systematics*, 3: 427–456, https://doi.org/10.1146/annurev.es.03.110172.002235
Farris, J.S. (1970) Methods for computing Wagner trees. *Systematic Zoology*, 19: 83–92, https://doi.org/10.2307/2412028
Farris, J.S. (1979) The information content of the phylogenetic System. *Systematic Zoology*, 28: 483–519, https://doi.org/10.2307/2412562
Favret, C. (2024) The 5 'D's of taxonomy: A user's guide. *The Quarterly Review of Biology*, 99, 131–156, https://doi.org/10.1086/732044
Goloboff, P. and Arias, J.S. (2019) Likelihood approximations of implied weights parsimony can be selected over the Mk model by the Akaike information criterion. *Cladistics*, 35: 695–716, https://doi.org/10.1111/cla.12380
Goloboff, P., Torres, A. and Arias, J. (2017) Weighted parsimony outperforms other methods of phylogenetic inference under models appropriate for morphology. *Cladistics*, 34: 407–437, https://doi.org/10.1111/cla.12205
Goloboff, P.A. (2015) Computer science and parsimony: A reappraisal, with discussion of methods for poorly structured datasets. *Cladistics*, 31: 210–225, https://doi.org/10.1111/cla.12082
Goloboff, P.A. (2022) *From Observations to Optimal Phylogenetic Trees: Phylogenetic Analysis of Morphological Data*. Volume 1, Taylor and Francis, CRC Press, Boca Raton.
Goloboff, P.A. and Wilkinson, M. (2018) On defining a unique phylogenetic tree with homoplastic characters. *Molecular Phylogenetics and Evolution*, 122: 95–101, https://doi.org/10.1016/j.ympev.2018.01.020
Grande, L. and Bemis, W.E. (1998) A comprehensive phylogenetic study of Amiid fishes (Amiidae) based on comparative skeletal anatomy. An empirical search for interconnected patterns of natural history. *Journal of Vertebrate Paleontology*, 18(1), Memoir 4: 1–696, https://doi.org/10.1080/02724634.1998.10011114
Groves, C.E. (1993) [Review of] *Principles of Systematic Zoology. Second Edition. Journal of Human Evolution*, 24: 251–253, https://doi.org/10.1006/jhev.1993.1020
Haeckel, E. (1864) *Über die Entwickelungstheorie Darwins. Vortrag auf der 38. Versammlung Deutscher Naturforscher und Ärzte in Stettin am 19. September 1863. Amtlicher Bericht über die acht und dreissigste Versammlung deutscher Naturforscher und Ärzte in Stettin im September 1863*. Hg. von den Geschäftsführern derselben Dr. C. A. Dohrn und Dr. Behm, H. Essenland's Buchdruckerei, Stettin, pp. 17–30.
Haeckel, E. (1866) *Generelle Morphologie der Organismen: Allgemeine Grundzüge der Organischen Formen-Wissenschaft, Mechanisch Begründet Durch Die Von C. Darwin Reformirte Decendenz-Theorie*. 2 volumes. G. Reimer, Berlin.

Haffer, J. (2007) *Ornithology, Evolution, and Philosophy. The Life and Science of Ernst Mayr 1904–2005.* Springer-Verlag, Berlin Heidelberg.

Hennig, W. (1950) *Grundzüge einer Theorie der phylogenetischen Systematik.* Deutsche Zentralverlag, Berlin [reprinted 1980 Otto Koeltz, Koenigstein].

Hennig, W. (1957) Systematik und Phylogenese, pp. 50–71. In *Bericht über die Hundertjahrfeier der Deutschen Entomologischen Gesellschaft, Berlin.* 30 September bis 5 Oktober 1956, (ed H. von Hannemann), Akademie-Verlag, Berlin.

Hennig, W. (1965) Phylogenetic systematics. *Annual Review of Entomology*, 10: 97–116, https://doi.org/10.1146/annurev.en.10.010165.000525

Hennig, W. (1966) *Phylogenetic Systematics.* University of Illinois Press, Urbana [Reprinted 1979, 1999].

Hilton, E.J. and Forey, P.L. (2005) Contributions of Walter G. Ridewood to systematic comparative anatomy, especially of the osteology of "lower" vertebrates. *Journal of Natural History*, 39: 641–655, https://doi.org/10.1080/00222930400001475

Hull, D.L. (1988) *Science as Process: An Evolutionary Account of the Social and Conceptual Development of Science.* University of Chicago Press, Chicago.

Janvier, P. (2003) Skeleton keys. *Nature*, 424: 493.

Jeiter, J. and Smets, E. (2023) Integrating comparative morphology and development into evolutionary research. *Taxon*, 72: 724–732, https://doi.org/10.1002/tax.12983

Jensen, R.J. (1993) The 25th International Numerical Taxonomy Conference. *Systematic Biology*, 42: 599–601, http://www.jstor.org/stable/2992496

Källersjö, M., Albert, V.A. and Farris, J.S. (1999) Homoplasy increases phylogenetic structure. *Cladistics*, 15: 91–93, https://doi.org/10.1111/j.1096-0031.1999.tb00400.x

Khakurel, B., Grigsby, C., Tran, T.D, Zariwala, J., Höhna, S. and Wright, A.M. (2024) The fundamental role of character coding in Bayesian morphological phylogenetics. *Systematic Biology*, 73: 861–871, https://doi.org/10.1093/sysbio/syae033

Kitching, I., Forey, P.L., Humphries, C.J. and Williams, D.M. (1998) *Cladistics: The Theory and Practice of Parsimony Analysis. Systematics Association Publications No. 11.* Oxford University Press, Oxford.

Kling, M.M., Mishler, B.D., Thornhill, A.H., Baldwin, B.G. and Ackerly, D.D. (2019) Facets of phylodiversity: Evolutionary diversification, divergence and survival as conservation targets. *Philosophical Transactions of the Royal Society B*, 374: 20170397, https://doi.org/10.1098/rstb.2017.0397

Kluge, A.G. (1971) Concepts and principles of morphologic and functional studies, pp. 3–41. In *Chordate Structure and Function* (ed A.J. Waterman). MacMillan, New York.

Kluge A.G. (1976) Phylogenetic relationships in the lizard family Pygopodidae: An evaluation of theory, methods and data. *Miscellaneous Publications of the Zoological Museum, University of Michigan*, 152: 1–72, https://hdl.handle.net/2027.42/56396

Kluge, A.G. (1977) Concepts and principles of morphologic and functional studies, pp. 3–41. In *Chordate structure and function*, 2nd ed. (ed A.G. Kluge). MacMillan, New York.

Kluge, A.G. (1989) A concern for evidence and a phylogenetic hypothesis of relationships among Epicrates (Boidae, Serpentes). *Systematic Zoology*, 38: 7–25, https://doi.org/10.1093/sysbio/38.1.7

Kluge, A.G. (2004) On total evidence: For the record. *Cladistics*, 20: 205–207, https://doi.org/10.1111/j.1096-0031.2004.00020.x

Kluge, A.G. and Farris, J.S. (1969) Quantitative phyletics and the evolution of Anurans. *Systematic Zoology*, 18: 1–32, https://doi.org/10.1093/sysbio/18.1.1

Larson, J. (1967) Linnaeus and the Natural Method. *Isis*, 58: 304–320, https://doi.org/10.1086/350265

Larson, J. (1971) *Reason and Experience; The Representation of Natural Order in the Work of Carl von Linne.* University of California Press, Berkeley.

Lewens, T. (2012) Pheneticism reconsidered. *Biology and Philosophy*, 27: 159–177, https://doi.org/10.1007/s10539-011-9302-2

Lewis, P.O. (2001) A likelihood approach to estimating phylogeny from discrete morphological character data. *Systematic Biology*, 50: 913–925, https://doi.org/10.1080/106351501753462876

Love, A.C. (2007) Functional homology and homology of function: Biological concepts and philosophical consequences. *Biology and Philosophy*, 22: 691–708, https://doi.org/10.1007/s10539-007-9093-7

Mabee, P.M., Balhoff, J.P., Dahdul, W.M., Lapp, H., Mungall, C.J. and Vision, T.J. (2020) A logical model of homology for comparative biology. *Systematic Biology*, 69, 345–362, https://doi.org/10.1093/sysbio/syz067

Mattox, G.M. (2011) Gareth Nelson: Os desafios atuais e o futuro da biogeografia. Gareth Nelson: Current challenges and the future of biogeography. *Revista da Biologia*, 7: 49–55, https://doi.org/10.7594/revbio.07.09

Mayr, E. and Ashlock, P. (1991) *Principles of Systematic Zoology*. McGraw-Hill, New York.

Michener, C.D. (2012) In Memoriam: Robert R. Sokal (1926–2012). *Journal of the Kansas Entomological Society*, 85: 173–175, http://dx.doi.org/10.2317/JKES120710.1

Michener, C.D. and Sokal, R.R. (1957) A quantitative approach to a problem in classification. *Evolution*, 11: 130–162, https://doi.org/10.1111/j.1558-5646.1957.tb02884.x

Michevich, M.F. (1982) Transformation series analysis. *Systematic Zoology*, 31: 461–478, https://doi.org/10.2307/2413369

Mitchell, P.C. (1901) On the intestinal tract of birds; with remarks on the valuation and nomenclature of zoological characters. *Transactions of the Linnean Society of London, Zoology*, Ser. 2, 8: 173–275, https://doi.org/10.1111/j.1096-3642.1901.tb00477.x

Mitchell, S.D. (2002) Integrative pluralism. *Biology and Philosophy*, 17: 55–70, https://doi.org/10.1023/A:1012990030867

Mitchell, S.D. and Dietrich, M.R. (2006) Integration without unification: An argument for pluralism in the biological sciences. *American Naturalist*, 168: 573–579, https://doi.org/10.1086/509050

Moonlight, P.W., Baldaszti, L., Cardoso, D., Elliott, A., Särkinen, T. and Knapp, S. (2024) Twenty years of big plant genera. *Proceedings of the Royal Society, B*, 291: 20240702, https://doi.org/10.1098/rspb.2024.0702

Morrone, J.J. and Carpenter, J.M. (1994) In search of a method for cladistic biogeography: An empirical comparison of component analysis, Brooks Parsimony Analysis, and three-area statements. *Cladistics*, 10: 99–153, https://doi.org/10.1111/j.1096-0031.1994.tb00169.x

Muñoz-Rodríguez, P., Carruthers, T., Wood, J.R.I. *et al.* (2019) A taxonomic monograph of *Ipomoea* integrated across phylogenetic scales. *Nature Plants*, 5: 1136–1144, https://doi.org/10.1038/s41477-019-0535-4

National Academy of Sciences (1988) *Biodiversity*. The National Academies Press, Washington, DC, https://doi.org/10.17226/989

Nelson, G. (1978a) Professor Michener on phenetics – old and new. *Systematic Zoology*, 27: 104–112, https://doi.org/10.2307/2412820

Nelson, G. (1978b) Ontogeny, phylogeny, paleontology and the biogenetic law. *Systematic Zoology*, 27: 324–345, https://doi.org/10.2307/2412883

Nelson, G. (1979) Cladistic analysis and synthesis: Principles and definitions, with a historical note on Adanson's *Familles des Plantes*. *Systematic Zoology*, 28: 1–21, https://doi.org/10.1093/sysbio/28.1.1

Nelson, G. (1989) Cladistics and evolutionary models. *Cladistics*, 5: 275–289, https://doi.org/10.1111/j.1096-0031.1989.tb00490.x

Nelson, G. (1991) 'Principles of systematic zoology: Second edition' (review of E. Mayr and P.D. Ashlock). *Cladistics*, 7: 305–307, https://doi.org/10.1111/j.1096-0031.1991.tb00040.x

Nelson, G.J. (2004) Cladistics: Its arrested development, pp. 127–147. In *Milestones in Systematics* (eds D.M. Williams and P.L. Forey), Systematics Association Special Volume 67. CRC Press, Boca Raton.

Nelson, G. (2011) Resemblance as evidence of ancestry. *Zootaxa*, 2946: 137–141, https://doi.org/10.11646/zootaxa.2946.1.18

Nelson, G.J. (2017) Donald Eric Rosen (1929–1986), pp. 5–13. In *Assumptions Inhibiting Progress in Comparative Biology* (eds B.I. Crother and L.R. Parenti). CRC Press, Boca Raton.

Nelson, G., Atz, J.W., Kallman, K.D. and Smith, C.L. (1987) Donn Eric Rosen, 1929–1986. *Copeia*, 1987(2): 541–547, http://www.jstor.org/stable/1445812

Nelson, G. and Ladiges, P.Y. (1990) Biodiversity and biogeography. *Journal of Biogeography*, 17: 559–560, http://www.jstor.org/stable/2845139

Nelson, G.J. and Ladiges, P.Y. (1991) Three-area statements: Standard assumptions for biogeographic analysis. *Systematic Zoology*, 40: 470–485, https://doi.org/10.2307/2992241

Nelson, G. and Ladiges, P.Y. ([1991] 1992) Three-area statements: Standard assumptions for biogeographic analysis. *Australian Systematic Botany* 4:41–58 [Amendment. *Australian Systematic Botany* 5:247].

Nelson, G.J. and Ladiges, P.Y. (1992) Information content and fractional weight of three-item statements. *Systematic Biology*, 41(4): 490–494, https://doi.org/10.2307/2992590

Nelson, G.J. and Ladiges, P.Y. (2009) Biogeography and the molecular dating game: A futile revival of phenetics? *Bulletin de la Société Géologique de France*, 180: 39–43, https://doi.org/10.2113/gssgfbull.180.1.39

Nelson, G.J. and Platnick, N.I. (1978) The perils of plesiomorphy: Widespread taxa, dispersal, and phenetic biogeography. *Systematic Zoology*, 27(4): 474–477, https://doi.org/10.2307/2412931

Nelson, G.J. and Platnick, N.I. (1981) *Systematics and Biogeography: Cladistics and Vicariance*. Columbia University Press, New York, xi+567 pp. [Extracts translated as: La biogéographie historique, de Linné à Darwin, In Systematics and Biogeography, Cladistics and Vicariance, Chapter 6, Biogeography History: Kind of questions, pp. 357–375, New York, Columbia University Press, in *Biosystema*, 7: 33–55, 1991].

Nelson, G.J. and Platnick, N.I. (1991) Three-taxon statements: A more precise use of parsimony? *Cladistics*, 7: 351–366, https://doi.org/10.1111/j.1096-0031.1991.tb00044.x

Nixon, K.C. and Carpenter, J.M. (1996a) On simultaneous analysis. *Cladistics*, 12: 221–241, https://doi.org/10.1006/clad.1996.0016

Nixon, K.C. and Carpenter, J.M. (1996b) On consensus, collapsibility, and clade concordance. *Cladistics*, 12: 305–321, https://doi.org/10.1111/j.1096-0031.1996.tb00017.x

Nixon, K.C. and Carpenter, J.M. (2012a) On homology. *Cladistics*, 28: 160–169, https://doi.org/10.1111/j.1096-0031.2011.00371.x

Nixon, K.C. and Carpenter, J.M. (2012b) More on homology. *Cladistics*, 28: 225–226, https://doi.org/10.1111/j.1096-0031.2011.00388.x

Olsson, L., Levit, G.S. and Hoßfeld, U. (2017) The "Biogenetic Law" in zoology: From Ernst Haeckel's formulation to current approaches. *Theory Biosciences*, 136: 19–29, https://doi.org/10.1007/s12064-017-0243-4

Olwen, M.G., Pérez-Escobar, O.A., Lucas, E.J., Vorontsova, M.S., Lewis, G.P., Walker, B.E., Lohmann, L.G., Knapp, S., Wilkie, P., Sarkinen, T., Darbyshire, I., Nic Lughadha, E., Monro, A., Woudstra, Y., Demissew, S., Muasya, A.M., Díaz, S., Baker, W.J. and Antonelli, A. (2021) Botanical monography in the Anthropocene. *Trends in Plant Science*, 26: 433–441, https://doi.org/10.1016/j.tplants.2020.12.018

Padial, J.M., Miralles, A., De la Riva, I. and Vences, M. (2010) The integrative future of taxonomy. *Frontiers in Zoology*, **7**, 16, https://doi.org/10.1186/1742-9994-7-16

Patterson, C. (1982) Morphological characters and homology, pp. 21–74. In *Morphological Characters and Homology* (eds K.A. Joysey and A.E. Friday). Academic Press, London.

Pavlinov, I.Y. (2020) Multiplicity of research programs in the biological systematics: A case for scientific pluralism. *Philosophies*, 5(2): 7, https://doi.org/10.3390/philosophies5020007

Peterson, A.S. (2020) Aristotelian explanation and homology in biology. *Ancient Philosophy Today*, 2(1):45–69, https://doi.org/10.3366/anph.2020.0024

Platnick, N.I. (1975) A revision of the Holarctic spider genus *Callilepis* (Araneae, Gnaphosidae). *American Museum Novitates*, 2573: 1–32.

Platnick, N.I. (1976) Drifting spiders or continents?: Vicariance biogeography of the spider subfamily Laroniinae (Araneae: Gnaphosidae). *Systematic Zoology*, 25: 101–109, https://doi.org/10.2307/2412735

Platnick, N.I. (1989) Cladistics and phylogenetic analysis today, pp. 17–24. In *The Hierarchy of Life* (eds B. Fernholm, K. Bremer and H. Jörnvall). Elsevier, Amsterdam.

Platnick, N.I. (1991) Patterns of biodiversity: Tropical vs temperate. *Journal of Natural History*, 25: 1083–1088, https://doi.org/10.1080/00222939100770701

Platnick, N.I. (2013) Reification, matrices, and the interrelationships of goblin spiders (Araneae, Oonopidae). *Zootaxa*, 3608(4): 278–280, https://doi.org/10.11646/zootaxa.3608.4.6

Platnick, N.I. (2016) Foreword: Willi Hennig and systematics: A personal view, pp. xi–xvi. In *The Future of Phylogenetic Systematics: The Legacy of Willi Hennig* (eds D.M. Williams, M. Schmitt and Q. Wheeler), Systematics Association Special Volume Series. Cambridge University Press, Cambridge.

Platnick, N.I. (2017) Foreword: Donn Rosen and inhibiting assumptions, pp. vii–ix. In *Assumptions Inhibiting Progress in Comparative Biology* (eds B.I. Crother and L.R. Parenti), CRC Press, Boca Raton.

Platnick, N.I. and Gaffney, E.S. (1977) Systematics: A Popperian perspective. *Systematic Zoology*, 26: 360–365, https://doi.org/10.2307/2412686

Platnick, N.I. and Gaffney, E.S. (1978) Evolutionary biology: A Popperian perspective. *Systematic Zoology*, 27: 137–141, https://doi.org/10.2307/2412832

Platnick, N.I. and Nelson, G.J. (1982) The purposes of biological classification. *Proceedings of the 1978 Biennial Meeting of the Philosophy of Science Association*, 2: 117–129, https://doi.org/10.1086/psaprocbienmeetp.1978.2.192465

Platnick, N.I. and Rosen, D.E. (1987) Popper and evolutionary novelties. *History and Author: Philosophy of the Life Sciences*, 9: 5–16, http://www.jstor.org/stable/23328766.

Platnick, N.I. and Shadab, M.U. (1974) A revision of the spider family Stenochilidae (Arachnida, Araneae). *American Museum Novitates*, 2556: 1–14.

Poe, S. and Wiens, J.J. (2000) Character selection and the methodology of morphological phylogenetics, pp. 20–36. In *Phylogenetic Analysis of Morphological Data* (ed J.J. Wiens). Smithsonian Institution Press, Washington, D.C.

Prendini, L. (2020a) Norman Ira Platnick (1951–2020). *Arachnology*, 18(5): 507–516, https://doi.org/10.13156/arac.2020.18.5.507

Prendini, L. (2020b) Biography of Norman I. Platnick. *Entomologica Americana*, 126(1–4): 1–21, https://doi.org/10.1664/1947-5136-126.1.1

Prendini, L. (2020c) Bibliography of Norman I. Platnick. *Entomologica Americana*, 126(1–4): 22–57, https://doi.org/10.1664/1947-5136-126.1.22

Prendini, L. (2020d) Arachnida (Araneae, Ricinulei) described by Norman I. Platnick. *Entomologica Americana*, 126(1–4): 58–100, https://doi.org/10.1664/1947-5136-126.1.58

Prendini, L. (2020e) Taxa dedicated to Norman I. Platnick. *Entomologica Americana*, 126(1–4): 101–107, https://doi.org/10.1664/1947-5136-126.1.101

Ray, G.C. (1996) Biodiversity is biogeography: Implications for conservation. *Oceanography* 9(1): 50–59, https://doi.org/10.5670/oceanog.1996.27

Rieppel, O. (2003) Popper and systematics. *Systematic Biology*, 52: 259–271, https://doi.org/10.1080/10635150390192762

Rieppel, O. (2004) What happens when the language of science threatens to break down in systematics – A Popperian perspective, pp. 57–100. In *Milestones in Systematics* (eds D. Williams and P. Forey). *Milestones in Systematics.* CRC Press, Boca Raton.

Rieppel, O. (2008) Re-writing Popper's philosophy of science for systematics. *History and Philosophy of the Life Sciences*, 30(3/4): 293–316, http://www.jstor.org/stable/23334453

Rieppel, O. (2023) "Living fossils" and the mosaic evolution of characters. *Frontiers in Ecology and Evolution*, 11: 1119418, https://doi.org/10.3389/fevo.2023.1119418

Rieppel, O. (2013) Homology: A philosophical and biological perspective. In *Handbook of Paleoanthropology* (eds W. Henke and I. Tattersall). Springer, Berlin, Heidelberg, https://doi.org/10.1007/978-3-642-27800-6_6-5

Ríos-Tamayo, D., Engelbrecht, I. and Goloboff, P.A. (2021) A revision of the genus *Hermacha* Simon, 1889 (Mygalomorphae: Entypesidae) in Southern Africa with revalidation of *Hermachola* Hewitt, 1915 and *Brachytheliscus* Pocock, 1902. *Argentina. American Museum Novitates*, 2021(3977): 1–80, https://doi.org/10.1206/3977.1

Ross, H.H. (1964) [Review of: *Principles of Numerical Taxonomy.* Robert R. Sokal; Peter H. A. Sneath; W. H. Freeman]. *Systematic Zoology*, 13: 106–108, https://doi.org/10.2307/2411834

Rozanov, A., Lopatin, A. and Rozhnov, S. (2010) On the road towards integrative paleontology. *Herald of the Russian Academy of Science*, 80: 466–473, https://doi.org/10.1134/S1019331610050072

Schuh, R.T. and Herman, L.H. (1988) Biography and bibliography. Petr Wolfgang Wygodzinsky (1967–1987). *Journal of the New York Entomological Society*, 96: 227–244.

Schultze, H.-P. (1981) Hennig und der Ursprung der Tetrapoda. *Palaontologische Zeitschrift*, 55: 71–86, https://doi.org/10.1007/BF02986038

Scotland, R.W., Olmstead, R.G. and Bennett, J.R. (2003) Phylogeny reconstruction: The role of morphology. *Systematic Biology*, 52: 539–548, http://www.jstor.org/stable/3651141

Simpson, G.G. (1975) Recent advances in methods of phylogenetic inference, pp. 3–19. In *Phylogeny of the Primates* (eds W.P. Luckett and F. Szalay). Plenum Press, New York.

Slater, J.A. and Polhemus, J.T. (1990) Obituary: Peter D. Ashlock 1929–1989. *Journal of The New York Entomological Society*, 98: 113–122.

Sneath, P.H.A. and Sokal, R.R. (1973) *Numerical Taxonomy.* W.H. Freeman, San Francisco.

Sokal, R.R. and Sneath, P.H.A. (1963) *Principles of Numerical Taxonomy.* W.H. Freeman, San Francisco.

Spade, D.A. (2020) An extended model for phylogenetic maximum likelihood based on discrete morphological characters. *Statistical Applications in Genetics and Molecular Biology*, 19: 20190029, https://doi.org/10.1515/sagmb-2019-0029

Srivastava, D.S., Cadotte, M.W., MacDonald, A.A.M., Marushia, R.G. and Mirotchnick, N. (2012) Phylogenetic diversity and the functioning of ecosystems. *Ecology Letters*, 15: 637–648, https://doi.org/10.1111/j.1461-0248.2012.01795.x

Stevens, P.F. (2002) Money, morphology and molecules. Some comments on Wortley, Bennett and Scotland (2002), Taxonomy and phylogeny reconstruction: Two distinct research agendas in systematics. (see p. 335 of this issue). *Edinburgh Journal of Botany*, 59(3): 451–457, https://doi.org/10.1017/S0960428602000288

Takhtajan, A. (1959) *Die Evolution der Angiospermen. VEB Gustav Fischer*, Jena, Germany.

Takhtajan, A. (1991) *Evolutionary Trends in Flowering Plants.* Columbia University Press, New York.

Tamborini, M. (2022) *The Architecture of Evolution: The Science of Form in Twentieth-Century Evolutionary Biology.* University of Pittsburgh Press, Pittsburgh.

Telford, M.J. and Littlewood, D.T.J. (2009) *Animal Evolution: Genomes, Fossils, and Trees.* Oxford University Press, Oxford.

Unzicker, J.D. and Wallace, B.J. (1979) Herbert Holdsworth Ross 1908–1978. *Journal of Economic Entomology*, 72(2): 309–310, https://doi.org/10.1093/jee/72.2.309

Vinarski, M.V. (2020) Roots of the taxonomic impediment: Is the 'integrativeness" a remedy? *Integrative Zoology*, 15: 2–15, https://doi.org/10.1111/1749-4877.12393

Vinarski, M.V. (2022) Pattern without process: Eugen Smirnov and the earliest project of numerical taxonomy (1923–1938). *Journal of the History of Biology*, 55(3): 559–583, https://doi.org/10.1007/s10739-022-09688-3

Wallace, A.R. (1908) [Acceptance Speech on Receiving the Darwin-Wallace Medal] In: *The Darwin-Wallace celebration held on Thursday, 1st July, 1908 by the Linnean society of London.* Printed for the Linnean Society, London.

Werneburg, I., Hossfeld, U. and Levit, G.S. (2022) Discovery of rare lecture notes from 1866 provides exceptional insights into the conceptualization and visualization of paleontology by Ernst Haeckel. *Vertebrate Zoology*, 72: 577–597, https://doi.org/10.3897/vz.72.e84983

West-Eberhard, M.J. (2009) Michener, Charles D. (b.1918), pp. 732–733. In *Evolution: The First Four Billion Years* (eds M. Ruse and J. Travis). Harvard University Press, Cambridge, USA.

West-Eberhard, M.J. and Engel, M.S. (2017) Charles D. Michener, 1918–2015. *Biographical Memoirs of the National Academy of Sciences*, 2017: 1–23.

Wheeler, Q.D. (2008) Introductory: Towards the New Taxonomy, pp. 1–17. In *The New Taxonomy* (ed Q.D. Wheeler). Taylor and Francis, Boca Raton.

Wiens, J.J. (2004) The role of morphological data in phylogeny reconstruction. *Systematic Biology*, 53: 653–661, https://doi.org/10.1080/10635150490472959

Will, K.W., Mishler, B.D. and Wheeler, Q.D. (2005) The perils of DNA barcoding and the need for integrative taxonomy. *Systematic Biology*, 54: 844–851, https://doi.org/10.1080/10635150500354878

Williams, D.M. and Ebach, M.C. (2004) The reform of palaeontology and the rise of biogeography – 25 years after 'Ontogeny, Phylogeny, Paleontology and the Biogenetic law' (Nelson 1978). *Journal of Biogeography*, 31: 685–712, http://onlinelibrary.wiley.com/doi/10.1111/j.1365-2699.2004.01063.x/pdf

Williams, D.M. and Ebach, M.C. (2008) *The Foundations of Systematics and Biogeography.* Springer Science, New York. [Preface by Professor Gareth Nelson].

Williams, D.M. and Ebach, M.C. (2020) *Cladistics: A Guide to Biological Classification.* Cambridge University Press, Cambridge (3rd edition of *Cladistics: A Practical Manual*)

Williams, D.M., Ebach, M.C. and Wheeler, Q.D. (2010) Beyond belief: The steady resurrection of phenetics, pp. 169–197. In *Beyond Cladistics* (eds D.M. Williams and S. Knapp). University of California Press, Berkeley, https://doi.org/10.1525/california/9780520267725.003.0010

Williams, R.F.G. (2009) Review article: Everyday sorrows are not mental disorders: The clash between psychiatry and western cultural habits. *Prometheus*, 27(1): 47–70, https://doi.org/10.1080/08109020802690942

Wortley, A. and Scotland, R. (2006) The effect of combining molecular and morphological data in published phylogenetic analyses. *Systematic Biology*, 55(4): 677–685, https://doi.org/10.1080/10635150600899798

Wortley, A., Bennett, J. and Scotland, R. (2002) Taxonomy and phylogeny reconstruction: Two distinct research agendas in systematics. *Edinburgh Journal of Botany*, 59(3): 335–349, http://dx.doi.org/10.1017/S0960428602000203

Yeates, D.K., Seago, A., Nelson, L., Cameron, S.L., Joseph, L. and Trueman, J.W.H. (2011) Integrative taxonomy, or iterative taxonomy? *Systematic Entomology*, 36: 209–217, https://doi.org/10.1111/j.1365-3113.2010.00558.x

# 3 Minimalist Species Descriptions

## *Are They the Answer? And if so, What Was the Question?*

*Gavin R. Broad*

## INTRODUCTION

How do we accelerate the enormous task of describing all of the world's species? If we are in a rush, against a ticking clock, then what is the minimum needed to partition life into diagnosable species? And how 'diagnosable' does a description need to be? In the era of Linnaean taxonomy, we have only described a small percentage of the planet's biota (Mora, 2011; Wiens, 2023), at a time when we are losing biodiversity at alarming rates (e.g., Díaz et al., 2019). This lack of knowledge, together with a shortfall in taxonomic expertise, has been named the 'taxonomic impediment' (Engel et al., 2021), and many papers have been written on how to overcome this impediment. Addressing the shortfall in taxonomic expertise has been the focus of various studies and official enquiries such as, in the UK, the House of Lords enquiries (House of Lords Science and Technology Committee, 2001–2002, 2002–2003, 2007–2008). This is the particular focus of Engel et al. (2021), who point out that the most serious impediment to taxonomy is a lack of skilled taxonomists. As there is no immediate sign of substantial increases in taxonomists or investment in taxonomy, many researchers have proposed initiatives to increase the rate at which species can be described. Understandably, these proposals have concentrated on methods and technology (e.g., Meierotto et al., 2019; Meier et al., 2024), which probably seem easier to implement than a reframing of the importance of taxonomy in the mindset of funders. Rather few studies have explicitly put these ideas to the test and described large numbers of species in a short space of time. Butcher et al. (2012) were among the first to label a rapid taxonomic revision as being conceptually different from a standard revision, when they coined the term 'turbo taxonomy', for a rapidly produced taxonomic revision using standard DNA barcodes (Cytochrome c oxidase 1 [*cox* 1, CO1]) and brief descriptions, concentrating on a few morphological characters, photographs and an identification key.

The more prominent examples of work on increasing the speed of mass species descriptions (e.g., Butcher et al., 2012, Meierotto et al., 2019) have involved descriptions of new species of parasitoid wasps (Hymenoptera). This assemblage of superfamilies has always been comparatively neglected. Yes, there are certainly a lot of beetles out there (Erwin, 1982) but the species richness of Hymenoptera, which are


DOI: 10.1201/b22822-3

mostly parasitoid wasps, almost certainly far exceeds Coleoptera, while the number of described species lags far behind (Forbes et al., 2018). In many parts of the world, hardly anybody has studied parasitoid wasps, despite their species richness and ecological importance being widely recognized (e.g., Hanson and Gauld, 1995). Butcher et al. (2012) recognized the need to at least partly describe a species-rich fauna, to raise one particular genus of parasitoid wasps from a metaphorical shrug of the shoulders to a big number. They described 179 species of *Aleiodes* (braconid wasps which mummify host caterpillars) where previously only three species were known from Thailand. True of Thailand, this high proportion of undescribed species is also the case for large areas of the world. The methods of Butcher et al. (2012) involved DNA barcoding as many specimens as possible, concentrating on recently collected specimens and not attempting a comprehensive revision of all available material. This work was in many respects 'traditional', in that the morphology of each species was (briefly) described and an identification key included. At the time, this 'turbo taxonomy' garnered a reasonable amount of press and discussion amongst taxonomists as to whether the speed of species description was much higher than in many 'unbranded', or traditional, taxonomic revisions.

Some of the most significant work to describe a tropical fauna of parasitoid wasps, a very neglected group which comprises a large proportion of insect diversity (Shaw and Hochberg, 2001), has come through efforts to monograph the fauna of Costa Rica (e.g., Gauld, 1991). There was an expectation that, given the relatively small size of Costa Rica, it should be possible to describe the vast majority of insect species within a relatively short period. Whether or not that expectation was naïve, there has been a trend towards shorter species descriptions and more rapid descriptions of species, extrapolated towards a vanishing point by a series of publications on ichneumonoid wasps resulting from efforts to describe the fauna of the Area de Conservación Guanacaste in Costa Rica (Meierotto et al., 2019; Sharkey et al., 2021a 2021b, 2023). These minimalist descriptions aim to satisfy the requirements of the International Code of Zoological Nomenclature (ICZN, 1999) and provide names to species, essentially as defined by DNA barcodes (Hebert et al., 2003). What does a minimalist species description look like? The description of *Mesochorus unocincoocho* Sharkey et al. (2023), the final species to be described in Sharkey et al. (2023), is reproduced here in full (Figure 3.1).

Meierotto et al. (2019) described their methods as 'revolutionary'. Are the revolutionary claims backed up by the data? Is there anything which sets apart minimalist species descriptions other than the brevity of the actual description and diagnosis? In this chapter, I am interested in (a) exploring whether minimalist species descriptions achieve what their proponents claim; (b) testing the efficiency of these methods against other taxonomic methods; (c) asking whether this is taxonomy; and (d) whether anybody needs this. Or, from another perspective, is life as a wasp taxonomist better now?

## METHODS

The proponents of minimalist species descriptions have been describing parasitoid wasps from Costa Rica. Therefore, to test whether minimalist descriptions result in species being described more rapidly, I have compared their species description

### *Mesochorus unocincoocho*

urn:lsid:zoobank.org:act:368E3662-B144-4FC8-8458-A9F80960BB55

**Diagnostics:** Fig. 187.
Holotype barcode.
TTTTATATTTTATTTTTGGCTTATGAGCAGGAATAATTGGGTCTTCAATAAGATTAATTATTCGCCTAGA
ATTAGGAAACCCAGGCTATTTAATCAATAATGATCAAATTTATAATTCATTTGTGACTGCTCATGCTTTTATTA
TAATTTTTTTTATAGTTATACCTATCATAATTGGTGGATTTGGAAATTGACTAATTCCTTTAATAATTGGAGCTCCT
GATATAGCATTTCCACGAATAAATAATATAAGATTTTGATTATTACCTCCCTCTCTTTTTTTATTAATTTTAGGA
AGTATTATTCATAAAGGTGTTGGAACTGGTTGAACTGTTTACCCTCCTTTATCTTTAAATTCTAGTCATGAA
GGTATATCTGTAGATTTATCAATCTTTTCTTTACATCTAGCAGGAATATCCTCAATTATAGGAGCAATTAATTTTAT
CACAACTATCTTAAATATGCGATGTTTAGGCACATCTTTAGATCAAATATCATTATTTACTTGATCAATAAAAATTAC
TACAATTTTATTATTATTAGCAGTTCCTGTACTTGCAGGTGCCATCACAATATTATTAGCAGATCGTAATTTAAATACTT
CATTTTTTGATCCTGCAGGAGGGGGAGACCCTATTTTATATCAACATTTA

**BOLD data:** BIN: BOLD:AEJ1909.
**Holotype** ♀: BIOUG64937-G06, Área de Conservación Guanacaste, Guanacaste, Sector Pailas, Pailas Dos, PL12-9, 10.76000, -85.33400, 809 m, Malaise trap, 10/01/2020 (CNC). GenBank accession code OM237697.
**Holotype host data:** None.
**Other host data:** None.

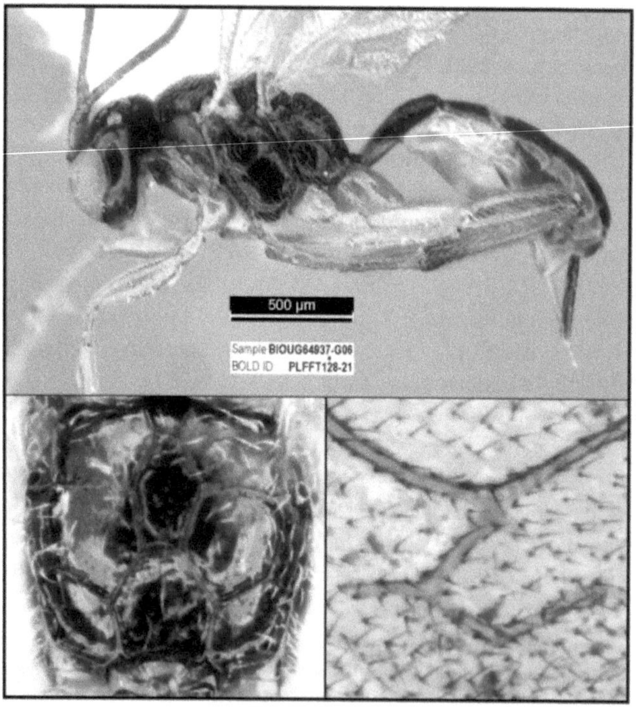


**Fig. 187.** *M. unocincoocho*, holotype female.

**FIGURE 3.1** Full description of Mesochorus *unocincoocho* Sharkey from Sharkey et al. (2023).

rates to other publications describing tropical parasitoid faunas, mostly wasps (Hymenoptera) from Costa Rica, with some exceptions. The exceptions are some papers describing parasitoid flies (Diptera: Tachinidae), resulting from the same caterpillar rearing programme as many of the wasps (Fleming et al., 2021, 2023), and Butcher et al. (2012), the 'turbo taxonomy' paper which described new species of *Aleiodes* wasps from Thailand. To compare the speed of description, data from type material (holotypes and paratypes) of species described from Costa Rica were gathered from 11 publications (Gauld, 1991; Gauld and Janzen, 2004; Butcher et al., 2012; Fernández-Triana et al., 2014a, 2014b; Meierotto et al., 2019; Fleming et al., 2021, 2023; Sharkey et al., 2021b; Noyes, 2023; Sharkey et al., 2023). The data gathered were the year of specimen collection and the year of publication. For publications dealing with the Costa Rican fauna (all except Butcher et al., 2012), the few non-Costa Rican specimens were excluded as the comparisons were between different approaches to broadly equivalent specimen bases. Occasionally, dates of collection were missing from some type of data in species descriptions, so these were excluded. For some publications (e.g., Noyes, 2023), there were simply too many specimens listed to database in the time available, so a large proportion was databased instead (e.g., 587 specimens from Noyes, 2023) in the expectation that these would be representative of the entire specimen base for the monograph. The interval between specimen collection and publication of the description was calculated for all 6,170 specimens and the intervals summarized as median and average numbers of years. Additionally, the proportion of species newly described was calculated for each publication. Authors were asked how long they had spent working on the publications.

Unfortunately, much of the specimen data behind Fernández-Triana's publications on Costa Rican Microgastrinae (Braconidae) could not be used as only holotype data are recorded in the publications. Data for paratypes are supposedly available via Janzen's rearing database but this database is currently unavailable online.

## RESULTS AND DISCUSSION

### How Efficient Are Minimalist Species Descriptions Compared to Other Methods?

In all the selected taxonomic revisions of tropical parasitoids, species have arguably been described relatively rapidly (Table 3.1 and Figure 3.2). This is perhaps unsurprising, given that intensive collecting in many tropical regions is a relatively new phenomenon. The minimalist studies sit around the middle of the spread of time lags between specimen collection and species description. The descriptions of 403 assorted species of Braconidae (Sharkey et al., 2021a) stand out for the number of species described, although it is noteworthy that several taxonomists contributed to this. Gauld (1991) stands out for the remarkable speed with which the first monograph of Costa Rican Ichneumonidae was produced, with a median lag of only three years separating species description from specimen collection. The lack of DNA barcodes and stacked images undoubtedly helped, although there are plenty of SEMs (scanning electron micrographs) in Gauld (1991).

**TABLE 3.1**
Selected taxonomic revisions describing tropical parasitoids, with the median time lag between specimen collection and publication of the species description. Publications in bold are 'minimalist revisions'. Note that the *Aleiodes* species descriptions are separated from the other descriptions of Braconidae (in square brackets) to compare with Butcher et al. (2012).

| Author | Study | Date | Median time lag |
|---|---|---|---|
| Gauld | Pimpliformes | 1991 | 3 |
| Areekul Butcher et al. | Thai *Aleiodes* | 2012 | 5 |
| Fernández-Triana et al. | *Pseudapanteles* | 2014 | 6 |
| Fernández-Triana et al. | *Apanteles* | 2014 | 8 |
| **Sharkey et al.** | **Costa Rican *Aleiodes*** | **2021** | **9** |
| **Sharkey et al.** | **Braconidae** | **2021** | **9** |
| **Meierotto et al.** | **Agathidinae** | **2019** | **11.5** |
| **Sharkey et al.** | ***Mesochorus*** | **2023** | **12** |
| Gauld & Janzen | *Thyreodon* | 2004 | 14 |
| Fleming et al. | *Belvosia* | 2023 | 21 |
| Fleming et al. | *Metaplagia* | 2021 | 22 |
| Noyes (2023) | Encyrtidae | 2023 | 23 |

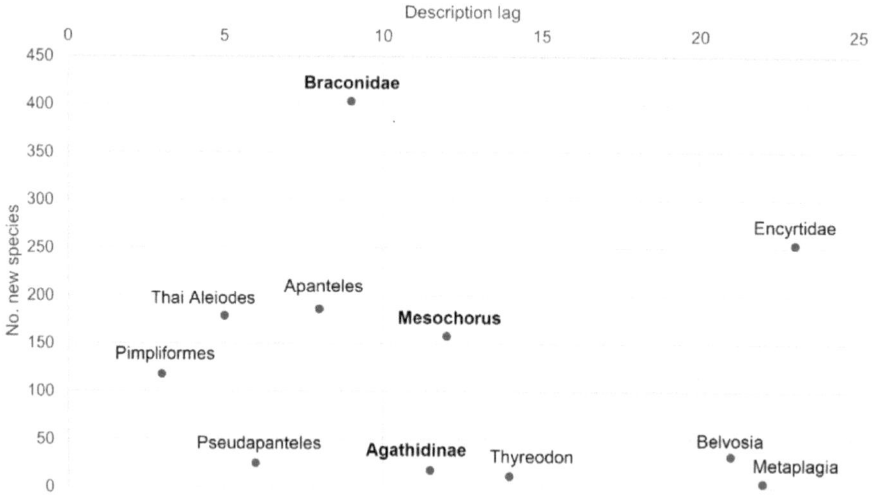


**FIGURE 3.2** Time lags, in years, between specimen collection and date of publication for selected publications describing new species of parasitoids, plotted against the number of new species described in that publication. Publications in bold are 'minimalist revisions'. Citations are listed in Table 3.1.

Clearly, 'turbo taxonomic' and 'minimalist' methods are comparatively speedy. I have spent about ten years on one small revision, describing a few species and synonymizing a few names, which is not an approach begging to be rolled out to the world's poorly known faunas. On the other hand, I have collaborated on revisions which have been finished within the timeframe of a part of a PhD programme (e.g., Veijalainen et al., 2014). These differing speeds are true of most taxonomists. One needs to account for the long list of authors for minimalist revisions and ask whether this is as efficient as one or two people putting the time in.

## Do Minimalist Species Descriptions Achieve What They Set Out to? Is This Taxonomy?

These descriptions are minimalist and they arguably fulfil the basic requirements of the Code for the descriptions to be valid. However, several authors have already pointed out that species described minimally might not be fully diagnosable. There is a problem with reliance on the BOLD BIN algorithm (Meier et al., 2022) and potential problems with species described from pseudogenes or Nuclear Mitochondrial DNA (numts). Although Sharkey et al. (2023) have disputed the examples given by Meier et al. (2022), it is noteworthy that numts have been found to be widespread in some Hymenoptera. As these nuclear copies of mitochondrial genes are under very little selection pressure, these sequences can confound phylogenetic reconstructions and species hypotheses, as demonstrated for some *Bombus* species (Williams et al., 2023). In the true sense of 'diagnosability', the *Mesochorus* species described by Sharkey et al. (2023) are not reliably diagnosable because the work ignores existing species descriptions, including the 30 species previously described or recorded from Costa Rica (Dasch, 1974), even species previously identified from the same caterpillar rearing programme (Janzen et al., 2003). It is also true that the Code (ICZN, 1999) is not particularly prescriptive when it comes to the definition of a diagnosis, stating that a name must (Art. 13.1.1) 'be accompanied by a description or definition that states in words characters that are purported to differentiate the taxon'. This is one reason for the large number of poor-quality descriptions, minimalist, maximalist and everything in between. Accurately and efficiently defining a taxon is a major part of the art and science of taxonomy. In the *Mesochorus* example, Sharkey et al. (2023) are relying on the words, 'Fig. [for example] 9', and 'Consensus barcode', to constitute a valid description as these are the parts of 'diagnostics' which are in words, rather than a string of nucleotides or pictures. Some species are (poorly) distinguished morphologically from nearest barcode 'neighbours'. Many have host data. Are these validly published while those introduced by just a barcode and photos are not?

Taxonomy is the science of classification, of the discovery of taxa (Williams and Ebach, 2020). There is not a lot of science in the minimalist approach. At their worst, minimalist descriptions simply validate a DNA barcode cluster. In the sense that this cluster of sequences, together with a photograph of what the species looks like, is a species hypothesis, then this can be considered taxonomy. However, there is no real test of this species concept and no account taken of previously described species. It is a 'year zero' approach which seems to state that the real work of taxonomy is too difficult (although many other taxonomists manage to do it). The Sharkey et al. papers are titled 'revisions' when they are manifestly not.

## Does Anybody Need Minimalist Species Descriptions?

There is no evidence that the taxonomic community needs these minimalist species descriptions. Almost all citations of the Sharkey *et al.* minimalist taxonomy papers are in publications discussing the drawbacks of minimalist species descriptions, or simply pointing out the existence of this approach. Is there a conservation need? It is clearly too early to tell yet but the idea that describing everything in Area de Conservación Guanacaste (ACG) will lead to greater conservation effort is untested and potentially untestable, unless funders state that they were persuaded to pay to protect forests now that a certain number of parasitoid wasps have been described.

The wealth of life history and ecology data associated with these specimens reared in Costa Rica are ill-served by a DNA barcode and a poor-quality picture. 'Minimalist' descriptions with maximum information content are also possible. For example, the descriptions of new species of ants (Hymenoptera: Formicidae) of the genus *Zasphinctus* by Hita Garcia et al. (2017). The species descriptions are essentially tables of measurements and images taken from three-dimensional Micro-CT scans. These morphological descriptions are highly portable and comparable. It is often difficult to sequence DNA barcodes from specimens, particularly if they are not fresh, or if funds for sequencing are unavailable, both of which are exceedingly common limitations.

## Perspectives

> 'Minimalism is a tool to rid yourself of life's excess in favor of focusing on what's important—so you can find happiness, fulfillment, and freedom'
>
> *The minimalists: https://www.theminimalists.com/minimalism/*

'Minimalism' as an art movement or as a lifestyle choice can lead to different perspectives. It has certainly led to great art. But as a form of taxonomy, does minimalism offer happiness, fulfilment and freedom? As a wasp taxonomist, do minimalist species descriptions improve my life?

The proponents of minimalist taxonomy make the rather uncontentious point that a step change in the rate of species descriptions is needed if we are to complete the task of describing all species. They then make the claim that their methods are the most appropriate to the task. Although a stated aim of minimalist taxonomy is that a DNA barcode-based approach to species identification will result in species being more readily identifiable to a general audience, there seems simultaneously to be the inherent assumption that nobody will ever use the product, that species will be described and then never become biologically meaningful entities. Ultimately, this shortcut to taxonomy is letting down the many people who have collected, reared, sorted and prepared these thousands of specimens. Those processes take time, and those data are invaluable. However, criticism from numerous directions has just prompted the proponents of minimalist taxonomy to double down and in fact reduce the useful content even further.

The minimalist descriptions of 158 species of *Mesochorus* (Sharkey et al., 2023) are noteworthy for featuring, among the 24 authors, no expertise on the taxonomy of *Mesochorus* and no attempt to hide the authors' contempt for traditional taxonomy. An extensive introduction probably took longer to write than some brief, useful descriptions would have, and the names of the new species are deliberately utilitarian, simply cod-Spanish transliterations of the numbers one through 158. This is the essence of minimalist taxonomy, simply giving names to barcode BINs. As Fernández-Triana (2022) has noted, various taxonomists are probably approaching an upper limit to the number of parasitoid wasp species which can be described in a year. Several colleagues who are prolific taxonomists have described, using different metrics, their maximum capacity for species descriptions:

José Fernández-Triana: 100 species of Microgastrinae (Braconidae) per year.

Niklas Johansson: 80 species of Ichneumonidae per year; working on the Palaearctic fauna in his spare time, this involves redescriptions and extensive checking of types.

John Noyes: 60–70 species per year; slide-mounting is standard for *Encyrtidae.*

Donald Quicke: 3–4 hours per species, 17 weeks for the description of 180 species of *Aleiodes*; does not account for the part-time barcoding effort over several years.

Ian Gauld, in his monograph on Costa Rican pimpliform Ichneumonidae (Gauld, 1991), took about three years to do all the work for a revision of 161 species, 73% of which (118 species) were described as new. Ian was employed as a taxonomist (at the Natural History Museum, London) and had a full-time research assistant, Sondra Ward, who really deserved co-authorship. A very rough approximation could be that useful taxonomic revisions, incorporating diverse sources of data, can be produced at the rate of about 100 species descriptions per year, sometimes more, sometimes fewer.

Species descriptions do not necessarily require lots and lots of images, as is claimed by Meierotto et al. (2019, p. 139): '[m]uch more could be included in the species treatments, e.g., extensive images of all body parts, genitalic images (if informative), or SEM images of microsculpture'; a taxonomic revision, and species descriptions, are rendered useful by an expert illustrating what is necessary and describing concisely what needs to be described. Ian Gauld's keys and descriptions are wonderful examples of this. As a side note, the habit of not listing paratype data properly is a hindrance to studies such as this, and a curatorial nuisance.

As Fernández-Triana (2022) pointed out, there have been many proposals to accelerate species description but very few practical examples. Meierotto et al. (2019) and Sharkey et al. (2021a, b, 2023) have provided us with examples of how accelerated species descriptions can work in practice. However, these are not the first attempts, they are simply the latest effort to reduce taxonomy to naming. 'Turbo taxonomy' is a convenient term for the effort to describe species within the shortest possible timeframe, as exemplified, for parasitoid wasps, by Butcher et al. (2012), Fernández-Triana et al. (2014a), and others. In fact, pioneering efforts such as Gauld (1991) could be termed 'turbo taxonomy'. The essential difference between 'turbo taxonomy' and minimalist descriptions is in the epithets: there is a taxonomic framework inherent in 'turbo taxonomy' which is lacking in minimalist descriptions. One major problem with minimalist descriptions is that the actual taxonomy is still needed, so the hard work is just put on a shelf for later. We now have a lot of names for parasitoids collected in ACG in Costa Rica, but the taxonomic work, of placing diagnosable species within an evolutionary framework, has been bypassed for the descriptions of 607 species described from DNA barcodes and a couple of photos each. Papers are already being published which include useful redescriptions of some of the names published by Sharkey et al. (2021b), for example, Jasso-Martínez et al. (2024). We appear to be back in the era many of us thought had passed. So many of us can rely on good quality photographs of old type specimens which are increasingly readily available. But for these ACG names, we will need to barcode match and hope for the best; and see those types.

## CONCLUSIONS

> 'With minimalism, no attempt is made to represent an outside reality, the artist wants the viewer to respond only to what is in front of them'
>
> *Tate (https://www.tate.org.uk/art/art-terms/m/minimalism#:~:text=With%20minimalism%2C%20no%20attempt%20is,the%20work%20is%20the%20reality)*

Fundamentally, there is a strong argument that minimalist species descriptions do not represent good science. Extreme minimalist descriptions disregard previous work; rely on a proprietary algorithm used on a nearly arbitrary selection of specimens for species delimitation and distinction; and offer nothing to communities lacking DNA barcode data. Minimalist species descriptions do not result in species being described faster; they do little to address the taxonomic impediment as taxonomy is not improved and the case for taxonomists is explicitly not made.

Minimalist species descriptions conflate methods with results: DNA barcodes are tools for species discovery, they are not species definitions. Do we need to describe the entire world's species? Or, to reframe this as a question of process, do we need to be quickly describing them? We know there are huge numbers of species in the world, the majority formally undescribed. Does the continuing lack of species names hinder their conservation? Initiatives such as iBOL (International Barcode of Life) and Bioscan (https://ibol.org/programs/bioscan/; Hobern, 2021) are revealing very large numbers of genetic clusters which are often equivalent to species. And these data are being used, in huge numbers of studies assessing community structure, invasive species, turnover of species, etc. Do these BINs need formal names urgently?

Where morphology-based systematics is probably missing a trick is in not surfacing the information in collections, the huge numbers of unidentified specimens and undescribed species which are, in effect, known unknowns. We can draw attention to the vast numbers of species in particular habitats and areas and then taxonomists can work on species hypotheses, constructing the tree of life and diagnosing species in a way which can hopefully be accessible to all. Giving formal names at the 'first pass' stage of the process is unnecessary. The drivers for minimalist taxonomy are less to do with the urgency in quickly describing threatened faunas and more to do with project-based expediency. Species can be described quickly as parts of revisionary studies, in ways which make those species concepts accessible. However, in ACG, there is a perceived need for these species to be described quickly by people who are not necessarily well placed to understand the taxonomy of the genera concerned. The assumption is that nobody will refer to these species again anyway. There are very instructive examples from the ongoing inventory of Costa Rican biodiversity of how to rapidly publish both exceedingly useful work and work that drags us back to the worst of 19th-century taxonomy.

There are many disciplines in which basic scientific exploration has attracted serious funding but, so far, taxonomy isn't one of them. We should be making the case that describing the world's biota is an urgent task and one which requires serious investment, as has been attempted at various times (e.g., Systematics Agenda 2000, 1994). Undercutting our discipline by describing species as quickly and cheaply as possible means that we are storing up problems for the future. Let's not kid ourselves that increased species description will solve the biodiversity crisis. There is, however, the very real possibility that enhanced knowledge of organisms will go some way to better characterizing the scale and nature of ecosystem dynamics, and the potential huge loss.

## ACKNOWLEDGMENTS

I am very grateful to Alan Fleming, Andrew Bennett, David Williams, Donald Quicke, John Noyes, Jose Fernández-Triana, and Niklas Johansson for interesting discussions and for data on work rates. David Williams and Quentin Wheeler organized a thought-provoking and wide-ranging symposium which forced me to write this chapter.

## REFERENCES

Butcher, B.A., Smith, M.A., Sharkey, M.J and Quicke, D.L.J. (2012) A turbo-taxonomic study of Thai *Aleiodes* (*Aleiodes*) and *Aleiodes* (*Arcaleiodes*) (Hymenoptera: Braconidae: Rogadinae) based largely on COI barcoded specimens, with rapid descriptions of 179 new species. *Zootaxa*, 3457: 1–232, https://doi.org/10.11646/zootaxa.3457.1.1

Cheng, R., Luo, A., Orr, M., Ge, D., Hou, Z., Qu, Y., Guo, B., Zhang, F., Sha, Z., Zhao, Z., Wang, M., Shi, X., Han, H., Zhou, Q., Li, Y., Liu, X., Shao, C., Zhang, A., Zhou, X. and Zhu, C. (2024) Cryptic diversity begets challenges and opportunities in biodiversity research. *Integrative Zoology*: 1–17, https://doi.org/10.1111/1749-4877.12809

Dasch C.E. (1974) Neotropic Mesochorinae (Hymenoptera: Ichneumonidae). *Memoirs of the American Entomological Institute*, 22: 1–509.

Díaz, S., Settele, J., Brondízio, E.S., Ngo, H.T., Agard, J., Arneth, A., Balvanera, P., Brauman, K.A., Butchart, S.H.M., Chan, K.M.A., Garibaldi, L.A., Ichii, K., Liu, J., Subramanian, S.M., Midgley, G.F., Miloslavich, P., Molnár, Z., Obura, D., Pfaff, A., Polasky, S., Purvis, A., Razzaque, J., Reyers, B., Chowdhury, R.R., Shin, Y.-J., Visseren-Hamakers, I., Willis, K.J. and Zayas, C.N. (2019) Pervasive human-driven decline of life on Earth points to the need for transformative change. *Science*, 366(6471): https://doi.org/10.1126/science.aax3100

Engel, M.S., Ceríiaco, L.M.P., Daniel, G.M., Dellapé, P.M., Löbl, I., Marinov, M., Reis, R.E., Young, M.T., Dubois, A., Agarwal, I., Lehmann, A. P., Alvarado, M., Alvarez, N., Andreone, F., Araujo-Vieira, K., Ascher, J.S., Baêta, D., Baldo, D., Bandeira, S.A., Barden, P., Barrasso, D.A., Bendifallah, L., Bockmann, F.A., Böhme, W., Borkent, A., Brandão, C.R.F., Busack, S.D., Bybee, S.M., Channing, A., Chatzimanolis, S., Christenhusz, M.J.M., Crisci, J.V., D'Elía, G., Da Costa, L.M., Davis, S.R., De Lucena, C.A.S., Deuve, T., Elizalde, S.F., Faivovich, J., Farooq, H., Ferguson, A.W., Gippoliti, S., Gonçalves, F.M.P., Gonzalez, V.H., Greenbaum, E., Hinojosa-Díaz, I.A., Ineich, I., Jiang, J., Kahono, S., Kury, A.B.,Lucinda, P.H.F., Lynch, J.D., Malécot, V., Marques, M.P., Marris, J.W.M., McKellar, R.C., Mendes, L.F., Nihei, S.S., Nishikawa, K., Ohler, A.,Orrico, V.G.D., Ota, H., Paiva, J., Parrinha, D., Pauwels, O.S.G., Pereyra, M.O., Pestana, L.B., Pinheiro, P.D.P., Prendini, L., Prokop, J., Rasmussen, C., Rödel, M.-O., Rodrigues, M.T., Rodríguez, S.M., Salatnaya, H., Sampaio, Í., Sánchez-García, A., Shebl, M.A., Santos, B.S., Solórzano-Kraemer, M.M., Sousa, A.C.A., Stoev, P., Teta, P., Trape, J.-F., Van-Dúnem Dos-Santos, C., Vasudevan, K., Vink, C.J., Vogel, G., Wagner, P., Wappler, T., Ware, J.L., Wedmann, S. and Zacharie, C.K. (2021) The taxonomic impediment: a shortage of taxonomists, not the lack of technical approaches. *Zoological Journal of the Linnean Society*, 193: 381–387, https://doi.org/10.1093/zoolinnean/zlab072

Erwin, T.L. (1982) Tropical forests: their richness in Coleoptera and other arthropod species. *The Coleopterists Bulletin*, 36: 74–75.

Fernández-Triana, J., Whitfield, J.B., Rodriguez, J.J., Smith, M.A., Janzen, D.H., Hallwachs, W.D., Hajibabaei, M., Burns, J.M., Solis, M.A., Brown, J., Cardinal, S., Goulet, H., Hebert, P.D.N. (2014a) Review of *Apanteles sensu stricto* (Hymenoptera, Braconidae, Microgastrinae) from Area de Conservación Guanacaste, northwestern Costa Rica, with keys to all described species from Mesoamerica. *ZooKeys*, 383: 1–565, https://doi.org/10.3897/zookeys.383.6418

Fernández-Triana, J., Whitfield, J.B., Smith, M.A., Braet, Y., Hallwachs, W. and Janzen, D.H. (2014b) Review of the neotropical genus *Prasmodon* (Hymenoptera, Braconidae, Microgastrinae), with emphasis on species from Area de Conservación Guanacaste, northwestern Costa Rica. *Journal of Hymenoptera Research*, 37: 1–52, https://doi.org/10.3897/jhr.37.6748

Fernández-Triana, J.L. (2022) Turbo taxonomy approaches: lessons from the past and recommendations for the future based on the experience with Braconidae (Hymenoptera) parasitoid wasps. *ZooKeys*, 1087: 199–220, https://doi.org/10.3897/zookeys.1087.76720

Fleming, A.J., Wood, D.M., Smith, M.A., Hallwachs, W. and Janzen, D.H. (2021) Revision of *Metaplagia* Coquillett (Diptera: Tachinidae) with description of five new species from Area de Conservación Guanacaste in northwestern Costa Rica. *Biodiversity Data Journal*, 9: e68598, https://doi.org/10.3897/BDJ.9.e68598

Fleming, A.J., Woodley, N., Smith, M.A., Hallwachs, W. and Janzen, D.H. (2023) Revision of *Belvosia* Robineau-Desvoidy (Diptera, Tachinidae) and 33 new species from Area de Conservación Guanacaste in northwestern Costa Rica with a key to known North and Mesoamerican species. *Biodiversity Data Journal*, 11: e103667, https://doi.org/10.3897/BDJ.11.e103667

Forbes, A.A., Bagley, R.K., Beer, M.A., Hippee, A.C. and Widmayer, H.A. (2018) Quantifying the unquantifiable: why Hymenoptera, not Coleoptera, is the most speciose animal order. *BMC Ecology*, 18(21):11 pp, https://doi.org/10.1186/s12898-018-0176-x

Gauld, I.D. (1991) The Ichneumonidae of Costa Rica, 1. *Memoirs of the American Entomological Institute*, 47: 1–589.

Gauld, I.D. and Janzen, D.H. (2004) The systematics and biology of the Costa Rican species of parasitic wasps in the *Thyreodon* genus-group (Hymenoptera: Ichneumonidae). *Zoological Journal of the Linnean Society*, 141: 297–351, https://doi.org/10.1111/j.1096-3642.2004.00116.x

Hanson, P.E. and Gauld, I.D (1995) *The Hymenoptera of Costa Rica*. Oxford University Press and The Natural History Museum, Oxford.

Hebert, P.D.N., Cywinska, A., Ball, S.L. and DeWaard, J.R. (2003) Biological identifications through DNA barcodes. *Proceedings of the Royal Society of London B*, 270: 313–321, https://doi.org/10.1098/rspb.2002.2218

Hita Garcia, F., Fischer, G., Liu, C., Audisio, T.L. and Economo, E.P. (2017) Next-generation morphological character discovery and evaluation: An X-ray micro-CT enhanced revision of the ant genus *Zasphinctus* Wheeler (Hymenoptera, Formicidae, Dorylinae) in the Afrotropics. *ZooKeys*, 693: 33–93, https://doi.org/10.3897/zookeys.693.13012

Hobern, D. (2021) BIOSCAN: DNA barcoding to accelerate taxonomy and biogeography for conservation and sustainability. *Genome*, 64: 161–164, https://doi.org/10.1139/gen-2020-0009

House of Lords Science and Technology Committee, Fourth Report, Session 2001–2002, *What on Earth? The threat to the science underpinning conservation*. HL Paper 118(i), https://publications.parliament.uk/pa/ld200102/ldselect/ldsctech/118/11801.htm

House of Lords Science and Technology Committee, Third Report, Session 2002–2003, *What on Earth? The threat to the science underpinning conservation: the government's response and the committee's commentary*. HL Paper 130(i), https://publications.parliament.uk/pa/ld200203/ldselect/ldsctech/130/13001.htm

House of Lords Science and Technology Committee, Fifth Report, Session 2007–2008, *Systematics and Taxonomy: Follow-up*. HL Paper 162, https://publications.parliament.uk/pa/ld200708/ldselect/ldsctech/162/162.pdf

ICZN (1999) *International Code of Zoological Nomenclature*. 4th ed. The International Trust for Zoological Nomenclature c/o The Natural History Museum, London.

Janzen, D.H., Walker, A.K., Whitfield, J.B., Delvare, G. and Gauld, I.D. (2003) Host specificity and hyperparasitoids of three new Costa Rican species of *Microplitis* Foerster (Hymenoptera: Braconidae: Microgastrinae), parasitoids of sphingid caterpillars. *Journal of Hymenoptera Research*, 12(1):42–76, https://biostor.org/reference/302

Jasso-Martínez, J.M., Martínez, J.J., Aguilar-Velasco, R.G. and Zaldívar-Riverón, A. (2024) Four new species of *Triraphis* Ruthe, 1855 (Braconidae, Rogadinae) from a Mexican tropical dry forest and morphological descriptions of *T. bradzlotnicki* Sharkey, 2021 and *T. davidwahli* Sharkey, 2021. *European Journal of Taxonomy*, 917: 50–73, https://doi.org/10.5852/ejt.2024.917.2387

Meier, R., Blaimer, B.B., Buenaventura, E., Hartop, E., vonRintelen, T., Srivathsan, A. and Yeo, D. (2022) A re-analysis of the data in Sharkey et al.'s (2021) minimalist revision reveals that BINs do not deserve names, but BOLD Systems needs a stronger commitment to open science. *Cladistics*, 38: 264–275, https://doi.org/10.1111/cla.12489

Meier, R., Hartop, E., Pylatiuk, C. and Srivathsan, A. (2024) Towards holistic insect monitoring: species discovery, description, identification and traits for all insects. *Philosophical Transactions of the Royal Society (B)*, 379: 20230120, https://doi.org/10.1098/rstb.2023.0120

Meierotto, S., Sharkey, M.J., Janzen, D.H., Hallwachs, W., Hebert, P.D.N., Chapman, E.G. and Smith, M.A. (2019) A revolutionary protocol to describe understudied hyperdiverse taxa and overcome the taxonomic impediment. *Deutsche Entomologische Zeitschrift*, 66(2): 119–145, https://doi.org/10.3897/dez.66.34683

Mora, C., Tittensor,D.P., Adl, S., Simpson, A.G.B. and Worm, B. (2011) How many species are there on Earth and in the ocean? *PLoS Biology*, 9(8): e1001127, https://doi.org/10.1371/journal.pbio.1001127

Noyes, J.S. (2023) Encyrtidae of Costa Rica (Hymenoptera: Chalcidoidea), 4. Subfamily Encyrtinae: Tribes Arrhenophagini, Habrolepidini, Cerapterocerini, Cheiloneurini, Trechnitini, Cercobelini, Polaszekiini, Protyndarichoidini, Gahaniellini and Syrphophagini (part), mainly primary parasitoids and hyperparasitoids of Coccoidea and Psylloidea (Hemiptera). *Taxonomic Monographs on Neotropical Hymenoptera*, 2: 1–923.

Sharkey, M.J., Baker, A., McCluskey, K., Smith, A., Naik, S., Ratnasingham, S., Manjunath, R., Perez, K., Sones, J., D'Souza, M., St. Jacques, B., Hebert, P., Hallwachs, W. and Janzen, D. (2021a) Addendum to a minimalist revision of Costa Rican Braconidae: 28 new species and 23 host records. *ZooKeys*, 1075: 77–136, https://doi.org/10.3897/zookeys.1075.72197

Sharkey, M.J., Janzen, D.H., Hallwachs, W., Chapman, E.G., Smith, M.A., Dapkey, T., Brown, A., Ratnasingham, S., Naik, S., Manjunath, R., Perez, K., Milton, M., Hebert, P., Shaw, S.R., Kittel, R.N., Solis, M.A., Metz, M.A., Goldstein, P.Z., Brown, J.W., Quicke, D.L.J., Achterberg, C. van, Brown, B.V. and Burns, J.M. (2021b) Minimalist revision and description of 403 new species in 11 subfamilies of Costa Rican braconid parasitoid wasps, including host records for 219 species. *ZooKeys*, 1013: 1–665, https://doi.org/10.3897/zookeys.1013.55600

Sharkey, M.J., Baker, A., McCluskey, K., Smith, A., Naik, S., Ratnasingham, S., Manjunath, R., Perez, K., Sones, J., D'Souza, M., St. Jacques, B., Hajibabaei, M., Whitfield, J., Arias, D., Solis, A., Metz, M., Burns, J., Zuñiga, R., Phillips-Rodriguez, E., Espinoza, B., Chacon, I., Hebert, P., Hallwachs, W. and Janzen, D. (2023) Minimalist revision of *Mesochorus* Gravenhorst, 1829 (Hymenoptera: Ichneumonidae: Mesochorinae) from Área de Conservación Guanacaste, Costa Rica, with 158 new species and host records for 129 species. *Revista de Biología Tropical*, 71 (Supplement 2): e53316.

Shaw, M.R. and Hochberg, M.E. (2001) The neglect of parasitic Hymenoptera in insect conservation strategies: The British fauna as a prime example. *Journal of Insect Conservation*, 5: 253–263, https://doi.org/10.1023/A:1013393229923

Systematics Agenda 2000 (1994) *Systematics Agenda 2000: Charting the Biosphere. Produced by Systematics Agenda 2000.* A Consortium of the American Society of Plant Taxonomists, the Society of Systematic Biologists, and the Willi Hennig Society, in cooperation with the Association of Systematics Collections, Washington, D.C.

Veijalainen, A., Broad, G.R. and Sääksjärvi, I.E. (2014) Twenty seven new species of *Orthocentrus* (Hymenoptera: Ichneumonidae; Orthocentrinae) with a key to the Neotropical species of the genus. *Zootaxa*, 3768 (3): 201–252, https://doi.org/10.11646/zootaxa.3768.3.1

Wiens, J.J. (2023) How many species are there on Earth? Progress and problems. *PLoS Biology*, 21(11): e3002388, https://doi.org/10.1371/journal.pbio.3002388

Williams, D.M. and Ebach, M.C. (2020) *Cladistics: A Guide to Biological Classification.* Systematics Association Special Volume Series. Cambridge University Press, Cambridge.

Williams, P.H., An, J., Dorji, P., Huang, J., Jaffar, S., Japoshvili, G., Narah, J., Ren, Z., Streinzer, M., Thanoosing, C., Tian, L. and Orr, M.C. (2023) Bumblebees with big teeth: Revising the subgenus *Alpigenobombus* with the good, the bad and the ugly of numts (Hymenoptera: Apidae). *European Journal of Taxonomy*, 892: 1–65, https://doi.org/10.5852/ejt.2023.892.2283

# 4 The Old, the New, and Lots of People

## *How Taxonomy Will Thrive*

*Frank-Thorsten Krell*

Changes in how systematists work and interact with other branches of science must take place in order to speed up research and re-establish the importance of Floras but only a substantial infusion of funds and personnel will really accelerate our progress

*Funk, 2006, p. 581*

Taxonomy has never had so many tools at its disposal and so many connections to the rest of biology. The tragedy is that there are too few taxonomists who can utilize these advances and tackle the challenging questions of our time

*Lane, 2009*

Our premise (and experience) is that meaningful taxonomy is constrained mainly by the limited supply of person-hours. The most efficient solution to speeding taxonomic production, ultimately, is to train and hire more taxonomists

*Deans et al., 2012, p. 81*

## INTRODUCTION

Taxonomy provides the foundation for all organismic and a large part of molecular biology by providing diagnoses and names for the entities of research. Despite its fundamental role in science, it has long struggled with its overwhelming task to get the biodiversity of the planet organized with little support—financial and otherwise (Löbl et al., 2023). But it continued with its task, due to a wealth of dedicated people. Nevertheless, taxonomic expertise is in decline and has been so for some time (Schminke, 1994, 1997; Buyck, 1999; Hopkins and Freckleton, 2002; Lyal, 2005; Drew, 2011; Frobel and Schlumprecht, 2016; Hochkirch et al., 2022; Zhu et al., 2022; Páll-Gergely et al., 2024). In our time of a biodiversity crisis, this is a critical situation, but all previous suggestions to mitigate the situation have had largely no consequences. The number of active taxonomists has remained stagnant or even decreased further.

Taxonomy is an old science, going back to the earliest societies that needed to name and classify animals, plants, and mushrooms of their environment to distinguish, for example, between edible and dangerous species. Naming and knowing the living creatures we share the world with was even considered a fundamental task in

DOI: 10.1201/b22822-4

the Abrahamic scriptures as it was one of the first exercises God assigned to Adam (Genesis 2: 19), indicating that communities thousands of years ago recognized the importance of doing so. In the Age of Enlightenment, after multiple attempts to name animals and plants with short descriptive sentences, Linnaeus formalized naming by introducing the binominal nomenclature that we still use today, almost 270 years later. From then on, humankind progressed rapidly in discovering, describing, naming and understanding the organismic world, but has only accomplished a fraction of this work. Despite being a fundamental part of human endeavor, taxonomy is not as respected or supported as one might expect. Still, taxonomy has produced an enormous amount of knowledge and continues to do so, increasingly under the criticism that it does not achieve enough fast enough, is 'only' descriptive, and is old-fashioned. Taxonomy has persisted all the same. Here I will try to outline how taxonomy can not only persist but might even flourish and accelerate the discovery of the yet to be known organisms on earth.

To preemptively address arguments repeatedly brought up in public and scientific discourse, it appears necessary to clarify what taxonomy is not:

- counting entities in samples of biodiversity, or
- pure identification of species or identifiable entities.

Taxonomy does encompass these two popular activities, but they are only minor parts of it: the first is the initial, technical phase of handling samples before taxonomic analysis; the second is an application of taxonomy. They are neither the essence nor the end goal of taxonomy.

## OLD WAYS

Ernst Mayr wrote in 1942:

> There is considerable doubt in the minds of some taxonomists and even more in the minds of most non-taxonomists as to what the real functions of the systematist are. Some laboratory men and ecologists seem to think that the taxonomist should content himself with identifying material and with devising keys. Beyond that, he should keep his collections in good order, describe new species, and have every specimen properly labeled. According to this view systematics is the mere pigeonholing of specimens. No taxonomist will deny that these particular tasks are part of his job, and the workers of the less-known groups may not be able to go beyond the cataloguing phase of taxonomic work. The systematist of the better-known groups, however, will not be satisfied with such a limitation: for him, systematics is more than an auxiliary science. He inquires not only the "What?" but also as to the "Why?".
>
> *Mayr, 1942, pp. 8–9*

Mayr expressed what is commonly considered the 'old ways' of taxonomy: descriptive, morphological work on preserved specimens. While the description of the morphology of an organism is an old method, it cannot be replaced. Describing the species on this planet is the central duty of taxonomy. Whether we describe morphology or a DNA-barcode, or both, or the whole genome, the work remains descriptive and is essential for all further analyses, for all further questions of 'Why?'.

The descriptive component has been and will be essential for taxonomy and other sciences (Grimaldi and Engel, 2007; Casadevall and Fang, 2008). But traditional taxonomy goes further, integrating data from distributions and ecology. This makes it even more challenging than just describing dead specimens. Cockerell muses about botanical taxonomy: 'the real reason for the neglect of taxonomy is that it is too difficult and takes too much time. In order to revise a genus of plants it will usually be necessary to consult or borrow from many herbaria and to read and interpret literature in many languages, most of which is obtainable in only a few libraries. When all this is done, it still remains very desirable to have a knowledge of the living plants, wild and in the botanic garden' (Cockerell, 1946). Such demands cannot compete with the relative ease of conducting controlled experiments in the laboratory or even experiments in the wild that can be carried out after reasonable technical training. Taxonomy is generally demanding in terms of training and experience. 'The main problem is that in its current form taxonomy cannot be easily automatized [...]. Worse still, not only does it need manpower, but it also needs it in the form of highly trained personnel' (Meier, 2008, p. 101). Traditional taxonomy might be seen as merely descriptive, but it is a serious, complex, and rather challenging task.

## NEW WAYS

Taxonomy has a great track record for adopting new technologies, if useful for its purposes, and also for jumping on bandwagons, e.g., DNA taxonomy (e.g., Tautz et al., 2003). It is no different from any other branch of science. There cannot be any reasonable doubt that taxonomists in general welcome new ways of analysis and exploration if they are an improvement over previous procedures and methods (Löbl et al., 2023). A large portion of current taxonomy integrates or relies on molecular data, sometimes opening up increased funding (Packer et al. 2009); even genomics caught on swiftly (Card et al., 2021). Deep learning can be applied (Kloster et al. 2020); big databases or apps, such as *iNaturalist* (https://www.inaturalist.org/), are regularly used, but expert checking remains necessary (Reverté et al., 2023; López-Guillén et al., 2024); -omics technologies are being explored (Raupach et al. 2016); and so forth.

## BOTH WAYS

The difference between taxonomy and many other branches of science is that (a) it has been practiced longer than most of them; and (b) it does not replace but complements its oldest method, morphological description, with newer approaches. This complementation is the key to taxonomy's success. Good taxonomy keeps all the useful old ways and adds new technologies, new ways of thinking, new infrastructure, and everything that helps to find and analyze new taxa and characters.

The idea of considering all available kinds of characters for taxonomical analysis is not new. Simpson wrote before World War II: 'Taxonomy is at the same time the most elementary and the most inclusive part of zoology, most elementary because animals cannot be discussed or treated in a scientific way until some taxonomy has been achieved, and most inclusive because taxonomy in its various guises and branches eventually gathers together, utilizes, summarizes, and implements everything that is known about animals, whether morphological, physiological, psychological, or ecological' (Simpson, 1945, p. 1). The many kinds of taxonomic characters listed by Mayr et al. (1953, p. 108) hint at considering or integrating information from morphology, physiology, ecology, ethology, and biogeography to achieve a reliable taxonomy and classification. Simpson summarizes it so: 'It is a corollary of the fact that *organisms* are to be classified, not their characters, that the *whole* organism should be considered in all its parts and all its aspects' (Simpson, 1961, p. 71). In the German-speaking world, Schilder (1952) called for integrating morphology, biology, chorology, and ecology to transform taxonomy into 'biotaxonomy'. Stafleu called it 'holotaxonomy' (Stafleu, 1965) or 'biosystematics' (Stafleu 1969), while Heywood (1973) considered taxonomy at its best to be the 'digestive system of biology': Taxonomy requires digesting and synthesizing data from all available sources to reach the best possible results. 'The integration of the old and the new has resulted in new insights into biological diversity and has often shown more clearly the pathways along which life evolves' (Stafleu, 1969). While some taxonomists in the past presented themselves as narrow-minded traditionalists (see Mollenhauer, 1975), taxonomy at large has long been integrative (see also Valdecasas et al., 2008). For a short period, morphological and molecular taxonomy began drifting apart (e.g., Tautz et al., 2003; but see Seberg et al., 2003 and Lipscomb et al., 2003) and efforts were supposedly needed to bring the two branches together again. The term 'Integrative Taxonomy' was formally introduced in 2005, independently by Dayrat (2005) and Will et al. (2005). Whether formalized or not, integrative taxonomy has been and is the way forward.

## RE-BRANDING AND RE-INVENTING TAXONOMY?

For a long time, and probably even to this day, taxonomy has been considered a dusty pastime from the 19th century (Duncker, 1992). It has been suggested repeatedly that taxonomy needs to re-invent itself if it is to survive and needs a 'major image revamp' if it is to attract sufficient funding (Godfray, 2002; Korf, 2005; Meier, 2008, p. 101; Wheeler, 2008, p. 12). And taxonomy has done so, as it always has throughout its history, by adopting new methods, new tools, higher quality documentation, and so forth. However, the demands for re-invention have not gone away. The reason might be that taxonomy cannot get rid of its descriptive component and that this descriptive component is old. The 'old-fashioned' or 'traditional' aspects of taxonomy are still the foundation for making taxa known and comparing them. It is purely and simply impossible to abandon species descriptions, just because they have been around for three centuries and are descriptive. They have been refined, increased in content and improved in quality (Sangster and Luksenburg, 2015), significantly more useful than the short phrase descriptions of Linnaean times, often being well-illustrated with scanning electron microscope (SEM) images, or digital, stacked images. But this massive improvement has not helped. The old (= tested and proven), if it is still essential, cannot be replaced by something new just because it is old. If people in the scientific community do not want to accept taxonomy as a worthwhile science, they will find ways to discredit it, and they do. Often advocates for taxonomy are taxonomy's own worst enemies as they confirm that we apparently cannot continue as we did, that we did not deliver, and must re-invent the field, then go on presenting their own agenda and silver bullets, just to start the cycle again.

Scientific disciplines are not centrally governed like the Catholic Church or corporations. They are not governed at all. They are diverse, global endeavors, which are often misunderstood. We still find statements like, 'Taxonomy is primarily governed by two branches of the International Union of Biological Sciences (IUBS): the International Commission on Zoological Nomenclature (ICZN) and the International Association for Plant Taxonomy (IAPT)' (Guedes et al., 2023; with prokaryotes being conveniently overlooked). No, taxonomy is not governed. Nomenclature is, but following nomenclatural rules is voluntary and by mutual agreement. How the science of taxonomy is done is up to the individual scientist, as it is in biochemistry, forestry, endocrinology, or any other branch of science. Good practices should guide every scientist, and this is as far as governing goes. With its strong amateur component, taxonomy is even less governable. We cannot 'change taxonomy' by decree. We can slowly change it by example and subsequent mutual agreement. This is how taxonomy has evolved.

The ongoing criticism of taxonomy might suggest that a re-branding exercise would be helpful, as already suggested by Engel et al. (2021). Re-branding, the change of name, logos, and identity of an entity (mostly a business, corporation, or a product line) became common in the commercial realm during the last thirty years—and also with natural history museums, from widely known name changes of large institutions in London or Leiden, to smaller ones in Denver or Lyon (Colwell-Chanthaphonh et al., 2013; Green, 2019). Modernizing or revamping their image is one of the most common reasons for rebranding in the corporate world (Muzellec et al., 2003). Changing a name can be an inspirational new start or indicate a broadened perspective. However, rebranding is more than just a name change. It is a complex process (Lomax et al., 2002) and can easily go wrong. To avoid naively suggesting strategies or solutions, we need to consult the relevant expertise, i.e., marketing and corporate research.

Rebranding efforts must consider two fundamental dimensions, brand identity change (expressed, e.g., in the brand name) and brand strategy change (Zhao et al., 2018). Should we re-name and re-define taxonomy to get rid of the impression of antiquatedness? Felis et al. (2010), and independently Reboleira and Enghoff (2016) proposed the term *taxonomics*, in line with metabolomics, genomics, or phenomics. And, in 2017, an established priority program of the Deutsche Forschungsgemeinschaft (German Research Council) bears the title 'Taxon-Omics, New approaches to discovering and naming biodiversity' (www.taxon-omics.com; Karbstein et al. 2024). The term taxonomics would stress that current taxonomy is a big data science whereby the data comprehensiveness might contribute more to the quality rather than just the quantity of data. This would match the concept of integrative taxonomy. While integrative taxonomy has been widely adopted and is increasingly practiced (Vinarski, 2020), the term taxonomics did not catch on widely.

Only a few papers have adopted it so far (Beermann et al., 2018; Koch et al., 2018; Ernst et al., 2021; Maletti et al., 2021; Cerri et al., 2022; Hodač et al., 2023; Karbstein et al., 2024), others have used it erroneously for the term taxonomy (Blaxter and Floyd, 2003; Owen, 2004; Peng et al., 2019). 'For a new name to be launched, however, the old name has to be abandoned' (Muzellec and Lambkin, 2006). This is unlikely to happen with taxonomy.

A brand strategy change will not be necessary because taxonomy has always been taxonomics. Taxonomy has always been, at its best performance, an integrative science, as we have seen above. We do need to convince our friends and colleagues in the scientific community and also science policy makers to be open to realizing what taxonomists are actually doing.

Like non-profit organizations (Glynn, 2000; Lee and Bourne, 2017) or social ventures (Moss et al., 2011), taxonomy has two identities: normative (mission-driven) and utilitarian (business or service driven). Taxonomy as a science produces hypotheses about distinct units of the organic world. Taxonomy as a service for other disciplines, intertwined with phylogenetics and nomenclature, produces classifications and identification tools that are based on scientific hypotheses but also consider user needs. This duality has been recognized for over a century (Hertwig, 1914, p. 79) and has caused issues and discussions ever since in terms of expectations from and opportunities for taxonomy (e.g., Thiele and Yeates, 2002; Thomson et al., 2018). 'Pressure to secure resources can lead nonprofit organizations to emphasize utilitarian identities in rebranding, and so surface hidden tensions among stakeholders reluctant to relinquish established normative identities' (Lee and Bourne, 2017). Taxonomy is in the same boat. Do we focus on the utilitarian aspect of taxonomy as an identification tool and get funded for, e.g., a barcoding initiative, or should we follow taxonomy's mission to discover, describe and analyze the biodiversity of our planet, accepting the lower probability of obtaining funding?

Going back to the business perspective: '[…] given brand repositioning is preferable, the use of evolutionary rebranding strategies is superior in enhancing consumer brand preference in the case of pleasant original brand name attitude, while the use of revolutionary strategies is superior when consumers hold less pleasant attitude' (Le et al., 2014). With taxonomy, we have both perspectives: Most taxonomists and natural historians are very loyal to taxonomy and are thoroughly convinced that taxonomy is a great science and a fundamental and useful undertaking; our peers outside taxonomy not so much. As taxonomy primarily depends on the often-unfunded work of taxonomists, an evolutionary strategy for enhancing taxonomy might be appropriate. Lomax et al. (2002) point to 'The difficulty of controlling the process, when powerful stakeholders are at work'. In our case, the stakeholders are independent taxonomists, and their power is their independence. Taxonomists are not a corporation, but an almost libertarian community of individualists (cf., Dubois, 2003), without any governance structure. Calls for rebranding or reinventing taxonomy are easy to make but are unrealistic. Without governing structure and with a big part of the 'old way' still being essential, reinventing or major rebranding is impossible to execute. Fortunately, rebranding is 'a continuum, from revitalizing a current brand, to a full name change involving alterations in brand value and promises' (Daly and Moloney 2004). Given the structure of taxonomic science,

I suggest revitalizing taxonomy in an evolutionary fashion should be the way forward. Taxonomy will remain taxonomy but will develop in time as it always has. If we work integratively and use as many data sources as possible, we might call what we are doing taxonomics and see if it will catch on over time. Being considered another -omics discipline cannot harm.

## PEOPLE

Science is done by people, is helped by people, is financed by people, and appreciated by people. People, not technologies, are the essential component of all sciences.

Artificial Intelligence (AI) is a big new technological development (Wang et al., 2023). If AI is to be replacing people in small parts of the scientific endeavor, people still must invest substantial work to refine and apply it. 'AI is not going to replace humans, but humans with AI are going to replace humans without AI' (Karim Lakhani, Harvard Business School, in Ignatius and Lakhani, 2023). New technology is of use only if people apply and control it, particularly in a creative field like science, and people will be most successful if they integrate useful new technology into their workflows. Easy tasks, well-structured and of low complexity, are at a risk of being taken over by AI, potentially threatening some citizen scientist involvement, which may require careful reassessment or task-allocation and re-structuring of projects (Ponti and Seredko, 2022). Complex, creative, and interdependent tasks are at a lesser risk.

At this point in time, and in the foreseeable future, science is largely a people-driven and people-performed activity as it has been before. It is not surprising that even in the realm of biodiversity databases, people are considered essential for linking data (Groom et al., 2020). At the same time, people are one of the costliest components of all sciences. For science administrators and decision-makers, replacing people with some less expensive mechanism is certainly tempting. Reducing the number of people involved is probably the easiest strategy but does not consider the consequences for science productivity and quality. Technology, including AI, augments and accelerates research, but cannot replace scientists' reasoning and creativity any time soon. And it shouldn't.

People have been, are, and will be the central component in scientific progress. Without enough people, progress is hampered. In 2001, an All Species Summit Meeting was held at Harvard, supported by the Gordon and Betty Moore Foundation. It was well-attended by 'leaders of organizations devoted to comprehensive taxonomic surveys on a global or continental scale' from around the globe. 'The attendees agreed that a complete or, more realistically, a nearly complete global biodiversity census is technically feasible within 25 years' (Wilson, 2003). This would be 2026. New technologies, particularly 'the more advanced revolutions in bioinformatics and genomics', had given rise to such optimism. A year later, Wilson (2004) added what was lacking in his 2003 report: 'Who will conduct these studies?' The crucial component for scientific progress entered the argument again. Wilson suggested that twice the number of existing taxonomists, 'with several technical assistants each' and using new technologies 'could move the global biodiversity survey to near completion within a single human generation'. In the real world, people were not added to the equation, and the global biodiversity survey has made little progress since.

Every analysis of the situation in taxonomy has come to the same conclusion, which has been heard over and over again for decades: If we want to accelerate taxonomy, we do not need fewer people with more technology, but *more* people (albeit with improved technology) (e.g., Sabrosky, 1950; Hartmann, 1965; Stafleu, 1969; Heywood, 1973; Holm, 1975; Wilson, 1985; Winston, 1988; Akingbohungbe, 1989; Bruton, 1989; Conert et al., 1990; Duncker, 1992; Wilson, 2004; Lyal, 2005; Funk, 2006; Lane, 2009; Pearson et al., 2011; Dar et al., 2012; Deans et al., 2012; Sluys, 2013; Nelson et al., 2015; Paknia et al., 2015; Wilson, 2017; Engel et al., 2021; Hochkirch et al., 2022; Zhu et al., 2022; Löbl et al., 2023; Rogers et al., 2023). The literature is vast and originates from different parts of the world. It seems to be a persistent and widespread issue. Adding more people to the taxonomic endeavor is not only straightforward but also widely unpopular and generally scoffed at as a solution to increase taxonomic output. People are not cheap, they need to be trained, paid, and retained, but this is the cost of doing business if we want to progress with a global biodiversity inventory, or, being less ambitious, if we want to save our knowledge of the species we share the planet with from disappearing into thin air.

## People Collecting for Taxonomy—Ecologists, Taxonomists, Students, Collectors, Everybody

Specimen collections have been and will be essential for reproducible biodiversity research, be it for morphology-based, integrative, or genomics-based research (Krell and Wheeler, 2014; Krell et al., 2016; Buckner et al., 2021; Nachman et al., 2023).

Natural history collections contain a wealth of material (Johnson et al., 2023) but overall, the exploration of the natural world is biased and shows many gaps (Troudet et al., 2017; Garcia-Rosello et al., 2023; Lo et al., 2024). Collecting new material from under-collected taxa and areas is urgently needed, considering the state of the planet, the reduction of natural habitats, the change of our climate and resulting assumed extinction rates. At the same time, additions to museum collections decline (Rohwer et al., 2023) and new, unhelpful ideas surface, such as compassionate collection (Byrne, 2023), i.e., non-lethal collecting that concentrates on photographs or minimally invasive sampling, such as tissue samples. This is a rather irresponsible and uninformed suggestion. Weller (2023) called it an 'assault on collections'. It was convincingly rebutted by Nachman et al. (2023) and Toussaint et al. (2024). We do need continuous species collecting, particularly in biodiversity-rich regions, but also in the northern hemisphere to document mega-biodiversity before it disappears and decline of known biodiversity in highly developed areas, respectively. In biodiversity-rich regions, biodiversity research is sometimes quite well-supported (Carvalho et al., 2005). Local taxonomists or ecologists, potentially with the help of knowledgeable local, often indigenous people, can drive such projects (Paknia et al., 2015; Copete et al., 2023), and they do so very successfully (e.g., Cupello et al., 2023).

Technological inventions can give a boost to discovery in areas where specimen collection and documentation have been notoriously difficult. Burns et al. (2024), for example, present a new technology that has great potential to facilitate the discovery and description of ocean life in previously unimaginable ways. While newly developed submergible vehicles are operated remotely, people are still needed to operate them.

## People Doing Taxonomy—Professional and Amateur Taxonomists

More than half a century ago, the Conference of Directors of Systematic Collections (1971, p. 11) stated: 'Maintenance of the enormous collections [of major Museums and Herbaria of the U.S.] [...] and the services connected with them [...] are performed by an extremely small total staff [...]. In many institutions, the demands of these combined curatorial tasks are so great that a trained scholar [= taxonomist] spends much of his time as a technician'. The problem of extra-scientific burdens has also been mentioned by Hennig (1971) but was certainly not an issue limited to the 1970s, neither is it specific to taxonomists. It is a general problem in science (Leshner 2008) that becomes more severe with fewer staff and is a persistent issue in many museums and university collections. The staffing numbers stated in the 1971 report are compared to the current numbers in Table 4.1. Wilson's (2004) call for doubling our taxonomic work force has apparently not been followed, but the trends in staffing numbers are mixed. Some institutions suffered from the 'thinning out' of the taxonomic work force, but most have remained stagnant with minor losses, and a few others increased staffing levels (Figure 4.1).

The people problem is not limited to professional (i.e., paid) taxonomists. The number of amateur collectors and amateur taxonomists is also declining (Hopkins and Freckleton, 2002; Fischer et al., 2021). Amateur taxonomists ['amateurs-éclairés' as Brygoo (1988) calls them] have been and are responsible for a significant share of taxonomic publications and of collecting specimens available for study (Fürsch, 1992; Richman, 1996; Köhler, 1997; Watling, 1998; Fontaine et al., 2012; Löbl et al., 2023). At the beginning of taxonomy, there were no professional biologists, and later in the nineteenth-century enthusiasts or members of the public interested and engaged in natural history were abundant (Toellner, 1986; Alberti, 2001; Loïs, 2014; Dawson et al., 2015), now living on as citizen scientists. It appears that nowadays, fewer evolve into serious amateur taxonomists, aggravating the problem of the decline in professional taxonomy.

While the number of authors of taxonomic papers has certainly risen in the last half-century, this does not indicate an increasing global capacity for species discovery (Bebber et al., 2013). Analysed for flowering plants, the number of authors describing only one or a few new species is up, but the number of taxonomically productive authors has fallen, and the overall number of described new species remains stagnant. The trend of more authors involved in taxonomic papers is most likely explained by a general trend of a rising number of authors per paper, which has increased fivefold over the last century (Abukhalil, 2014). We now involve more colleagues as authors than decades ago, when, e.g., technicians or students were not generally included.

There is certainly no surge of new taxonomists. Every analysis and metric, apart from the number of authors of taxonomic papers, indicates the contrary. The task of discovering and describing the undescribed part of biodiversity remains the same: a staggering, intimidating workload. With a shortage of people, we could end up with

**TABLE 4.1**
Staffing of research and collections in selected US Institutions, excluding anthropology and other non-taxonomic fields. 1971 Figures from Conference of Directors of Systematic Collections (1971) and Anonymous (1972). 2024 figures were reported by staff of the respective institutions. 'Before 2023' figures are from Johnson et al. (2023). Current numbers from self-reporting are crude, not independently verified and do not consider temporary or grant funded positions.

| Institution | Research Staff 1971 | Support Staff 1971 | Research Staff 2024 | Support Staff 2024 |
|---|---|---|---|---|
| Academy of Natural Sciences, Philadelphia | 20 | 27 | 7 (1990: 14) | 15 |
| American Museum of Natural History | 37.5 | 61 | 34 (before 2023) | 54 (before 2023) |
| Bernice P. Bishop Museum | 18 | 21 | 6 (1992: 19) | 7 |
| California Academy of Sciences | 19 | 8.5 | 15 | 19.75 |
| Carnegie Museum of Natural History | 13 | 7 | 5* | 7 |
| Denver Museum of Nature & Science | 4 | 5 | 11 | 12 |
| Field Museum of Natural History | 21 | 41 | 30 | 52 |
| Florida Museum of Natural History | 7 | 6 | 24 | 21 |
| Illinois State Natural History Survey | 12 | 4 | 5–6 | 4 |
| Los Angeles County Museum of Natural History | 17 | 23 | 16 | 17.5 |
| Missouri Botanical Garden | 11 | 6 | 27 (+7 abroad) | 31 |
| New York Botanical Garden | 14 | 15 | 7 | 16 |
| Smithsonian Institution, National Museum of Natural History | 73 | 94 | 42 (+29 USDA, NOAA, DOD) | 71 (+16 USDA, NOAA, DOD) |
| University of Nebraska State Museum | 10 | 9 | 9 | 7 |

* The Carnegie Museum has an additional 4 curators and 3 collection managers who are not taxonomists but ecologists.

**FIGURE 4.1** Entomology volunteer, Alex Leede, at the Denver Museum of Nature & Science doing an excellent job pinning samples of necrophagous beetles from Côte d'Ivoire, West Africa (with permission).

solutions such as those presented by Sharkey et al. (2021, 2023): a minimalist treatment, with bare-bone descriptions, or no diagnoses at all (Ahrens et al., 2021, see Broad, this Volume, Chapter 3). Taxonomy becomes restricted to naming countable entities and as a consequence, loses its status as a deeply intellectual science exploring, documenting, and explaining the earth's biodiversity.

Technical and infrastructure developments help accelerate the process of taxonomy in many ways. Deep learning and improvement in computer vision might soon help to sort large samples and identify species (Høye et al., 2021; Wührl et al., 2022). People will still be needed to deal with the interesting fraction in a sample that the automated system has problems with, such as species not yet described, or with diagnostic characters not yet defined (Valdecasas and Wheeler, 2010). Automated systems can suggest novelties, but researchers need to take their studies from there. While this is an extremely fast developing and improving field, I doubt that the creative part of scientific research will be, to any serious extent, taken over by automated systems. In the best case, experts will be relieved from routine sorting and identification (Wührl et al., 2022; Vasilita et al., 2024).

Online databases shorten the time for taxonomists by magnitudes to find specimens in a vast array of collections, as it was never possible before. If non-experts mine online databases for data, taxonomic experts are still needed because of outdated taxonomy and issues in data quality (Mesibov, 2013; Hjarding et al., 2014; López-Guillén et al., 2024). Misidentifications are common in museum collections and are regularly translated into online databases (Goodwin et al., 2015; Nekola et al., 2019). 'Knowledge from expert taxonomists and ecologists was especially critical to evaluate the data retrieved from online databases, as the latter includes a substantial proportion of potentially erroneous occurrence records uploaded by non-experts, so the validation of data coming from this type of sources is necessary' (Reverté et al., 2023) but neglected in many cases when taxonomists are not involved in projects. This relates, even more severely, to the species identifications in BOLD and GenBank (Meiklejohn et al., 2019). Digitizing museum collections can attract windfall funding, such as the $200 million project the Natural History Museum attracted to digitize Britain's collections (Pennisi, 2024), but taxonomists are not given a role in such projects.

We cannot expect the number and quality of taxonomists to maintain itself without teaching and subsequent promotion of young talent. Malicky (1980) reminded us that even professional taxonomists are largely self-taught, similar to amateurs, because taxonomy as such has not been taught at universities. This fits with my personal experience. Relying on the self-motivation of a few is a risky strategy forward, although it has been the dominating strategy in most universities globally. Where are the chairs of taxonomy in the university systems? Taxonomists are needed in university faculty positions to teach taxonomy and raise a new generation of young biodiversity researchers (Britz et al., 2020). What Malicky (1980) and Schminke (1994) demanded a generation ago for German-speaking countries should nowadays be demanded everywhere: additional professorships for integrative taxonomy and systematics. Particularly in biodiversity-rich countries, investment in taxonomists is a great opportunity to promote biodiversity discovery regionally (Brygoo, 1988; Paknia et al., 2015), and it sometimes has happened (Carvalho et al., 2005). According to Schminke (1990), it takes seven to ten years for taxonomists to establish themselves as an international expert. This is a long-term strategy, difficult to defend when short-term return of investment is expected, but the alternative would be a further decline of taxonomy. With modern infrastructure, the time for taxonomists to achieve internationally recognized qualification might be halved, but it is still a long-term investment.

## People Helping to Do Taxonomy—Support Staff, Citizen Scientists, Volunteers

Even if taxonomists often are doing their work alone, support staff, paid or unpaid, increase their efficiency and output enormously. Paid support staff, like curatorial assistants or collection assistants, and particularly professional collection management is a lifeline for taxonomists. Wilson (2004) clearly and readily recognized this need when he stated that a taxonomist should have several technical assistants each. Ted Edwards described a setting from bygone times: 'The Lepidoptera unit settles into a three person team [scientist, experimental officer and assistant], which achieved an efficiency now no longer possible when scientists have to grovel for funding and assistants are seen as short-term and dispensable' (Edwards, 2008). Technical help is no less than essential to perform taxonomy efficiently. This technical help could come as dedicated staff or, in a more imaginative approach, in the form of a taxonomical support service as currently suggested by Senckenberg in Germany. Sigwart et al. state: 'The real intellectual, skilled and creative part of taxonomy is the process of recognizing an organism as new to science, and formulating the hypothesis that it is an independent evolutionary lineage from all other described species. The additional laborious work that comes associated with publishing the description – producing images, drawings and sequence data as evidence for the new hypothesis – does not always require a researcher and was historically the remit of technicians' (Sigwart et al., 2023, p. 2).

While many technical components of taxonomic work are now much faster and cheaper, such as genetic analyses or photographic documentation, they still need to be done. The production of publication-ready stacked images, a once half-hour long process, is now achievable in a few seconds. People are still needed to prepare and clean the specimen and operate the microscope. Three years ago, a group in Stanford sequenced a human genome for medical diagnostics in 7–8 hours (Gorzynski et al., 2022). Times and costs for genetic analyses, be it whole genomes or barcodes, decrease rapidly (e.g., Vasilita et al., 2024). Whole genome sequencing cost $300 million in 2003 and is now available for several hundred dollars. Vasilita et al. (2024) produced 250 barcodes for 285 specimens in 6 hours, including specimen identification through BLAST. Still, somebody needs to operate the system and check and interpret the results, particularly when barcodes turn out to be ineffective (Quek et al., 2024; Nymoen et al., 2024).

It is well known, and a common experience, that sorting quantitative ecological samples to useful entities, e.g., to species, is time-consuming with the few personnel that is normally assigned to such projects. Blair et al. call it 'prohibitively time-consuming and expensive, limiting how much data a project can afford to process' (Blair et al., 2023, p. 147). This is very true. However, instead of choosing the easiest and most efficient solution, namely employing more people to do the sorting, a technological approach often seems to be preferable, or cheaper (Wührl et al., 2022; Vasilita et al., 2024). Admittedly it can produce fascinating stories for a general audience, or funding bodies, or benefactors, for that matter, combining interesting technology with biodiversity themes.

Like gentlemen or amateur scientists, citizen scientists are by no means a new phenomenon (e.g., Clavero and Revilla, 2014). While the delimitations between amateur and citizen scientists are blurry, both had a significant share of biodiversity research since its beginnings. For many decades, citizen science has flourished, be it through large projects (Silvertown, 2009; Deacon et al., 2023), online infrastructure (Skvarla and Fisher, 2023), in museums such as the Denver Museum of Nature & Science with approaching 2,000 volunteers in pre-Covid times, or opportunistic or one-off opportunities, such as my Japanese Beetle mapping project in Colorado (Krell, 2018). Citizen scientists, volunteers (Figure 4.1), and amateur experts have been and will be an essential component of the taxonomy endeavor.

## People Supporting Taxonomy—Administrators, Colleagues, Donors

Taxonomy needs support in both society and in the scientific community. The general public appears to appreciate natural history stories and exhibitions, given the large, sometimes exorbitant numbers of visitors to natural history museums. In 2022, for example, the Natural History Museum, in London was the most visited museum in the country. In the UK, we have a large number of superb, highly interesting museums, but the Natural History Museum made it to first place, nevertheless. Natural history narratives are often easier to understand than stories from other sciences, and authentic, tangible objects enhance the experience and enjoyment of visitors substantially (Schwan and Dutz, 2020). The success of citizen science gives an additional indication that the general public is supportive of biodiversity research.

Colleagues are a different story. In a survey on systematic biology training and personnel, the perceived need for positions in revisionary taxonomy was mostly lower than for positions for survey work and often lower than positions for phylogenetic analysis. Only 21.0–42.1% of the respondents saw a need for taxonomic positions (Celebuski and Farris, 1990). This was thirty-four years ago, but the general attitude in biological circles has not changed much. Taxonomists are in the lower or lowest position in the academic pecking order where they have traditionally been for a long time (Malicky, 1980; Slater, 1991; Wägele et al., 2012), it seems we need to do a better, and probably more aggressive, job presenting to our colleagues in our institutions and universities fascinating stories about the nature of taxonomic discovery and why it is essential. Despite all the increasingly harsh competition in academia, taxonomy does need the support of our colleagues and administrations in our institutions but is all too often still sidelined.

## People Appreciating Taxonomy—Lawmakers, Policy Makers, Science Policy Makers

Decisions are made by decision-makers. Taxonomists rarely are decision-makers. Taxonomists are scientists arguing rationally and with facts (one would hope). Decision-makers increasingly get informed by ideology and beliefs, and are often less interested in facts and reality, at least in my country (see O'Brien, 2011). Strong arguments for the revival of taxonomy have been made for over half a century, without too much success. To let people outside our field of competence see value and excitement in species discovery and documentation, we can present exciting, inspiring, and touching stories about the natural world and its exploration. We can, for example, share the wonder of discovering a new species that nobody has ever seen before or show micro-photographs of structures nobody has ever expected to see. Let's get a little bit out of our comfort zone, even if all this will be a shot in the dark.

We need sympathetic decision-makers who are willing to invest in biodiversity research and are willing to create supporting legislation. As Prathapan et al. (2018),

Britz et al. (2020), Bouchet et al. (2023), and Löbl et al. (2023) have explained, one of the major and growing obstacles to our endeavors to discover undescribed biodiversity is the legal framework that regulates the collection of organisms. Legislation that was initially introduced to regulate the trade of threatened species is now generally applied to non-commercial biodiversity research and requires permitting of research, collecting, genetic study, exporting, and importing. Paperwork is rising, the process can be lengthy and expensive, stipulations can decrease the efficiency of research substantially, and applications are not always successful. In times of a biodiversity crisis, with rapid loss of habitat and species, this will lead to irrevocable loss of knowledge. We need decision-makers to understand that taxonomy does not need to be feared like biopiracy. Taxonomy will not lead to substantial financial gain for taxonomists, their institutions or their countries, but will provide knowledge for us all and help everybody, independent of which country or institution in which it is produced. A separate paper discussing potential solutions for this issue is in preparation.

## What Next?

If we have a shortage of teachers, would we refuse to employ additional staff and instead propagate the development of AI-robot-teachers that would make the employment of additional human teachers unnecessary? Or dentists? Or administrators? Society would most likely find ways to employ additional people.

Yet, it is unlikely that taxonomy will get more funding, more positions, more respect in the short term. 'Funding bodies fund opportunities, they don't fund need', Richard Lane, former Science Director of the Natural History Museum London stated in a talk at Naturalis Biodiversity Center in 2010 (https://www.youtube.com/watch?v=YJYk9BEdpeA). While this seems a realistic, albeit in some ways depressing assessment, we should not give up, present the extraordinary achievements of taxonomy and develop visions of what we could achieve in the future—and what needs to change or evolve to make these visions a reality, but also what needs to stay the same. For instance, we will continue to need people to do taxonomy. Still, taxonomy is not entitled to thrive or even to survive. Neither is biodiversity. Neither is humankind. We must work on it, explain why it matters and present understandable narratives that excite and inspire people.

It is easy and cheap to continue complaining that we know so little about the natural world and that taxonomy has not been delivering, to find excuses why we cannot afford exploration and enhancing our knowledge of the natural world. It is our decision as humanity to do so. Alternatively, humanity could decide to not let taxonomists go extinct and take steps toward preserving and enhancing the taxonomic workforce. It is our decision.

A Nature editorial stated 20 years ago: 'So far, taxonomists and ecologists have failed to muster a comparable response to the galloping loss of our planet's biodiversity. It's time that they did'. (Anonymous, 2004). Yes, please, let us do it! But it won't happen by closing herbaria or cutting curator positions, and I doubt that creating another acronym will do the trick either. Taxonomy is a vast endeavor performed by a distributed army of highly motivated, individualistic enthusiasts. This has been and is taxonomy's structure and strength (Dubois, 2003). Whether we keep this army strong will determine if we win the battle against disappearing knowledge and disappearing biodiversity, or not.

## ACKNOWLEDGMENTS

I am grateful to David Williams, Natural History Museum London, and Quentin Wheeler, State University of New York College of Environmental Science and Forestry, for the invitation to contribute to this symposium. Brian Brown, Natural History Museum of Los Angeles County, Los Angeles; Dmitry A. Dmitriev, Illinois Natural History Survey, Champaign; Neal Evenhuis, Bernice Pauahi Bishop Museum, Honolulu; Thorsten Lumbsch, Field Museum, Chicago; Scott Miller, Smithsonian Institution, National Museum of Natural History, Washington, DC; Robert Naczi, New York Botanical Garden; Gary Rosenberg and Mark Sabaj Perez, Academy of Natural Sciences of Drexel University, Philadelphia; Scott Sampson and Shannon Bennett, California Academy of Sciences, San Francisco; Ainsley Seago, Carnegie Museum of Natural History, Pittsburgh; Keith R. Willmott, Florida Museum of Natural History, Gainsville; and Susan Weller, University of Nebraska State Museum provided information about their institutions. My daughter, Victoria Krell, took on the proofreading.

## REFERENCES

Abukhalil, R. (2014) The rising trend in authorship. *The Winnower*, 2: e141832.26907 (5 pp.).

Ahrens, D., Ahyong, S.T., Ballerio, A., Barclay, M.V.L., Eberle, J., Espeland, M., Huber, B.A., Mengual, X., Pacheco, T.L., Peters, R.S., Rulik, B., Vaz-de-Mello, F., Wesener, T. and Krell, F.-T. (2021) Is it time to describe new species without diagnoses?—A comment on Sharkey et al. (2021). *Zootaxa*, 5027: 151–159, https://doi.org/10.11646/zootaxa.5027.2.1

Akingbohungbe, A.E. (1989) Six-Legged Science in Nigeria and Its Development. Inaugural Lecture Delivered at the University of Ife (Now Obafemi Awolowo University) on 12th March, 1985. *Inaugural Lecture Series 72*. Obafemi Awolowo University Press, Ile-Ife, Nigeria. 37 pp.

Alberti, S.J.M.M. (2001) Amateurs and professionals in one county: Biology and natural history in late Victorian Yorkshire. *Journal of the History of Biology*, 34: 115–147, https://doi.org/10.1023/A:1010373912743

Anonymous (1972) *Annual Report for 1971 of the Denver Museum of Natural History*. Museum, Denver, Colorado. 44 pp.

Anonymous (2004) Ignorance is not bliss. *Nature*, 430: 385, https://doi.org/10.1038/430385a

Bebber, D.P., Wood, J.R.I., Barker, C. and Scotland, R.W. (2013) Author inflation masks global capacity for species discovery in flowering plants. *New Phytologist*, 201: 700–706, https://doi.org/10.1111/nph.12522

Beermann, J., Westbury, M.V., Hofreiter, M., Hilgers, L., Deister, F., Neumann, H. and Raupach, M.J. (2018) Cryptic species in a well-known habitat: Applying taxonomics to the amphipod genus *Epimeria* (Crustacea, Peracarida). *Scientific Reports*, 8: 68793 (26 pp.), https://doi.org/10.1038/s41598-018-25225-x

Blair, J.D., Gaynor, K.M., Palmer, M.S. and Marshall, K.E. (2023) A gentle introduction to computer vision-based specimen classification in ecological datasets. *Journal of Animal Ecology*, 93: 147–158, https://doi.org/10.1111/1365-2656.14042

Blaxter, M. and Floyd, R. (2003) Molecular taxonomics for biodiversity surveys: Already a reality. *Trends in Ecology and Evolution*, 18: 268–269, https://doi.org/10.1016/S0169-5347(03)00102-2

Bouchet, P., Decock, W., Lonneville, B., Vanhoorne, B. and Vandepitte, L. (2023) Marine biodiversity discovery: The metrics of new species descriptions. *Frontiers in Marine Science*, 10: 929989 (14 pp.), https://doi.org/10.3389/fmars.2023.929989

Britz, R., Hundsdörfer, A.K. and Fritz, U. (2020) Funding, training, permits—The three big challenges of taxonomy. *Megataxa,* 1: 49–52, https://doi.org/10.11646/megataxa.1.1.10

Bruton, M.N. (1989) Does animal systematics have a future in South Africa? *South Africa Journal of Science*, 85: 348–350.

Brygoo, E.R. (1988) Défense et illustration de la taxinomie. *Archives de l'Institut Pasteur de Madagascar*, 54: 11–40.

Buckner, J.C., Sanders, R.S., Faircloth, B.C. and Chakrabarty, P. (2021) The critical importance of vouchers in genomics. *ELife*, 10: e68264 (7 pp.), https://doi.org/10.7554/elife.68264

Burns, A., Becker, K.P., Casagrande, D., Daniels, J., Roberts, P., Orenstein, E., Vogt, D.M., Teoh, J.E., Wood, R., Yin, A.H., Genot, B., Gruber, D.F., Katija, K., Wood, R.J. and Phillips, B.T. (2024) An in situ digital synthesis strategy for the discovery and description of ocean life. *Science Advances*, 10: eadj4960 (13 pp.), https://doi.org/10.1126/sciadv.adj4960

Buyck, B. (1999) Taxonomists are an endangered species in Europe. *Nature*, 401: 321, https://doi.org/10.1038/43762

Byrne, A.Q. (2023) Reimagining the future of natural history museums with compassionate collection. *PLoS Biology*, 21(5): e3002101 (4 pp.), https://doi.org/10.1371/journal.pbio.3002101

Card, D.C., Shapiro, B., Giribet, G., Moritz, C. and Edwards, S.V. (2021) Museum genomics. *Annual Review of Genetics*, 55: 633–659, https://doi.org/10.1146/annurev-genet-071719-020506

Carvalho, M.R. de, Bockmann, F.A., Amorim, D.S., Vivo, M. de, Toledo-Piza, M. de, Menezes, N.A., Figueiredo, J.L. de, Castro, R.M.C., Gill, A.C., McEachran, J.D., Compagno, L.J.V., Schelly, R.C., Britz, R., Lundberg, J.G., Vari, R.P. and Nelson, G. (2005) Revisiting the taxonomic impediment. *Science*, 307: 353, https://doi.org/10.1126/science.307.5708.353b

Casadevall, A. and Fang, F.C. (2008) Descriptive science. *Infection and Immunity*, 76: 3835–3836, https://doi.org/10.1128/iai.00743-08

Celebuski, C.A. and Farris, E. (1990) Systematic biology training and personnel. *Higher Education Surveys Report, Survey Number 10.* Westat, Rockville, Maryland. 57 pp.

Cerri, F., Saliu, F., Maggioni, D., Montano, S., Seveso, D., Lavorano, S., Zoia, L., Gusetti, F., Lasagni, M., Orlandi, M., Taglialatela-Scafati, O. and Galli, P. (2022) Cytotoxic compounds from Alcyoniidae: An overview of the last 30 years. *Marine Drugs*, 20: 134 (45 pp.), https://doi.org/10.3390/md20020134

Clavero, M. and Revilla, E. (2014) Mine centuries-old citizen science. *Nature*, 510: 35, https://doi.org/10.1038/510035c

Cockerell, T.D.A. (1946) Botanical taxonomy. *Science*, 104: 146, https://doi.org/10.1126/science.104.2693.146.a

Colwell-Chanthaphonh, C., Haglund, K.A., Stucky, R.K. & Wineman, P. (2013) "A museum here founded", a summative history, pp. 11–63. In *Denver's Natural History Museum: A History* (eds Johnson, K., Armstrong, B., Colwell-Chanthaphonh, C., Kruger, F., Haglund, K.A. and Krell, F.-T.). *Denver Museum of Nature & Science Annals*, 4: 1–427.

Conert, H.J., Gutmann, W.F., Janssen, R., Mollenhauer, D., Peters, D.S., Storch, G., Struve, W., Türkay, M. and Ziegler, W. (1990) Grundlagen und Anwendungsbezug von Taxonomie und Systematik. *Aufsätze und Reden der Senckenbergischen Naturforschenden Gesellschaft*, 36: 1–44.

Conference of Directors of Systematic Collections (1971) *The Systematic Biology Collections of the United States: An Essential Resource. Part II. The Great Collections: Statistical Information. A Report to the National Science Foundation.* New York Botanical Garden, Bronx, New York. IV, 52 pp.

Copete, J.C., Kir, A., Novotny, V. and Cámara-Leret, R. (2023) The importance of indigenous and local people for cataloging biodiversity. *Trends in Ecology & Evolution*, 38: 1112–1114, https://doi.org/10.1016/j.tree.2023.08.017

Cupello, M., Silva, F.A.B. and Vaz-de-Mello, F.Z. (2023) The taxonomic revolution of New World dung beetles (Coleoptera: Scarabaeidae: Scarabaeinae). *Frontiers in Ecology and Evolution*, 11: 1168754 (42 pp.), https://doi.org/10.3389/fevo.2023.1168754

Daly, A. and Moloney, D. (2004) Managing corporate rebranding. *Irish Marketing Review*, 17: 30–36.

Dar, G.H., Khuroo, A.A., Reddy, C.S. and Malik, A.H. (2012) Impediment to taxonomy and its impact on biodiversity science: An Indian perspective. *Proceedings of the National Academy of Science, India, Section B Biological Sciences*, 82: 235–240, https://doi.org/10.1007/s40011-012-0031-3

Dawson, G., Lintott, C. and Shuttleworth, S. (2015) Constructing scientific communities: Citizen science in the nineteenth and twenty-first centuries. *Journal of Victorian Culture,* 20: 246–254, https://doi.org/10.1080/13555502.2015.1022053

Dayrat, B. (2005) Towards integrative taxonomy. *Biological Journal of the Linnean Society,* 85: 407–415, https://doi.org/10.1111/j.1095-8312.2005.00503.x

Deacon, C., Govender, S. and Samways, M.J. (2023) Overcoming biases and identifying opportunities for citizen science to contribute more to global macroinvertebrate conservation. *Biodiversity and Conservation*, 32: 1789–1806, https://doi.org/10.1007/s10531-023-02595-x

Deans, A.R., Yoder, M.J. and Balhoff, J.P. (2012) Time to change how we describe biodiversity. *Trends in Ecology and Evolution*, 27: 78–84, https://doi.org/10.1016/j.tree.2011.11.007

Drew, L.W. (2011) Are we losing the science of taxonomy? *BioScience*, 61: 942–946, https://doi.org/10.1525/bio.2011.61.12.4

Dubois, A. (2003) The relationships between taxonomy and conservation biology in the century of extinctions. *Comptes Rendus Biologies*, 326: S9–S21, https://doi.org/10.1016/s1631-0691(03)00022-2

Duncker, H.-R. (1992) Schwerpunkte in der zoologischen Forschung, pp. 34–41. In *Jahrhundertwissenschaft Biologie?! Aktueller Stand der Biowissenschaften in Deutschland* (ed P. Präve), VCH, Weinheim.

Edwards, E.D. (2008) Ian Francis Bell Common (23 June 1917 to 3 June 2006). *Journal of the Lepidopterists' Society*, 62: 111–113.

Engel, M.S., Ceríaco, L.M.P., Daniel, G.M., Dellapé, P.M., Löbl, I., Marinov, M., Reis, R.E., Young, M.T., Dubois, A., Agarwal, I., Lehmann, A.P., Alvarado, M., Alvarez, N., Andreone, F., Araujo-Vieira, K., Ascher, J.S., Baêta, D., Baldo, D., Bandeira, S.A., Barden, P., Barrasso, D.A., Bendifallah, L., Bockmann, F.A., Böhme, W., Borkent, A., Brandão, C.R.F., Busack, S.D., Bybee, S.M., Channing, A., Chatzimanolis, S., Christenhusz, M.J.M., Crisci, J.V., D'Elía, G., Costa, L.M. da, Davis, S.R., Lucena, C.A.S. de, Deuve, T., Fernandes Elizalde, S., Faivovich, J., Farooq, H., Ferguson, A.W., Gippoliti, S., Gonçalves, F.M.P., Gonzalez, V.H., Greenbaum, E., Hinojosa-Díaz, I.A., Ineich, I., Jiang, J., Kahono, S., Kury, A.B., Lucinda, P.H.F., Lynch, J.D., Malécot, V., Marques, M.P., Marris, J.W.M., Mckellar, R.C., Mendes, L.F., Nihei, S.S., Nishikawa, K., Ohler, A., Orrico, V.G.D., Ota, H., Paiva, J., Diogo Parrinha, D., Pauwels, O.S.G., Pereyra, M.O., Pestana, L.B., Pinheiro, P.D.P., Prendini, L., Prokop, J., Rasmussen, C., Rödel, M.-O., Rodrigues, M.T., Rodríguez, S.M., Salatnaya, H., Sampaio, Í., Alba Sánchez-García, A., Shebl, M.A., Bruna S. Santos, B.S., Solórzano-Kraemer, M.M., Sousa, A.C.A., Stoev, P., Teta, P., Trape, J.-F., Van-Dúnem Dos Santos, C., Vasudevan, K., Vink, C.J., Vogel, G., Wagner, P., Wappler, T., Ware, J.L., Wedmann, S. and

Kusamba, C.Z. (2021) The taxonomic impediment: A shortage of taxonomists, not the lack of technical approaches. *Zoological Journal of the Linnean Society*, 22: 381–387, https://doi.org/10.1093/zoolinnean/zlab072

Ernst, R., Kehlmaier, C., Baptista, N., Pinto, P.V., Branquima, M.F., Dewynter, M., Fouquet, A., Ohler, A. and Schmitz, A. (2021) Filling the gaps: The mitogenomes of Afrotropical egg-guarding frogs based on historical type material and a re-assessment of the nomenclatural status of *Alexteroon* Perret, 1988 (Hyperoliidae). *Zoologischer Anzeiger*, 293: 215–224, https://doi.org/10.1016/j.jcz.2021.06.002

Felis, G., Del Casale, A., Salvetti, E., Fracchetti F. and Torriani, S. (2010) TAXON-OMICS: For a renaissance of taxonomy as a fundamental resource towards biodiversity-based innovation in industrial biotechnology. *14th International Biotechnology Symposium and Exhibition, Biotechnology for the Sustainability of Human Society*, 1: 133, https://iris.univr.it/handle/11562/749971?mode=full

Fischer, E.E., Cobb, N.S., Kawahara, A.Y., Zaspel, J.M. and Cognato, A.I. (2021) Decline of amateur Lepidoptera collectors threatens the future of specimen-based research. *BioScience*, 71: 396–404, https://doi.org/10.1093/biosci/biaa152

Fontaine, B., Achterberg, K. van, Alonso-Zarazaga, M.A., Araujo, R., Asche, M., Aspöck, H., Aspöck, U., Audisio, P., Aukema, B., Bailly, N., Balsamo, M., Bank, R.A., Belfiore, C., Bogdanowicz, W., Boxshall, G., Burckhardt, D., Chylarecki, P., Dearveng, L., Dubois, A., Enghoff, H., Fochetti, R., Fontaine, C., Gargominy, O., Gomez Lopez, M.S., Goujet, D., Harvey, M.S., Heller, K.-G., Helsdingen, P. van, Hoch, H., De Jong, Y., Karsholt, O., Los, W., Magowski, W., Massard, J.A., McInnes, S.J., Mendes, L.F., Mey, E., Michelsen, V., Minelli, A., Nieto Nafrıa, J.M., Nieukerken, E.J. van, Pape, T., De Prins, W., Ramos, M., Ricci, C., Roselaar, C., Rota, E., Segers, H., Timm, T., Tol, J. van and Bouchet, P. (2012) New species in the Old World: Europe as a frontier in biodiversity exploration, a test bed for 21st century taxonomy. *PloS ONE*, 7: e36881 (7 pp.), https://doi.org/10.1371/journal.pone.0036881

Frobel, K. and Schlumprecht, H. (2016) Erosion der Artenkenner. Ergebnisse einer Befragung und notwendige Reaktionen. *Naturschutz und Landschaftsplanung*, 48(4): 105–113.

Funk, V.A. (2006) Floras: A model for biodiversity studies or a thing of the past? *Taxon*, 55: 581–588, https://doi.org/10.2307/25065635

Fürsch, H. (1992) Amateur und Museum: Eine Symbiose zwischen freischaffenden Faunisten und Forschungsinstitutionen. *Spixiana, Suppl.*, 17: 223–226.

Garcia-Rosello, E., Gonzalez-Dacosta, J., Guisande, C. and Lobo, J.M. (2023) GBIF falls short of providing a representative picture of the global distribution of insects. *Systematic Entomology*, 48: 489–497, https://doi.org/10.1111/syen.12589

Glynn, M.A. (2000) When cymbals become symbols: Conflict over organizational identity within a symphony orchestra. *Organization Science*, 11: 285–298, https://doi.org/10.1287/orsc.11.3.285.12496

Godfray, H.C.J. (2002) Challenges for taxonomy. The discipline will have to reinvent itself if it is to survive and flourish. *Nature*, 417: 17–19, https://doi.org/10.1038/417017a

Goodwin, Z.A., Harris, D.J., Filer, D., Wood, J.R.I. and Scotland, R.W. (2015) Widespread mistaken identity in tropical plant collections. *Current Biology*, 25: R1066–R1067, https://doi.org/10.1016/j.cub.2015.10.002

Gorzynski, J.E., Goenka, S.D., Shafin, K., Jensen, T.D., Fisk, D.G., Grove, M.E., Spiteri, E., Pesout, T., Monlong, J., Baid, G., Bernstein, J.A., Ceresnak, S., Chang, P.C., Christle, J.W., Chubb, H., Dalton, K.P., Dunn, K., Garalde, D.R., Guillory, J., Knowles, J.W., Kolesnikov, A., Ma, M., Moscarello, T., Nattestad, M., Perez, M., Ruzhnikov, M.R., Samadi, M., Setia, A., Wright, C., Wusthoff, C.J., Xiong, K., Zhu, T., Jain, M., Sedlazeck, F.J., Carroll, A., Paten, B. and Ashley, E.A. (2022) Ultrarapid nanopore genome sequencing in a critical care setting. *New England Journal of Medicine*, 386: 700–702, https://www.nejm.org/doi/full/10.1056/NEJMc2112090

Green, C.D. (2019) Natural history disavowed: Confronting colonial legacies in the Musée des Confluences. *Museological Review*, 23: 25–36.

Grimaldi, D.A. and Engel, M.S. (2007) Why descriptive science still matters. *BioScience*, 57: 646–647, http://www.dx.doi.org/10.1641/B570802

Groom, Q., Güntsch, A., Huybrechts, P., Kearney, N., Leachman, S., Nicolson, N., Page, R.D.M., Shorthouse, D.P., Thessen, A.E. and Haston, E. (2020) People are essential to linking biodiversity data. *Database*, 2020: baaa072 (4 pp.), https://doi.org/10.1093/database/baaa072

Guedes, P., Alves-Martins, F., Arribas, J.M., Chatterjee, S., Santos, A.M.C., Lewin, A., Bako, L., Webala, P.W., Correia, R.A., Rocha, R. and Ladle, R.J. (2023) Eponyms have no place in 21st-century biological nomenclature. *Nature Ecology & Evolution*, 7: 1157–1160, https://doi.org/10.1038/s41559-023-02022-y

Hartmann, G. (1965) Zur gegenwärtigen Situation der Systematik in der Zoologie (Mit kritischen Bemerkungen zur Organisation der Zoologie an einigen alten und neuen Universitäten). *Zoologischer Anzeiger*, 175: 92–100.

Hennig, W. (1971) Zur Situation der biologischen Systematik. *Erlanger Forschungen B*, 4: 7–15.

Hertwig, R. (1914) Die Abstammungslehre, pp. 1–91. In *Die Kultur der Gegenwart, Ihre Entwicklung und Ihre Ziele. Dritter Teil Mathematik, Naturwissenschaften, Medizin. Vierte Abteilung. Organische Naturwissenschaften. Vierter Band. Abstammungslehre, Systematik, Paläontologie, Biogeographie* (eds R. Hertwig and R.V. Wettstein). Teubner, Leipzig.

Heywood, V.H. (1973) Taxonomy in crisis? Or taxonomy is the digestive system of biology. *Acta Botanica Academiae Scientiarum Hungaricae*, 19: 139–146.

Hjarding, A., Tolley, K.A. and Burgess, N.D. (2014) Red list assessments of East African chameleons: A case study of why we need experts. *Oryx*, 49: 652–658, http://dx.doi.org/10.1017/S0030605313001427

Hochkirch, A., Casino, A., Penev, L., Allen, D., Lilley, L., Georgiev, T., Gospodinov, K. and Barov, B. (2022) *European Red List of Insect Taxonomists*. Publication Office of the European Union, Luxembourg. viii, 32 pp.

Hodač, L., Karbstein, K., Tomasello, S., Wäldchen, J., Bradican, J.P. and Hörandl, E. (2023) Geometric morphometric versus genomic patterns in a large polyploid plant species complex. *Biology*, 12: 418 (36 pp.), https://doi.org/10.3390/biology12030418

Holm, E. (1975) Insect taxonomy in South Africa: How can we cope? *Proceedings of the 1st Congress of the Entomological Society of Southern Africa*, The Society, Pretoria, pp. 35–40.

Hopkins, G.W. and Freckleton, R.P. (2002) Declines in the numbers of amateur and professional taxonomists: Implications for conservation. *Animal Conservation*, 5: 245–249, https://doi.org/10.1017/S1367943002002299

Høye, T.T., Ärje, J., Bjerge, K., Hansen, O.L.P., Iosifidis, A., Leese, F., Mann, H.M.R., Meissner, K., Melvad, C. and Raitoharju, J. (2021) Deep learning and computer vision will transform entomology. *Proceedings of the National Academy of Sciences of the United States of America*, 118: e2002545117 (10 pp.), https://doi.org/10.1073/pnas.2002545117

Ignatius, A. and Lakhani, K. (2023) AI won't replace humans — but humans with AI will replace humans without AI. *Harvard Business Review*, https://hbr.org/2023/08/ai-wont-replace-humans-but-humans-with-ai-will-replace-humans-without-ai [accessed 28 Feb. 2024].

Johnson, K.R., Owens, I.F.P. and the Global Collections Group (2023) A global approach for natural history museum collections. *Science*, 379: 1192–1194, suppl. https://doi.org/10.1126/science.adf6434

Karbstein, K., Kösters, L., Hodač, L., Hofmann, M., Hörandl, E., Tomasello, S., Wagner, N.D., Emerson, B.C., Albach, D.C., Scheu, S., Bradler, S., Vries, J. de, Irisarri, I., Li, H., Soltis, P., Mäder, P. and Wäldchen, J. (2024) Species delimitation 4.0: Integrative taxonomy meets artificial intelligence. *Trends in Ecology & Evolution*, 39: 771–784, https://doi.org/10.1016/j.tree.2023.11.002

Kloster, M., Langenkämper, D., Zurovietz, M., Beszteri, B. and Nattkemper, T.W. (2020) Deep learning-based diatom taxonomy on virtual slides. *Scientific Reports*, 10: 14416 (13 pp.), https://doi.org/10.1038/s41598-020-71165-w

Koch, M.A., German, D.A., Kiefer, M. and Franzke, A. (2018) Database taxonomics as key to modern plant biology. *Trends in Plant Science*, 23: 4–6, https://doi.org/10.1016/j.tplants.2017.10.005

Köhler, F. (1997) Amateurwissenschaft: Entwicklung, Beschreibung und wissenschaftssoziologische Analyse am Beispiel der Koleopterologie. *Decheniana Beihefte*, 36: 351–420.

Korf, R.P. (2005) Reinventing taxonomy: A curmudgeon's view of 250 years of fungal taxonomy, the crisis in biodiversity, and the pitfalls of the phylogenetic age. *Mycotaxon*, 93: 407–415.

Krell, F.-T. (2018) Japanese beetles make Colorado home. *YourHub (Denver Post), 12 July 2018,* 6T+9T. Available online: www.researchgate.net/publication/326366522_Japanese_beetles_make_Colorado_home (accessed on 4 March 2024).

Krell, F.-T. and Wheeler, Q. (2014) Specimen collection: Plan for the future. *Science*, 344: 815–816, http://dx.doi.org/10.1126/science.344.6186.815

Krell, F.-T., Klimeš, P., Rocha, L.A., Fikáček, M. and Miller, S.A. (2016) Preserve specimens for reproducibility. *Nature*, 539: 168, http://dx.doi.org/10.1038/539168b

Lane, R. (2009) Taxonomy comes of age. *Nature*, 461: 597, https://doi.org/10.1038/461597a

Le, A.N.H., Cheng, J.M.S., Kuntjara, H. and Lin, C.T.-J. (2014) Corporate rebranding and brand preference, brand name attitude and product expertise as moderators. *Asia Pacific Journal of Marketing and Logistics*, 26: 602–620.

Lee, Z. and Bourne, H. (2017) Managing dual identities in nonprofit rebranding: An exploratory study. *Nonprofit and Voluntary Sector Quarterly*, 46: 794–816, https://doi.org/10.1177/0899764017703705

Leshner, A.I. (2008) Reduce administrative burden. *Science*, 322: 1609, https://doi.org/10.1126/science.1168345

Lipscomb, D., Platnick, N. and Wheeler, Q. (2003) The intellectual content of taxonomy: A comment on DNA taxonomy. *Trends in Ecology and Systematics*, 18: 65–66, http://dx.doi.org/10.1016/S0169-5347(02)00060-5

Lo, E., Nie, R.-E. and Vogler, A. (2024) The geographic and phylogenetic structure of public DNA barcode databases: An assessment using Chrysomelidae (leaf beetles). *Frontiers in Ecology and Evolution*, 12: 1395898 (13 pp.), https://doi.org/10.3389/fevo.2024.1305898

Löbl, I., Klausnitzer, B., Hartmann, M. and Krell, F.-T. (2023) The silent extinction of species and taxonomists—an appeal to science policymakers and legislators. *Diversity*, 15: 1053 (17 pp.), https://doi.org/10.3390/d15101053

Loïs, G. (2014) La contribution de l'observation citoyenne à la recherche scientifique. *Pour*, 2014/3: 43–51.

Lomax, W., Mador, M. and Fitzhenry, A. (2002) Corporate rebranding: Learning from experience. *Kingston Business School Occasional Paper Series*, 48: ii + 10 pp, Kingston upon Thames, U.K.: Kingston Business School, Kingston University.

López-Guillén, E., Herrera, I., Bensid, B, Gómez-Bellver, C., Ibáñez, N., Jiménez-Mejías, P., Mairal, M., Mena-García, L., Nualart, N., Utjés-Mascó, M. and López-Pujol, J. (2024) Strengths and challenges of using iNaturalist in plant research with focus on data quality. *Diversity*, 16: 42 (22 pp.), https://doi.org/10.3390/d16010042

Lyal, C.H.C. (2005) Crumbling foundations: The erosion of the taxonomic base of biology. *School Science Review*, 87: 65–73.

Maletti, S., Niehuis, O., Mayer, C., Sann, M., Klopfstein, S., Nottebrock, G., Baur, H. and Peters, R.S. (2021) Phylogeny, taxonomics, and ovipositor length variation of the

*Pteromalus albipennis* species group (Hymenoptera: Chalcidoidea: Pteromalidae: Pteromalinae). *Journal of Zoological Systematics and Evolutionary Research*, 59: 349–358, https://doi.org/10.1111/jzs.12433

Malicky, H. (1980) Betrachtungen über die Lage der Zootaxonomie. *Naturwissenschaftliche Rundschau*, 33: 179–182.

Mayr, E. (1942) *Systematics and the Origin of Species from the Viewpoint of a Zoologist.* Columbia University Press, New York. xiv, 334 pp.

Mayr, E., Linsley, E.G. and Usinger, R.L. (1953) *Methods and Principles of Systematic Zoology.* McGraw-Hill, New York. ix, 336 pp.

Meier, R. (2008) DNA sequences in taxonomy, opportunities and challenges, pp. 95–127. In Wheeler, Q.D. (ed.) *The New Taxonomy.* CRC Press, Boca Raton.

Meiklejohn, K.A., Damaso, N. and Robertson, J.M. (2019) Assessment of BOLD and GenBank – their accuracy and reliability for the identification of biological materials. *PLoS ONE*, 14: e0217084 (14 pp.), https://doi.org/10.1371/journal.pone.0217084

Mesibov, R. (2013) A specialist's audit of aggregated occurrence records. *ZooKeys*, 293: 1–18, https://doi.org/10.3897/zookeys.293.5111

Mollenhauer, D. (1975) Biologie ohne Systematik — oder Renaissance von Systematik und Taxonomie? *Botanische Jahrbücher für Systematik, Pflanzengeschichte und Pflanzengeographie*, 96: 276–289.

Moss, T.W., Short, J.C., Payne, G.T. and Lumpkin, G.T. (2011) Dual identities in social ventures: An exploratory study. *Entrepreneurship Theory and Practice*, 35: 805–830, https://doi.org/10.1111/j.1540-6520.2010.00372.x

Muzellec, L., Doogan, M. and Lambkin, M. (2003) Corporate rebranding – an explorative review. *Irish Marketing Review*, 16: 31–40.

Muzellec, L. and Lambkin, M. (2006) Corporate rebranding: Destroying, transferring or creating brand equity? *European Journal of Marketing*, 40: 803–824, http://dx.doi.org/10.1108/03090560610670007

Nachman, M.W., Beckman, E.J., Bowie, R.C.K., Cicero, C., Conroy, C.J., Dudley, R., Hayes, T.B., Koo, M.S., Lacey, E.A., Martin, C.H., McGuire, J.A., Patton, J.L., Spencer, C.L., Tarvin, R.D., Wake, M.H., Wang, I.J., Achmadi, A., Álvarez-Castañeda, S.T., Andersen, M.J., Arroyave, J., Austin, C.C., Barker, F.K., Barrow, L.N., Barrowclough, G.F., Bates, J., Bauer, A.M., Bell, K.C., Bell, R.C., Bronson, A.W., Brown, R.M., Burbrink, F.T., Burns, K.J., Cadena, C.D., Cannatella, D.C., Castoe, T.A., Chakrabarty, P., Colella, J.P., Cook, J.A., Cracraft, J.L., Davis, D.R., Davis Rabosky, A.R., D'Elía, G.,. Dumbacher, J.P., Dunnum, J.L., Edwards, S.V., Esselstyn, J.A., Faivovich, J., Fjeldså, J., Flores-Villela, O.A., Ford, K., Fuchs, J., Fujita, M.K., Good, J.M., Greenbaum, E., Greene, H.W., Hackett, S., Hamidy, A., Hanken, J., Haryoko, T., Hawkins, M.T.R., Heaney, L.R., Hillis, D.M., Hollingsworth, B.D., Hornsby, A.D., Hosner, P.A., Irham, M., Jansa, S., Jiménez, R.A., Joseph, L., Kirchman, J.J., LaDuc, T.J., Leaché, A.D., Lessa, E.P., López-Fernández, H., Mason, N.A., McCormack, J.E., McMahan, C.D., Moyle, R.G., Ojeda, R.A., Olson, L.E., Onn, C.K., Parenti, L.R., Parra-Olea, G., Patterson, B.D., Pauly, G.B., Pavan, S.E., Peterson, A.T., Poe, S., Rabosky, D.L., Raxworthy, C.J., Reddy, S., Rico-Guevara, A., Riyanto, A., Rocha, L.A., Ron, S.R., Rovito, S.M., Rowe, K.C., Rowley, J., Ruane, S., Salazar-Valenzuela, D., Shultz, A.J., Sidlauskas, B., Sikes, D.S., Simmons, N.B., Stiassny, M.L.J., Streicher, J.W., Stuart, B.L., Summers, A.P., Tavera, J., Teta, P., Thompson, C.W., Timm, R.M., Torres-Carvajal, O., Voelker, G., Voss, R.S., Winker, K., Witt, C., Wommack, E.A. and Zink, R.M. (2023) Specimen collection is essential for modern science. *PLOS Biology*, 21: e3002318 (6 pp.), https://doi.org/10.1371/journal.pbio.3002318

Nekola, J.C., Hutchins, B.T., Schofield, A., Najev, B. and Perez, K.E. (2019) Caveat consumptor notitia museo: Let the museum data user beware. *Global Ecology and Biogeography*, 28: 1722–1734, https://doi.org/10.1111/geb.12995

Nelson, W., Breitwieser, I., Fordyce, E., Bradford-Grieve, J., Penman, D., Roskruge, N., Trnski, T., Waugh, S. and Webb, C. (2015) *National Taxonomic Collections in New Zealand*. Royal Society of New Zealand, Wellington. 63 pp.

Nymoen, A.R., Kongsrud, J.A., Willassen, E. and Bakken, T. (2024) When standard DNA barcodes do not work for species identification: Intermixed mitochondrial haplotypes in the *Jaera albifrons* complex. (Crustacea: Isopoda). *Marine Biodiversity*, 54: 43 (17 pp.), https://doi.org/10.1007/s12526-024-01435-7

O'Brien, P. (2011) Reaching the U.S. Public through their patriotism, pastors, and pockets. *Conservation Biology*, 25: 1108–1111, https://doi.org/10.1111/j.1523-1739.2011.01781.x

Owen, R.J. (2004) Bacterial taxonomics. Finding the wood through the phylogenetic trees, pp. 353–383. In *Genomics, Proteomics, and Clinical Bacteriology. Methods and Reviews* (eds N. Woodford and A.P. Johnson). Methods in Molecular Biology 266. Humana Press, Totowa, New Jersey.

Packer, L., Grixti, J.C., Roughley, R.E. and Hanner, R. (2009) The status of taxonomy in Canada and the impact of DNA barcoding. *Canadian Journal of Zoology*, 87: 1097–1110, https://doi.org/10.1139/Z09-100

Paknia, O., Rajaei Sh., H. and Koch, A. (2015) Lack of well-maintained natural history collections and taxonomists in megadiverse developing countries hampers global biodiversity exploration. *Organisms Diversity Evolution*, 15: 619–629, https://doi.org/10.1007/s13127-015-0202-1

Páll-Gergely, B., Krell, F.-T., Ábrahám, L., Bajomi, B., Balog, L.E., Boda, P., Csuzdi, C., Dányi, L., Fehér, Z., Hornok, S., Horváth, A., Kóbor, P., Koczor, S., Kontschán, J., Kovács, T., Majoros, G., Murányi, D., Németh, T., Pernecker, B., Puskás, G., Rózsa, L., Soltész, Z., Szita, É., Szűts, T., Tóth, B., Tőke, A., Vas, Z., Zsuga, K., Zsupos, V., Csabai, Z. and Móra, A. (2024) Identification crisis: A fauna-wide estimate of biodiversity expertise shows massive decline in a Central European country. *Biodiversity and Conservation*, 33: 3871–3903, https://doi.org/10.1007/s10531-024-02934-6

Pearson, D.L., Hamilton, A.L. and Erwin, T.L. (2011) Recovery plan for the endangered taxonomy profession. *BioScience*, 61: 58–63, https://doi.org/10.1525/bio.2011.61.1.11

Peng, Y.-R., Shekhar, K., Yan, W., Herrmann, D., Sappington, A., Bryman, G.S., van Zyl, T., Tri, M., Do, H., Regev, A. and Sanes, J.R. (2019) Molecular classification and comparative taxonomics of foveal and peripheral cells in primate retina. *Cell*, 176, 1222–1237, https://doi.org/10.1016/j.cell.2019.01.004

Pennisi, E. (2024) U.K. boosts digitization of museum specimens, London's Natural History Museum gets nearly $200 million to scan country's collections. *Science*, 384: 17, https://doi.org/10.1126/science.adp6028

Ponti, M. and Seredko, A. (2022) Human-machine-learning integration and task allocation in citizen science. *Humanities & Social Sciences Communications*, 9: 48 (15 pp.), https://doi.org/10.1057/s41599-022-01049-z

Prathapan, K.D., Pethiyagoda, R., Bawa, K.S., Raven, P.H., Rajan, P.D., Acosta, L.E., Adams, B., Adl, S., Ahyong, S.T., Anderson, R., Arango, C.P., Arnedo, M.A., Armbruster, J.W., Avila, L.J., Azevedo, C.O., Baldo, D., Barclay, M.V.L., Baron-Szabo, R., Bauer, A.M., Bentlage, B., Bezděk, A., Bird, G., Blagoderov, V., Bocak, L., Bonaldo, A., Bond, J.E., Borkent, C.J., Branham, M.A., Carranza, S., Carreno, R., Carvalho, M.R. de, Castroviejo-Fisher, S., Chiba, H., Čiampor, F., Clarke, D.J., Collins, A.G., Constantino, R., Crespo, F.A., Daly, M., Dominiak, P., Dronen, N., Dubois, A., Duda, T.F., Eléaume, M., Erlacher, S., Estrela, P.C., Evenhuis, N., Fehlauer-Ale, K.H., Fery, H., Fritz, U., Gaimari, S.D., Garrison, R., Gaubert, P., Geiger, D.L., Gill, A.C., Gimmel, M.L., Goldschmidt, T., Goswami, R., González, A.P., Gonzalez, V.H., Gordon, D., Gower, D.J., Greenslade, P., Gusarov, V.I., Hajdu, E., Harms, D., Heinicke, M.P., Hilton, E.J., Hodgson, C.J., Hormiga, G., Hughes, L.E., Hutchings, P., Jäger, P., Jennings, J.T., Kadej, M., Kaile, L., Kamiński, M.J., Karaman, G.S., Karanovic, T., Kathirithamby,

J., Kerr, P.R., Kirkendall, L.R., Kitahara, M.V., Klautau, M., Kondratieff, B.C., Kroh, A., Labarque, F.M., Leavengood, J.M., Letardi, A., Liang, A.-P., Lima, F.C.T., Liu, Z., Löbl, I., Lohrmann, V., Malchus, N., Malipatil, M.B., Marques, A.C., Matzke-Karasz, R., Mayer, G., Mayoral, J.G., McInnes, S.J., Minelli, A., Moir, M.L., Monks, S., Morrone J.J., Muster, C., Nagy, Z.T., Narayanan, K.S., Nearns, E.H., Nekola, J., Nihei, S.S., Nützel, A., Ohler, A., Orrico, V.G.D., Padial, J.M., Page, L.M., Passos, P., Paulson, D., Perkins, P.D., Pfingstl, T., Prieto, C., Pinheiro, L.R., Pinto-da-Rocha, R., Prendini, L., Price, B., De Prins, J., Ramirez, M., Rasmussen, C., Rasmussen, P., Redei, D., Ribera, I., Ricarte, A., Rivera, A., Rix, M.G., Rossaro, B., Roy, A.D., Ruiz, G.R.S., Salles, F.F., Sanborn, A.F., Sartori, M., Schöller, M., Schmelz, R.M., Schrödl, M., Segniagbeto, G.H., Serrano, J., Shimano, S., Shin, M.K., Sidorchuk, E., Siler, C.D., Sket, B., Smith, A.D., Smith, A.B.T., Smith, R., Smith-Pardo, A.H., Sparks, J., Sterrer, W.E., Stroiński, A., Svavarsson, J., Toledo, M., Twomey, E., Vasudevan, K., Vences, M., Voogt, N. de, Wang, Q., Watson, G.W., Weiner, W.M., Weksler, M., Wesener, T., Whitmore, D., Wiklund, H., Williams, P.H., Winterton, S.L., Wood, T.S., Yen, S.-H., Zaher, H., Zhang, Z.-Q. and Zhou, H.-Z. (2018) When the cure kills—CBD limits biodiversity research. *Science*, 360: 1405–1406, https://doi.org/10.1126/science.aat9844

Quek, Z.B.R., Yip, Z.T., Jain, S.S., Wong, H.X.V., Tan, Z., Joseph, A.R. and Huang, D. (2024) DNA barcodes are ineffective for species identification of *Acropora* corals from the aquarium trade. *Biodiversity Data Journal*, 12: e125914 (14 pp.), https://doi.org/10.3897/BDJ.12.e125914

Raupach, M.J., Amann, R., Wheeler, Q.D. and Roos, C. (2016) The application of "-omics" technologies for the classification and identification of animals. *Organisms Diversity & Evolution,* 16: 1–12, https://doi.org/10.1007/s13127-015-0234-6

Reboleira, A.S.P.S. and Enghoff, H. (2016) Taxonomics—next-generation taxonomists. *Organisms, Diversity and Evolution*, 16: 679–680, https://doi.org/10.1007/s13127-016-0287-1

Reverté, S., Miličić, M., Ačanski, J., Andrić, A., Aracil, A., Aubert, M., Balzan, M.V., Bartomeus, I., Bogusch, P., Bosch, J., Budrys, E., Cantú-Salazar, L., Castro, S., Cornalba, M., Demeter, I., Devalez, J., Dorchin, A., Dufrêne, E., Đorđević, A., Fisler, L., Fitzpatrick, Ú., Flaminio, S., Földesi, R., Gaspar, H., Genoud, D., Geslin, B., Ghisbain, G., Francis Gilbert, F., Gogala, A., Grković, A., Heimburg, H., Herrera-Mesías, F., Jacobs, M., Janković Milosavljević, M., Janssen, K., Jensen, J.-K., Ješovnik, A., Józan, Z., Karlis, G., Kasparek, M., Kovács-Hostyánszki, A., Kuhlmann, M., Le Divelec, R., Leclercq, N., Likov, L., Litman, J., Ljubomirov, T., Madsen, H.B., Marshall, L., Mazánek, L., Milić, D., Mignot, M., Mudri-Stojnić, S., Müller, A., Nedeljković, Z., Nikolić, P., Ødegaard, F., Patiny, S., Paukkunen, J., Pennards, G., Pérez-Bañón, C., Perrard, A., Petanidou, T., Pettersson, L.B., Popov, G., Popov, S., Praz, C., Prokhorov, A., Quaranta, M., Radchenko, V.G., Radenković, S., Rasmont, P., Rasmussen, C., Reemer, M., Ricarte, A., Risch, S., Roberts, S.P.M., Rojo, S., Ropars, L., Rosa, P., Ruiz, C., Sentil, A., Shparyk, V., Smit, J., Sommaggio, D., Soon, V., Ssymank, A., Ståhls, G., Stavrinides, M., Straka, J., Tarlap, T., Terzo, M., Tomozii, B., Tot, T., van der Ent, L.-J., Steenis, J. van, Steenis, W. van, Varnava, A.I., Vereecken, N.J., Veselić, S., Vesnić, A., Weigand, A., Wisniowski, B., Wood, T.J., Zimmermann, D., Michez, D. and Vujić, A. (2023) National records of 3000 European bee and hoverfly species: A contribution to pollinator conservation. *Insect Conservation and Diversity*, 16: 758–775, https://doi.org/10.1111/icad.12680

Richman, D.B. (1996) The contribution of amateurs to entomology. *Backyard Bug Watching*, 17: 11–13.

Rogers, A.D., Appiah-Madson, H., Ardron, J.A., Bax, N.J., Bhadury, P., Brandt, A., Buttigieg, P.-L., De Clerck, O., Delgado, C., Distel, D.L., Glover, A., Gobin, J., Guilhon, M., Hampton, S., Harden-Davies, H., Hebert, P., Hynes, L., Lowe, M., MacIntyre, S.,

Madduppa, H., Mazzuco, A.C. de A., McCallum, A., McOwen, C., Nattkemper, T.W., Odido, M., O'Hara, T., Osborn, K., Pouponneau, A., Provoost, P., Rabone, M., Ramirez-Llodra, E., Scott, L., Sink, K.J., Turk, D., Watanabe, H.K., Weatherdon, L.V., Wernberg, T., Williams, S., Woodall, L., Wright, D.J., Zeppilli, D. and Steeds, O. (2023) Accelerating ocean species discovery and laying the foundations for the future of marine biodiversity research and monitoring. *Frontiers in Marine Science*, 10: 1224471 (21 pp.), https://doi.org/10.3389/fmars.2023.1224471

Rohwer, V.G., Rohwer, Y. and Dillman, C.B. (2023) Declining growth of natural history collections fails future generations. *PLoS Biology*, 20(4): e3001613 (4 pp.), https://doi.org/10.1371/journal.pbio.3001613

Sabrosky, C.W. (1950) Taxonomy and ecology. *Ecology*, 31: 151–152.

Sangster, G. and Luksenburg, J.A. (2015) Declining rates of species described per taxonomist: Slowdown of progress of a side-effect of improved quality in taxonomy? *Systematic Biology*, 64: 144–151, https://doi.org/10.1093/sysbio/syu069

Schilder, F.A. (1952) *Einführung in die Biotaxonomie (Formenkreislehre). Die Entstehung der Arten durch räumliche Sonderung.* Gustav Fischer, Jena. IV, 162 pp.

Schminke, H.K. (1990) Bedeutung und Probleme praxisorientierter Taxonomie. *Verhandlungen der Gesellschaft für Ökologie,* 19(2): 236–244.

Schminke, H.K. (1994) Wiederaufbau von Forschung und Lehre in Zoosystematik – eine nationale Aufgabe. *Spektrum der Wissenschaft*, 9/1994: 114–116.

Schminke, H.K. (1997) Naturschutzarbeit und Biodiversitätsforschung ohne systematische Zoologie? *Biologie in unserer Zeit*, 27: 340–345.

Schwan, S. and Dutz, S. (2020) How do visitors perceive the role of authentic objects in museums? *Curator: The Museum Journal*, 63: 217–237, https://doi.org/10.1111/cura.12365

Seberg, O., Humphries, C.J., Knapp, S., Stevenson, D.W., Petersen, G., Scharff, N. and Andersen, N.M. (2003) Shortcuts in systematics? A commentary on DNA-based taxonomy. *Trends in Ecology and Systematics*, 18: 63–65, http://dx.doi.org/10.1016/S0169-5347(02)00059-9

Sharkey, M.J., Baker, A., McCluskey, K., Smith, A., Naik, S., Ratnasingham, S., Manjunath, R., Perez, K., Sones, J., D'Souza, M., St. Jacques, B., Hajibabaei, M., Whitfield, J., Arias, D., Solis, A., Metz, M., Burns, J., Zuñiga, R., Phillips-Rodriguez, E., Espinoza, B., Chacon, I., Hebert, P., Hallwachs, W. and Janzen, D. (2023) Minimalist revision of *Mesochorus* Gravenhorst, 1829 (Hymenoptera: Ichneumonidae: Mesochorinae) from Área de Conservación Guanacaste, Costa Rica, with 158 new species and host records for 129 species. *Revista de Biología Tropical,* 71: e53316 (174 pp.), https://doi.org/10.15517/rev.biol.trop..v71iS2.56316

Sharkey, M.J., Janzen, D.H., Hallwachs, W., Chapman, E.G., Smith, M.A., Dapkey, T., Brown, A., Ratnasingham, S., Naik, S., Manjunath, R., Perez, K., Milton, M., Hebert, P., Shaw, S.R., Kittel, R.N., Solis, M.A., Metz, M.A., Goldstein, P.Z., Brown, J.W., Quicke, D.L.J., Achterberg, C. van, Brown, B.V. and Burns, J.M. (2021) Minimalist revision and description of 403 new species in 11 subfamilies of Costa Rican braconid parasitoid wasps, including host records for 219 species. *ZooKeys*, 1013: 1–665, https://doi.org/10.3897/zookeys.1013.55600

Sigwart, J.D., Chen, C., Tilic, E., Vences, M. and Riehl, T. (2023) Why is there no service to support taxonomy? *BioEssays*, 45: 2300070 (4 pp.), https://doi.org/10.1002/bies.202300070

Silvertown, J. (2009) A new dawn for citizen science. *Trends in Ecology and Evolution*, 24: 467–471, https://doi.org/10.1016/j.tree.2009.03.017

Simpson, G.G. (1945) The principles of classification and a classification of mammals. *Bulletin of the American Museum of Natural History*, 85: i–xvi, 1–350.

Simpson, G.G. (1961) *Principles of Animal Taxonomy.* Columbia University Press, New York. x, 247 pp.

Skvarla, M.J. and Fisher, J.R. (2023) Online community photo-sharing in entomology: A large-scale review with suggestions on best practices. *Annals of the Entomological Society of America*, 116: 276–304, http://dx.doi.org/10.1093/aesa/saad021

Slater, J.A. (1991) Taxonomist! Defend thyself! *Antenna*, 15: 2–3.

Sluys, R. (2013) The unappreciated, fundamentally analytical nature of taxonomy and the implications for the inventory of biodiversity. *Biodiversity and Conservation*, 22: 1095–1105, https://doi.org/10.1007/s10531-013-0472-x

Stafleu, F.A. (1965) Chemotaxonomy on the march [Review of *Chemotaxonomie der Pflanzen, eine Übersicht über die Verbreitung und die systematische Bedeutung der Pflanzenstoffe; Taxonomic Biochemistry and Serology*, by R. Hegnauer & C. A. Leone]. *Taxon*, 14: 69–71, https://doi.org/10.2307/1217254

Stafleu, F.A. (1969) Biosystematic pathways anno 1969. *Taxon*, 18: 485–500, https://doi.org/10.2307/1218377

Tautz, D., Arctander, P., Minelli, A., Thomas, R.H. and Vogler, A.P. (2003) A plea for DNA taxonomy. *Trends in Ecology & Evolution*, 18: 70–74, https://doi.org/10.1016/S0169-5347(02)00041-1

Thiele, K. and Yeates, D. (2002) Tension arises from the duality at the heart of taxonomy. *Nature*, 419: 337, https://doi.org/10.1038/419337a

Thomson, S.A., Pyle, R.L., Ahyong, S., Alonso-Zarazaga, M., Ammirati, J., Araya, J.F., Ascher, J.S., Audisio, T.L., Azevedo-Santos, V.M., Bailly, N., Baker, W.J., Balke, M., Barclay, M.V.L., Barrett, R.L., Benine, R.C., Bickerstaff, J.R.M., Bouchard, P., Bour, R., Bourgoin, T., Boyko, C.B., Breure, A.S.H., Brothers, D.J., Byng, J.W., Campbell, D., Ceríaco, L.M.P., Cernák, I.,Cerretti, P., Chang, C.-H., Cho, S., Copus, J.M., Costello, M.J., Cseh, A., Csuzdi C., Culham, A., D'Elía, G., d'Udekem d'Acoz, C., Daneliya, M.E., Dekker, R., Dickinson, E.C., Dickinson, T.A., van Dijk, P.P., Dijkstra, K.-D.B., Dima, B., Dmitriev, D.A., Duistermaat, L., Dumbacher, J.P., Eiserhardt, W.L., Ekrem, T., Evenhuis, N.L., Faille, A., Fernández-Triana, J.L., Fiesler, E., Fishbein, M., Fordham, B.G., Freitas, A.V.L., Friol, N.R., Fritz, U., Frøslev, T., Funk, V.A., Gaimari, S.D., Garbino, G.S.T., Garraffoni, A.R.S., Geml, J., Gill, A.C., Gray, A., Grazziotin, F.G., Greenslade, P., Gutiérrez, E.E., Harvey, M.S., Hazevoet, C.J., He, K., He, X., Helfer, S., Helgen, K.M., van Heteren, A.H., Hita Garcia, F., Holstein, N., Horváth, M.K., Hovenkamp, P.H., Hwang, W.S., Hyvönen, J., Islam, M.B., Iverson, J.B., Ivie, M.A., Jaafar, J., Jackson, M.D., Jayat, J.P., Johnson, N.F., Kaiser, K., Klitgård, B.B., Knapp, D.G., Kojima, J.-i., Kõljalg U., Kontschán, J., Krell, F.-T., Krisai-Greilhuber, I., Kullander, S., Latella, L., Lattke, J.E., Lencioni, V., Lewis, G.P., Lhano, M.G., Lujan, N.K., Luksenburg, J.A., Mariaux, J., Marinho-Filho, J., Marshall, C.J., Mate, J.F., McDonough, M.M., Michel, E., Miranda, V.F.O., Mitroiu, M.-D., Molinari, J., Monks, S., Moore, A.J., Moratelli, R., Murányi, D., Nakano, T., Nikolaeva, S., Noyes, J., Ohl, M., Oleas, N.H., Orrell, T., Páll-Gergely, B., Pape, T., Papp, V., Parenti, L.R., Patterson, D., Pavlinov, I.Ya., Pine, R.H., Poczai, P., Prado, J., Prathapan, D., Rabeler, R.K., Randall, J.E., Rheindt, F.E., Rhodin, A.G.J., Rodríguez, S.M., Rogers, D.C., Roque, F. de O., Rowe, K.C., Ruedas, L.A., Salazar-Bravo, J., Salvador, R.B., Sangster, G., Sarmiento, C.E., Schigel, D.S., Schmidt, S., Schueler, F.W., Segers, H., Snow, N., Souza-Dias, P.G.B., Stals, R., Stenroos, S., Stone, R.D., Sturm, C.F., Štys, P., Teta, P., Thomas, D.C., Timm, R.M., Tindall, B.J., Todd, J.A., Triebel, D., Valdecasas, A.G., Vizzini, A., Vorontsova, M.S., Vos, J.M. de, Wagner, P., Watling, L., Weakley, A., Welter-Schultes, F., Whitmore, D., Wilding, N., Will, K., Williams, J., Wilson, K., Winston, J.E., Wüster, W., Yanega, D., Yeates, D.K., Zaher, H., Zhang, G., Zhang, Z.-Q. and Zhou, H.-Z. (2018) Taxonomy based on science is necessary for global conservation. *PLoS Biology*, 16: e2005075 (12 pp.), https://doi.org/10.1371/journal.pbio.2005075

Toellner, R. (1986) Liebhaber und Wissenschaft, zur Rolle des Amateurs in der Geschichte der Wissenschaften (XXIII. Symposium der Gesellschaft für Wissenschaftsgeschichte, 16.–18. 5. 1985 in Münster). *Berichte zur Wissenschaftsgeschichte*, 9: 137–145.

Toussaint, E.F.A., Gillett, C.P.D.T. and Löbl, I. (2024) The fate of natural history museums in the face of good intention. *Biological Journal of the Linnean Society*, 143: blae012, (4 pp.) https://doi.org/10.1093/biolinnean/blae012

Troudet, J., Grandcolas, P., Blin, A., Vignes-Lebbe, R. and Legendre, F. (2017) Taxonomic bias in biodiversity data and societal preferences. *Scientific Reports*, 7(9132): 14 pp, https://doi.org/10.1038/s41598-017-09084-6

Valdecasas, A.G. and Wheeler, Q.D. (2010) Taxonomy: Add a human touch too. *Nature*, 467: 788, https://doi.org/10.1038/467788a

Valdecasas, A.G., Williams, D. and Wheeler, Q.D. (2008) 'Integrative taxonomy' then and now: A response to Dayrat (2005). *Biological Journal of the Linnean Society*, 93: 211–216, https://doi.org/10.1111/j.1095-8312.2007.00919.x

Vasilita, C., Feng, V., Hansen, A.K., Hartop, E., Srivathsan, A., Struijk, R. and Meier, R. (2024) Express barcoding with NextGenPCR and MinION for species-level sorting of ecological samples. *Molecular Ecology Resources*, 24: e13922 (7 pp.), https://doi.org/10.1111/1755-0998.13922

Vinarski, M.V. (2020) Roots of the taxonomic impediment: Is the 'integrativeness" a remedy? *Integrative Zoology*, 15: 2–15, https://doi.org/10.1111/1749-4877.12393

Wägele, J.W., Astrin, J.J., Balke, M., Hausmann, A., Krogmann, L., Hendrich, L., Pietsch, S., Raupach, M., Schmidt, S., Segerer, A.H. and Haszprunar, G. (2012) Taxonomie am Scheideweg? *Studia Dipterologica*, 18: 105–117.

Wang, H., Fu, T., Du, Y., Gao, W., Huang, K., Liu, Z., Chandak, P., Liu, S., Van Katwyk, P., Deac, A., Anandkumar, A., Bergen, K., Gomes, C.P., Ho, S., Kohli, P., Lasenby, J., Leskovec, J., Liu, T.-Y., Manrai, A., Marks, D., Ramsundar, B., Song, L., Sun, J., Tang, J., Veličković, P., Welling, M., Zhang, L., Coley, C.W., Bengio, Y. and Zitnik, Z. (2023) Scientific discovery in the age of artificial intelligence. *Nature*, 620: 47–60, https://doi.org/10.1038/s41586-023-06221-2

Watling, R. (1998) The role of the amateur in mycology – what would we do without them! *Mycoscience*, 39: 513–522.

Weller, S. (2023) Toward a brighter future for entomological collections. *Annals of the Entomological Society of America*, 116: 329–330, https://doi.org/10.1093/aesa/saad026

Wheeler, Q.D. (2008) Introductory. Toward the New Taxonomy, pp. 1–17. In *The New Taxonomy* (ed Q.D. Wheeler). The Systematics Association Special Volume Series 76. CRC Press, Boca Raton.

Will, K.W., Mishler, B.D. and Wheeler, Q.D. (2005) The perils of DNA barcoding and the need for an integrative taxonomy. *Systematic Biology*, 54: 844–851, https://doi.org/10.1080/10635150500354878

Wilson, E.O. (1985) Time to revive systematics. *Science*, 230: 1227, https://doi.org/10.1126/science.230.4731.1227

Wilson, E.O. (2003) The encyclopedia of life. *Trends in Ecology and Evolution*, 18: 77–80, https://doi.org/10.1016/S0169-5347(02)00040-X

Wilson, E.O. (2004) Taxonomy as a fundamental discipline. *Philosophical Transactions of the Royal Society London B*, 359: 739, https://doi.org/10.1098/rstb.2003.1440

Wilson, E.O. (2017) Biodiversity research requires more boots on the ground. *Nature Ecology & Evolution*, 1: 1590–1591, https://doi.org/10.1038/s41559-017-0360-y

Winston, J.E. (1988) The systematists' perspective, pp. 1–6. In *Biomedical Importance of Marine Organisms* (ed D.G. Fautin). Memoirs of the California Academy of Sciences 13. California Academy of Sciences, San Francisco.

Wührl, L., Pylatiuk, C., Giersch, M., Lapp, F., Rintelen, T. von, Balke, M., Schmidt, S., Cerretti, P. and Meier, R. (2022) DiversityScanner: Robotic handling of small invertebrates with machine learning methods. *Molecular Ecology Resources*, 22: 1626–1638, https://doi.org/10.1111/1755-0998.13567

Zhao, Y., Calantone, R.J. and Voorhees, C.M. (2018) Identity change vs. strategy change: The effects of rebranding announcements on stock returns. *Journal of the Academy of Marketing Science*, 46: 795–812.

Zhu, C., Luo, A., Bai, M., Orr, M.C., Hou, Z., Ge, S., Chen, J., Hu, Y., Zhou, X., Qiao, G., Kong, H., Lu, L., Jin, X., Cai, L., Wei, X., Zhao, R., Miao, W., Wang, Q., Sha, Y., Lin, Q., Qu, M., Jiang, J., Li, J., Che, J., Jiang, X., Chen, X., Gao, L., Ren, Z., Xiang, C., Luo, S., Wu, D., Liu, D., Peng, Y., Su, T., Cai, C., Zhu, T., Cai, W., Liu, X., Li, H., Xue, H., Ye, Z., Chen, X., Tang, P., Wei, S., Pang, H., Xie, Q., Zhang, F., Zhang, F., Peng, X., Zhang, A., Gao, T., Zhou, C., Shao, C., Ma, L., Wei, Z., Luan, X., Yin, Z., Dai, W., Wei, C., Huang, X., Liu, J., Chen, X., Yi, T., Zhang, Z., Aishan, Z., Li, Q. and Hu, H. (2022) A joint call for actions to advance taxonomy in China. *Zoological Systematics*, 47: 188–197, http://dx.doi.org/10.11865/zs.2022302

# 5 Databases
## *Juggling Nomenclature and Taxonomy*

*Michael D. Guiry*

## INTRODUCTION

A major goal of systematics is to provide a synthesis of knowledge on the diversity of a group of organisms. As knowledge of a group grows, the data associated with it, and the literature on it, can become unmanageable. One solution is to store such data in a database, preferably making it and the data therein freely available to all workers in that group via the World Wide Web.

The word 'database' is of very recent origin, as words go, but rather older than the word 'internet.' The Oxford English Dictionary (OED)—itself a database these days—gives a year of origin of 1953 (as 'data base') and offers a definition of a '... structured set of data held in computer storage and typically accessed or manipulated by means of specialized software'. The word 'internet' is even more recent—1974—and is described by the OED as 'a computer network comprising or connecting a number of smaller networks'. The World Wide Web (WWW), 1990, is a '[...] widely used multimedia information system on the internet'. In the last 30 years, multimedia information systems have developed exponentially, and now pervade every aspect of modern life, mostly for good, but not always, and like dark matter, few of us know anything of the 'dark web'. The pervasiveness of the wired world is such that most electronic devices are now interconnected and data are instantly available, sometimes in indigestible quantities, and sometimes in warped and misleading forms.

Taxonomy, with its strong hierarchical structures, at first sight appears to lend itself to 'databasing' (also a new word, 1978, '[...] the process or practice of using databases; the addition of data to a database.') and with the various nomenclatural codes to superintend the introduction and use of names, a taxonomic database would seem to be an ideal tool for the practicing taxonomist. While nomenclature is governed by a set of rules agreed upon by several taxonomic communities,[1] taxonomy is not, and a clash was inevitable between the two in attempting to impose a database structure with a single solution or opinion as to the correct name for an organism, and its place in a hierarchy.

The taxonomist's task is an enormous one with at least 8.7 (± 1.3) million species thought to be extant (Mora et al., 2011); but even that number is an estimate, with, for example, less than 10% of bacteria having been described, probably including the Cyanobacteria (Nabout et al., 2013). In 2022, the International Union for Conservation of Nature (IUCN) estimated that 2.16 million species have been

DOI: 10.1201/b22822-5

described.[2] Some insect experts claim that there are more than 5 million species of insects and at least 7 million arthropods (Stork, 2018).

## COMPILATIONS AND DATABASES

While John Ray (1627–1705) and Carl Linnaeus (1707–1778) never heard of the word, their floras and systems were in fact databases of a kind: reference summations of existing knowledge for the use of readers. Before printed books, knowledge was simply passed down by word of mouth from master to pupil, but we all know what 'Chinese Whispers' can result in. The old WWI adage of an order being sent down the line, 'Send reinforcements, we're going to advance' arriving at headquarters as 'Send three-and-four-pence, we're going to a dance'. As floras, monographs, treatises, theses, and journals proliferated in the 19th century, it became necessary to produce further summary works to provide an '*aide mémoire*' to these works.

For most systematic biologists, it became increasingly necessary to travel to locations with comprehensive libraries, mostly in large cities that were often the capitals of great European colonial powers. Most taxonomists had to do this in any case as most original material, including types, were also held in such cities. This was a laborious business, particularly prior to the advent of air travel and the photocopier as a commonly available tool (about 1960), entailing the taking of notes and often copying figures by hand or waiting for photographs. The practice of exchanging 'preprints', 'reprints' or 'offprints', now nearly defunct, helped a lot, but taxonomy was a slow and labor-intensive business of trying to source literature and absorb its contents. Nevertheless, some extraordinary works of reference were produced by the likes of Carl Ernst Otto Kuntze (1843–1907; *Revisio generum plantarum*, 1891–1898), Carl Fredrik Otto Nordstedt (1838–1924; *Index Desmidiacearum*; 1896–1908); Giovanni Batista De Toni (1864–1924; *Sylloge Algarum*, 1889–1924), and later compilations of names for particular groups of algae such as the diatoms by Frederick William Mills (1868–1949; *An index to the genera and species of the Diatomaceae and their Synonyms* 1933–1934) and Samuel Leighton VanLandingham (1935–2013, *Catalogue of the fossil and Recent genera and species of diatoms and their synonyms*, 1967–1979). The disadvantage of these printed works was, of course, that they went out of date almost the instant they were published, but they were extraordinary achievements by dedicated and perhaps obsessed individuals.

In 1949, Paul Claude Silva (1922–2014) initiated *Index Nominum Algarum* (INA) with the '*...goal of producing and maintaining an index of scientific names of algae, both living and fossil, at all ranks*' (Silva and Moe, 1999), which was initially stored on typed index cards. This was complemented by *Bibliographia Phycologica Universalis*, a bibliography of publications on algae. Computerization was initiated in 1986 and is still ongoing, and all the card images and computerized entries are available online (https://ucjeps.berkeley.edu/ina/) currently containing nearly 200,000 names of algae 'in the broad sense'. While the online version of INA is not a database in that it is not held in a custom-built commercially available data structure, it serves the function of a database in that the data can be accessed via the World Wide Web. INA is essentially a *nomenclator*,[3] a source of information on the details of the publication of names of all groups of algae and as such is a unique and invaluable resource.

More recently (1999–2011), the On-Line Catalogue of Diatom Names (CDN), a compilation of names of diatom genera, species, and taxa at infraspecific ranks (more than 62,000 names), was assembled at the California Academy of Sciences and edited by Elisabeth Fourtanier and J. Patrick Kociolek (Fourtanier and Kociolek, 2009). The VanLandingham Catalogue was used for the initial set of names. CDN included all scientific names of diatom genera, species, and taxa at infraspecific ranks, with authorship, date, place of publication, page of description, basionym or replaced name (if applicable), nomenclatural status (valid or invalid), and some type information. The CDN data were held in a Filemaker Pro® database and are still available online https://researcharchive.calacademy.org/research/diatoms/names/index.asp) although updating ceased in 2011. The CDN data were later transferred to DiatomBase, now held at https://diatombase.org, which is being updated continuously. A card-image-based 'Diatom New Taxon File' is also maintained at the Academy of Natural Sciences, Philadelphia (http://symbiont.ansp.org/dntf/), which includes descriptions and original illustrations extracted from the literature, and is also being updated continuously (Potapova et al., 2024).

An online database of cyanobacteria genera is available at http://www.cyanodb.cz. This is a comprehensive data source with a full taxonomic hierarchy and descriptions of 430 valid genera including some 1650 species, but has a notice recently (April 2023) that it is not being curated.

There is no doubt that all of these resources have been of immense value to the various psychological communities. Printed catalogues such as Mills and VanLandingham have been less used as physical and virtual copies are not widely available. Also, with the partial exception of DiatomBase and Cyano DB, most of the available catalogues include little or no updated taxonomic information, something to which most did not aspire.

## ALGAEBASE

In 1996, fascinated by the possibilities presented by the burgeoning Internet, I started to compile in a FileMaker® relational database a taxonomic database of seaweed names including nomenclatural, taxonomic and distributional information. It was my intention to use this database to produce 'fact-sheets' for the marine algae of the NE Atlantic, but it quickly became apparent that what was needed was a taxonomic database of all seaweeds (Nic Dhonncha and Guiry, 2002) and later of all algae (Guiry, 2012; Guiry et al., 2014). This was intended to be a counterpart to INA since this database did not (perhaps sensibly) provide taxonomic data.

AlgaeBase quickly built up to a database of more than 200,000 names and about 50,000 species of living algae (Guiry, 2024) and about 9,000 fossils in a relational, hierarchical structure that quickly outgrew the capabilities of Filemaker Pro® and in 2004 the data were restructured into a MySQL® database on my own servers, but recently AlgaeBase employs Amazon Web Services (AWS), which enhanced the speed, security and availability of the data worldwide.

## JUGGLING NOMENCLATURE AND TAXONOMY

While INA is a remarkable achievement, it was always intended to publish the Index. The information on the cards was sometimes incomplete. All algae and organisms treated at one time or another as algae come under the aegis of the International Code of Nomenclature for algae, fungi, and plants, currently the Shenzhen Code[4] (ICNafp; Turland et al., 2018 eds.) for purposes of naming and priority. The ICNafp is essentially a lexicon of agreed procedures or rules for establishing the nomenclaturally correct name for a taxon. It is most certainly not for the faint of heart—like any set of laws—and it was considered necessary to provide an explanatory document longer than the ICNafp to explain its usage (*The Code Decoded,* ed. 2. Turland 2019, https://ab.pensoft.net/article/38075/). Many of us have spent our lives working on taxonomy and still do not fully comprehend all the nomenclatural provisions of the Code.

In compiling a taxonomic database, however, one cannot avoid applying the nomenclatural rules, so one must apply three basic criteria:

1. *Effective publication.* Does the publication meet the requirements of the ICNafp for effective publication (Arts 29–31)? These are mostly obvious, but publication as a PDF has certain conditions that have to be met. Supplementary files[5] (increasingly a feature of publications) are not effectively published unless they are in a PDF format. A name that does not meet the requirements for effective publication is not valid.
2. *Validity.* A name, in order to be valid, in addition to being effectively published, has to meet the conditions set out in Arts. 32–45. It seems at first sight to be easy: decide on a name, describe your taxon, designate a type and an institution where the type has been lodged (for species and below), and give an illustration (for species and below) that reflects the characters of the type in some way. Even so, many names are still invalidly introduced.[6]
3. *Legitimacy.* A valid name must be legitimate when the principle of priority is applied. In other words, it must not have been used previously and must not include in synonymy a name that is valid and legitimate.

This all seems straightforward, but there are many other considerations. For example, for certain groups of algae, there are Later Starting Points (Art. 13e): prior to specified dates, names are invalid (and thus do not exist), something often ignored by authors.

Invalid names effectively do not exist under the ICN and are known officially as ‘designations’, a term not familiar to most phycologists. A designation is ‘the term used for what appears to be a name but that (1) has not been validly published and hence is not a name in the sense of the *Code* or (2) is not to be regarded as a name.’ (Glossary of ICNafp, Turland et al., 2018). A designation is also known as a ‘nomen’—somewhat contradictory since this is the Latin for ‘name’. AlgaeBase and other databases of algal names include these designations for information, merely labelling them as invalid. INA includes these designations and, if they are validated, the validated name has separate entries. AlgaeBase includes a designation initially but replaces it with the valid name when it is validated to avoid duplicate entries (and potential confusion).

The inclusion of such designations for reasons of nomenclature brings us to the matter of dealing with both nomenclature and taxonomy in one database. Such designations are likely to be taxonomically valid and should be included even if only from the point of view of counting potential taxa or notifying that the designation exists and needs validation.

Classification is another matter. If there is contradictory taxonomic information from different authors, how can this be accommodated? Currently, it cannot as the structure of our database does not permit multiple taxonomies. All we can do is accept a monograph, revision, or review of the group, or if nonesuch exists, simply choose the most recent opinion. Generally, molecular data tend to resolve such issues but not always and in some groups there can be a clash between morphological and genetic data. It is not always possible to be certain of molecular data if the analyses are not competently carried out and related to type material. All one can do is note any such contradictions and 'carry on.' We juggle as effectively and usefully as we possibly can, but at the end of the day the user must be aware that AlgaeBase is simply an *'aide mémoire'* and the user is responsible for checking all information provided.

Even higher-level taxonomies can be difficult to adapt to databases. Molecular data can be contradictory (and several genes are usually more informative than one, as many workers have found recently) and often does not conform to the defined hierarchical categories of the ICN as laid out in Art. 16. While the ICN is quite flexible in that new categories can be intercalated, molecular data (particularly when the monophyly demanded by cladistics is critical) often do not lend themselves to such rigid structures.

A further issue for the erstwhile data compiler is that the principle of priority does not apply to names above the level of family and even classes with the same circumspection can have several different names, even different names on separate continents. For example, the *Conjugatophyceae*, known as such in Europe, is more widely known in North America as the *Zygnematophyceae* (Guiry, 2013) even though both are typified as opposed to descriptive names.[7]

## PERSISTENCE OF DATABASES

A major problem in the future will be the survival of databases. In general, the most persistent algal databases will be those that are driven by one or two persons. In the case of INA, the longest-surviving data source, this was initially Paul Silva and, more recently, Richard Moe. The copyright of the INA data is vested in the Regents of the University of California, and eventually, some decisions will have to be made by them for the survival of this invaluable nomenclator. In the case of other databases, the ownership of the compilation and/or of the data is less clear and may cause difficulties in the future. In the case of AlgaeBase a large amount of data has been copied into Wikipedia and Wikispecies, but with little or no curation or updating. These pages have been translated into numerous languages. The links in both online encyclopaedias to AlgaeBase are critical for the proper assessment of the data but obviously would not survive its demise.

## CONCLUSIONS

While online databases have contributed hugely to taxonomy and nomenclature worldwide, their basic function should be recognized as stated clearly on the AlgaeBase website: 'AlgaeBase is often a compromise of taxonomic opinions that may or may not reflect your conclusions. It is also a work in progress and is thus incomplete. Please note that AlgaeBase is purely meant as an aid to taxonomic studies and not as a definitive source. You should always check the information included prior to use.' All databases are essentially an *'aide mémorie'* for taxonomists and nomenclaturists. Their existence came about merely because even the best human minds cannot possibly recall all the data and need help and a reference point.

A particular concern for the future is the lack of training in taxonomy and nomenclature in universities. In Britain and Ireland, most departments of zoology and botany have disappeared, and departments, schools and faculties mostly do not wish to employ taxonomists and even less nomenclaturalists, preferring to engage people with 'earning power.' And this is not just for algae; it applies to all biological groups.

Regarding nomenclature Silva and Hegewald (1988) summarized the situation then pertaining to the ICN:

> Compared to higher-plant taxonomists, phycologists include more than their fair share of workers who are either unaware of certain provisions or choose to ignore them.

Things have only gotten worse since.

## NOTES

1 The main Codes are: The International Code of Nomenclature for algae, fungi, and plants, The International Code of Zoological Nomenclature, and The International Code of Nomenclature of Prokaryotes (formerly the Bacteriological Code).
2 ICUN give 25,000 species of algae, which is only half the number of extant species documented in AlgaeBase (Guiry, 2024).
3 The original *nomenclator* in Rome was a person, often a slave, who discretely reminded politicians and other notables of people's names. Also used for someone who announced people by name.
4 The Madrid Code is currently in force as of July 2024 but the on-line and printed versions are not available at the time of writing (August 2024).
5 These 'supplementary files' seem quite unnecessary now that many journals are no longer printed. A worry is that with the inevitable demise of a publisher these files may not be available. Most people do not download them and they are not stored independently.
6 Under the ICZN the equivalent to a valid name in the ICN is an 'available name'. There, a valid name of a taxon is its correct scientific name.
7 The ICN permits descriptive names (i.e. names not based on included types) for classes and typified names, which can result in several different names being applied to the same class.

## REFERENCES

Fourtanier, E. and Kociolek, J.P. (2009) Catalogue of diatom names. *Part 1. Introduction and Bibliography. Occasional Papers of the California Academy of Sciences*, 156: 1–168.

Guiry, M.D. (2012) How many species of algae are there? *Journal of Phycology*, 48: 1057–1063, https://doi.org/10.1111/j.1529-8817.2012.01222.x

Guiry, M.D. (2013) Taxonomy and nomenclature of the Conjugatophyceae (=Zygnematophyceae). *Algae. An International Journal of Algal Research*, 28: 1–29, https://doi.org/10.4490/algae.2013.28.1.001

Guiry, M.D. (2024) How many species of algae are there? A reprise. Four kingdoms, 14 phyla, 63 classes and still growing. *Journal of Phycology*, 60: 214–228, https://doi.org/10.1111/jpy.13431

Guiry, M.D., Guiry, G.M., Morrison, L., Rindi, F., Valenzuela Miranda, S., Mathieson, A.C., Parker, B.C., Langangen, A., John, D.M., Bárbara, I., Carter, C.F., Kuipers, P. and Garbary, D.J. (2014) AlgaeBase: An on-line resource for algae. *Cryptogamie Algologie*, 35: 105–115, https://doi.org/10.7872/crya.v35.iss2.2014.105

Hegewald, E. and Silva, P.C. (1988) Annotated catalogue of *Scenedesmus* and nomenclaturally related genera: Including original descriptions and figures. *Bibliotheca Phycologica*, 80: i–xii, 1–586.

Mora, C., Tittensor, D.P., Adl, S., Simpson, A.G.B. and Worm, B. (2011) How many species are there on earth and in the ocean? *PLoS Biology*, 9(8 e1001127): 1–8, https://doi.org/10.1371/journal.pbio.1001127

Nabout, J.C., da Silva Rocha, B., Carneiro, F.M. and Sant'Anna, C. (2013) How many species of Cyanobacteria are there? Using a discovery curve to predict the species number. *Biodiversity and Conservation*, 22: 2907–2918, https://doi.org/10.1007/s10531-013-0561-x

Nic Dhonncha, E. and Guiry, M.D. (2002) AlgaeBase: Documenting seaweed biodiversity in Ireland and the world. *Biology and Environment: Proceedings of the Royal Irish Academy*, 102B (3): 185–188, https://www.jstor.org/stable/20500166

Potapova, M., Veselá, J., Smith, C., Minerovic, A. and Aycock, L., eds (2024) *Diatom New Taxon File at the Academy of Natural Sciences (DNTF-ANS)*, Philadelphia, http://dh.ansp.org/dntf

Silva, P.C. and Moe, R.L. (1999) The Index Nominum Algarum. *Taxon*, 48: 351–353, https://doi.org/10.2307/1224441

Stork, N.E. (2018) How many species of insects and other terrestrial arthropods are there on Earth? *Annual Review of Entomology*, 63: 31–45, https://doi.org/10.1146/annurev-ento-020117-043348

Turland, N. (2019) *The Code Decoded. A User's Guide to the International Code of Nomenclature for Algae, Fungi, and Plants.* 2nd ed., 196 pp. Pensoft Publishers, Sofia. https://doi.org/10.3897/ab.e38075

Turland, N.J., Wiersema, J.H., Barrie, F.R., Greuter, W., Hawksworth, D.L., Herendeen, P.S., Knapp, S., Kusber, W.-H., Li, D.-Z., Marhold, K., May, T.W., McNeill, J., Monro, A.M., Prado, J., Price, M.J. and Smith, G.F., eds (2018) *International Code of Nomenclature for Algae, Fungi, and Plants (Shenzhen Code)* Adopted by the Nineteenth International Botanical Congress Shenzhen, China, July 2017. *Regnum Vegetabile*, vol. 159, pp. [i]–xxxviii, 1–253, Koeltz Botanical Books, Glashütten.

# 6 Zootaxonomy in the Century of Extinctions
## *Time for Field Work and Collections*

*Alain Dubois*

## INTRODUCTION

Millions of animal species are disappearing from our planet, mainly as a consequence of man's activity. Whatever our efforts, we will not impede most extinctions. A vast majority of species are as yet unknown to science, and reducing significantly this taxonomic knowledge gap would require drastic decisions, before it is too late. Each species carries unique and unpredictable information on organisms, evolution and adaptations. In order to save as much of this irreplaceable information as possible, an unprecedented effort should be undertaken for the taxonomic survey of all natural habitats, including the collection of specimens, tissues and data, and expansion of natural history collections. Collections will be inexhaustible mines of knowledge of biodiversity, far more informative than photos, descriptions or molecular sequences. Models and trees are just hypotheses. Their fate is to be challenged and replaced by better ones – hence the inappropriateness of 'taxonomic stability'. In contrast, voucher specimens are endless information sources and remain precious under any classification or theory of evolution. Studying diversity can bring answers to many questions about the way the biological world has evolved and functions, and an infinite number of applications useful to man. However, it is not justified only by these utilitarian aims. Simple intellectual curiosity and aesthetic appreciation about biodiversity are pertinent motivations for study. 'Taxonomic urgency' is a new paradigm for biology, which should be acknowledged by the scientific community. In order to address it efficiently, two major biological disciplines, taxonomy and conservation biology, should act as partners, not as 'adversaries'. Such a program would require funds and salaries for exploration, collections and laboratory work all around the planet. The time has come for the community of taxonomists to build a powerful international organisation to promote their discipline.

## THE NEED TO STUDY THE DIVERSITY OF SPECIES ON EARTH

Ever since people started to employ science to explore and study the world, the inventory of earth's living species has been a constant preoccupation. Yet the scientific discipline in charge of this work, ***taxonomy***, has been in crisis for decades.


DOI: 10.1201/b22822-6

Several misconceptions about taxonomy are widespread in the scientific community. Among them are the ideas that, as it uses so-called 'type specimens', it is typological and therefore non-evolutionary, that taxonomists study only morphology, and that their work is ancillary, merely providing identification of specimens for other biological studies, such as molecular phylogeny, ecology or conservation biology (Engel et al., 2021). So-called 'type specimens' are in fact simply vouchers that serve as 'name-bearers', allowing objective allocation of names to taxa and are not meant to be 'typical' of taxa. Taxonomy, like every science, is based on explicit concepts and the formulation and testing of hypotheses. It requires an approach wherein scientists synthesise data derived from morphology, behaviour, life history, molecules and other fields of inquiry, making it perhaps the most integrative of all biological disciplines. As for the identification of specimens, this is just a practical outcome of taxonomy, not the core of the discipline which consists of taxonomic revisions of the delimitations and relationships of taxa. Taxonomy is a descriptive science that tests explicit hypotheses in the form of hierarchies of taxa.

Today, these have become burning issues as the ecosystems of our planet face major anthropogenic crisis and biodiversity is threatened by the sixth extinction, by far the most rapid, and therefore most severe, such episode in history.

## THE CENTURY OF EXTINCTIONS

### Extinctions and Threats

Although field naturalists identified it decades ago, many became aware of the 'biodiversity crisis' only when it received this designation (Wilson, 1985). Despite discussions regarding the causes, rates and magnitudes of extinctions, it is now clear that our planet has entered the ***sixth mass extinction*** (Cowie et al., 2022), which can be stressed by calling our epoch the 'century of extinctions' (Dubois, 2003). Millions of animal species are disappearing forever from our planet for a number of reasons, the most important of which are by far the perturbation, spoiling or destruction of their habitats by man, as well as their direct destruction by hunting and fishing (IPBES, 2019).

The *International Union for Conservation of Nature's Red List of Threatened Species* (Anonymous, 2023) is the most comprehensive resource detailing the global conservation status of plants and animals. It is a compilation of data from the scientific literature and various other sources. As such, it qualifies as a 'semi-scientific' document, which does not mean it is not useful, especially as it derives from the implementation of a data-driven protocol, which leads to consistent classifications of species into threat categories. The main problem with this document is its unbalanced coverage, which is vertebrate-centred and 'anthropocentric'. It is therefore hardly useful for an overall picture of the threats to, and extinctions of, animal species occurring on earth since the evolution and spread of destructive species of the genus *Homo*.

The last available version of the *Red List* mentions 993 named animal species extinct in historical times. This represents 0.06% of the 1,555,313 animal species considered valid by the IUCN in 2023. These figures have nothing to do with other recent estimates of extinctions. For example, Cowie et al. (2022) gave a range of 150,000–260,000 extinctions of species since 1500, i.e., 9.6–16.7% of the species considered valid. These authors also discussed man-induced extinctions previous to 1500, which were numerous, especially among large vertebrates in prehistoric times, either through hunting for food or deliberate destruction of dangerous or harmful species, or through changes in the environment.

The situation is no better if one considers the species categorised as 'threatened'. In 2023, only 89,856 species had been 'assessed' by the IUCN according to its criteria, and 17,416 of them were considered 'threatened'. According to these figures, the IUCN recognises as threatened only 1.1% of the species considered valid by taxonomists, which is doubtless dramatically undervalued. Besides, there is a large discrepancy between the two groups of animals distinguished by the IUCN.

In 2023, the number of vertebrate species considered valid by taxonomists was estimated by the IUCN as 74,962. More than four-fifths of them (62,493, i.e., 83.4%) had been assessed and referred to a category of threat. Among the latter, 11,195 (17.9%) were considered threatened. If these figures are adopted as an acceptable estimate, extending them to the whole group would give an estimate of about 13,429 threatened vertebrate species on earth. Although higher, this is of the same order of magnitude as the estimate (12,671) based on the IUCN data of 2010 (63,161 valid species, 6,714 threatened).

The situation is drastically different in the paraphyletic assemblage of 'invertebrates'. The 2023 *Red List* estimated the total number of species considered valid as 1,480,351, among which only 27,363 (1.8%) were assessed and, among these, 6,221 (22.7%) were considered threatened. If these values are seen as acceptable estimates, the number of 'threatened invertebrates' on earth would be about 336,559. Although lower, this is also of the same order of magnitude as the estimate (397,905) based on the IUCN data of 2010 (1,305,250 valid species, 2,904 threatened).

Such extrapolations are unwarranted, especially as it is likely that the species of invertebrates assessed by IUCN 'experts' were not randomly chosen. Yet, it is beyond doubt that the number of 17,416 threatened animal species on earth is a gross underestimation. It is even more so when one considers that only a small proportion of all species on earth is taken into account due to the gross incompleteness of our inventory as discussed below. Such warnings should be made clearer by the IUCN when publishing *Red Lists*. This would avoid the ridiculous appearance each year in the media of misleading statements like '17,416 animal species are currently threatened with extinction on earth', whereas it is doubtless that species at risk number in the hundreds of thousands, even millions.

Using data from 25,780 vertebrate species categorised by the IUCN, Hoffmann et al. (2010) analysed the status of the world's vertebrates. They stated that not only are one-fifth of the species assessed classified as threatened, but that this figure is increasing. Although they argued that this rate of deterioration would have been worse in the absence of conservation actions targeted at vertebrate species, they admitted that 'current conservation efforts remain insufficient to offset the main drivers of biodiversity loss in these groups: agricultural expansion, logging, overexploitation, and invasive alien species'; they concluded that 'conservation responses will need to be substantially scaled up to combat the extinction crisis'. In other words, although we may be able to slow extinctions, at least for a while, there is no evidence that we can stop them.

According to data of the 2011 and 2023 *Red Lists*, the number of species considered valid by taxonomists has increased by 186,902 in the period 2010–2023, which represents 14,377 species per year. A frequent conservative estimate is 18,000 new species each year (Wheeler, 2024). Even if we adopt the higher number of 20,000, it is ludicrous compared with the 2 million species already described and the very conservative estimate of 10 million species currently living on earth: if the current pace of discovery, collection, study, description and naming is not considerably and quickly increased, it would take 400 years to discover and describe the 8 remaining million. As the real number of species might well be much underestimated, this number may increase (see González-Oreja, 2008). Given the current acceleration of extinctions, accepting this perspective would mean that we consider as inevitable that most species currently living on earth will vanish – without any reaction from the scientific community – joining the massive number of extinctions prior to the onset of science, only a tiny proportion of which are known by fossils. In the meantime, efforts, budgets and personnel are devoted to 'biodiversity' and the (largely hopeless) 'conservation' of the minute portion which is 'close to man'; as well as the search for 'life on other planets', another hopeless endeavour with no urgency but great cost (Diamond, 2005). No coordinated worldwide effort exists to collect billions of specimens and tissues of condemned species in permanent collections, to at least save their memory and characters for future studies as 'recent fossils'. Perpetuating this reckless attitude would be a defeat of science as a search for knowledge about our planet.

## Conservation Biology and Taxonomy

The growing awareness of the effects of our globalised civilisation on the ecosystems and living species of the planet has had a justified and positive impact on the attitude of many towards nature. The discipline of conservation biology has developed as a response to this problem. It is justified for scientific, practical and ethical reasons. In particular, the recognition that each living species is unique and plays a role in functioning ecosystems has supported the idea that species should be protected against extinction.

A landmark in this process was the 1992 'Earth Summit' in Rio de Janeiro, 'United Nations Conference on Environment and Development', which adopted many benevolent resolutions, some of which had concrete consequences – not always positive for biodiversity. A major problem with this conference is that it did not produce a statement that the biosphere is a unique system and that biodiversity should not be considered the private property of individuals, corporations or states, but should be the responsibility of mankind as a whole. As long as this basic postulate is not made, people or countries will feel entitled to do what they wish with 'their' biodiversity and ecosystems, including exploiting, selling, modifying, impoverishing, spoiling or destroying them in the name of 'economic imperatives' (i.e., quite often, for the profit of a few persons or groups, rather than the 'common good').[1] From the viewpoint of scientists studying biodiversity, one of the few practical consequences of this 'Earth Summit' was to make the fieldwork of taxonomists increasingly difficult, often impossible. This work requires the collection of specimens from the field in order to study them in the laboratory. The idea, fostered by some 'conservationists', that photographs or a drop of blood for DNA sequencing would be enough to study, 'know' and describe a species, is based on a misunderstanding of the work of taxonomists, which requires fine and detailed studies of morphology and anatomy, and careful comparisons with other specimens belonging to other species considered closely related (see Winker, 1996; Dubois and Nemésio, 2017).

Carried to its logical extreme, biological conservation, so understood, may lead to a confrontation between 'conservation' and knowledge, such as questioning whether what is at stake is a scientific discipline or a blind, sectarian and aggressive lobby of sentimental 'protectors of animals' (as individuals, not as members of possibly threatened taxa[2]), having little concern with the growth of knowledge about living species, or the conservation of species and habitats. With biodiversity poorly known, to face the crisis of mass extinction, it is indispensable that conservation biology and taxonomy work as allies, not as adversaries (Dubois, 2003, 2009). The recent 'Buffon Declaration' on 'Natural history institutions and the environmental crisis', adopted in 2007 in Paris by representatives of 93 natural history institutions from 36 countries and four continents (Anonymous, 2008; Dubois et al., 2021), is an important step toward a solution for this problem. In any case, the solution to these problems should be sought at a higher level, i.e., that of institutions, states or international organisations, and should not remain a matter of 'personal arrangements' between taxonomists or other field naturalists and local populations and authorities.

In recent years, the focus of research and publications in conservation biology has been on the study of factors that lead to threats and extinctions, their importance in various zoological groups and in different environments, and in proposing interventions to limit them. The discipline is concentrating on proximal factors of threat and extinction (e.g., deforestation, pollution, translocation, hunting or fishing), not on ultimate factors such as the political and economic organisation and functioning of human societies. The interests and decisions of powerful actors such as industries, banks and shareholders, supported by influential lobbies, exert more influence than decisions taken by states for the good of the environment and citizens (see Dubois, 2003, p. S11). As most scientists take great care to appear 'objective' and 'disinterested' they, as a group, do not interfere with politics and concentrate their efforts on the Sisyphean task of addressing proximal factors.

All things considered, it is doubtless that 'an important proportion [...] of species extinctions are *inescapable*, that *we shall not be able* to prevent them, whatever the large international programs of conservation of biodiversity can be' (Dubois, 2003), that 'biodiversity will continue to decline over the 21st century' (Pereira et al., 2010) and that 'we are pessimistic about the fate of most of the Earth's biodiversity, much of which is going to vanish without us ever knowing of its existence' (Cowie et al., 2022). Such statements are more realistic than deceptive announcements like those of the Convention on Biological Diversity (CBD, http://www.cbd.int/). In April 2002, the parties of the CBD adopted the '2010 Biodiversity Target' in which they 'committed themselves to achieve by 2010 a significant reduction of the current rate of biodiversity loss at the global, regional and national level as a contribution to poverty alleviation and to the benefit of all life on Earth'. In October 2010, the Nagoya Summit adopted the 'Strategic Plan for Biodiversity 2011–2020, including Aichi Biodiversity Targets', Target 12 of which was: 'By 2020, the extinction of known[3] threatened species has been prevented and their conservation status, particularly of those most in decline, has been improved and sustained'. Although neither target was reached (see, e.g., IPBES, 2019), no perceptible change in the way of working of such international bodies is evident.

In contrast, it seems strange that the international community of biologists has not yet elaborated an action plan for this century that takes into consideration continuing biodiversity decline, as if ignoring scenarios of extinction would be enough to avoid them. Against all evidence, most biologists proceed as though actions undertaken or proposed for limiting biodiversity losses were likely to succeed. This is more or less imposed on them by 'untold injunctions' of a dominant ideology that favours 'positive', 'constructive' and 'optimistic' predictions or aims over rational and realistic goals (see Dubois, 2003), explaining the resistance to evidence of the biodiversity crisis and global climate change. Such a 'denial of reality' is a naive attitude based on the idea that 'mass extinctions cannot occur because this would be too sad and dreadful'; a position antithetical to science, as stressed by Jean Rostand (1939): 'there is no proof by the horrible'.

An illustration of this attitude is the currently fashionable idea of the 'de-extinction' of species, 'resurrected' from DNA samples (Dubois, 2010c; Dubois and Testart, 2013). Besides the technical difficulties of such projects, which could lead at best to hybrids, they could be carried out only for a few vertebrate species for which functional and complete DNA is available. Most importantly, this idea carries the counterproductive message that extinctions are not irreversible and therefore acceptable. It is in line with ideas of conservation biology regarding reintroductions of species or, worse, so-called 'repopulations'. These projects have roots in the idea that species are black boxes containing identical individuals designated by the same name,[4] ignoring intraspecific and intrapopulational variation. They lead to ***genetic pollution*** which may cause a collapse of the receiving population and obscure evolutionary relationships (Dubois, 2006; Dufresnes et al., 2024).

In contrast, the scientific community has shown its inability to impede or modify the many laws and regulations, derived from decisions at international conferences like those of Rio and Nagoya that have raised legal and administrative barriers to the collection and export of specimens for taxonomic research, thus acting as obstacles to the scientific study of biodiversity. Worst in this situation is that part of the scientific community involved in conservation biology supports these regulations, without understanding that most collections for research have no impact on the populations concerned, and above all that it is impossible to protect things that we do not yet know.

The process of mass extinction currently taking place on earth will not be stopped, despite optimistic statements to the contrary, because its major causes are deforestation and other ecosystem destruction that will not stop in the foreseeable future. Consequently, taxonomic urgency should be a main parameter of comparative biology for decades to come. As evolution is not teleological and predictable, it will be more useful for its proper description and understanding if species have been thoroughly inventoried, are represented in collections, and have been at least superficially studied. Time will exist for studies of phylogenetic relationships and evolutionary mechanisms after most species are extinct, assuming that we have collected voucher specimens, tissues and data for the 80% of species not yet known.

## TAXONOMIC IMPEDIMENT, TAXONOMIC GAP, AND TAXONOMIC URGENCY

The ***taxonomic gap*** (Dubois, 2010a) points to the fact that our inventory of the living species of our planet is largely incomplete. Less than 2 million animal species have been described so far. This is perhaps 20%, possibly much less, of the actual number of species which can only be guessed at using various estimation methods in the absence of a reliable, exhaustive list of species validated by taxonomic revisions. How incomplete is our inventory? Whichever estimate we accept, most species remain unknown and many, perhaps half or more, will be extinct before the end of this century. Rather than devoting efforts to obtain a 'better estimate', should we not first address this incompleteness itself and try to reduce it?

Above all, the taxonomic gap is due to the lack of professional taxonomists and expertise worldwide, i.e. the ***taxonomic impediment*** (Anonymous, 1994). The gap is difficult to overcome because, since the middle of the last century, taxonomy has been in a period of crisis. Although clearly identified decades ago, the lack of professional positions and appropriate funding for taxonomy has not been seriously addressed (Engel et al., 2021).

The combination of factors mentioned above (biodiversity crisis, taxonomic gap and crisis of taxonomy) creates a situation unprecedented in the history of science: ***taxonomic urgency*** (Dubois, 2010a). It calls for an acceleration of the exploration, collection, study, description and naming of the species of the globe.

Compared to a few decades ago, it may seem to outsiders that the current situation of taxonomy is improving. The term 'taxonomy' appears more and more in general declarations and scientific programs. However, a closer look reveals that the change is more in language than action, and the situation for the discipline remains substantially unchanged. Although not as much as 'biodiversity', the ***word*** 'taxonomy' has become more fashionable. But it is often seen as a tool to support other purposes, as stressed in 2004 by Wheeler et al.: 'Funds nominally allocated to taxonomy go largely to reconstruct molecular phylogenies, while thousands of species are threatened by imminent extinction' (Wheeler et al., 2004). Except in a few countries, which have voluntarily increased their ***taxonomic involvement*** in recent years, most students interested in biodiversity, if they wish to get a job, are discouraged from specialising in taxonomy and urged to go into disciplines like molecular phylogeny, conservation biology or bioinformatics.

In recent years, a number of publications have been devoted to the methodological problems of taxonomy, whether conceptual (e.g., the quest for the 'unified species concept', or nomenclatural questions) or technological (molecular techniques, computerisation of research, databases, electronic communication, georeferencing, etc.), and the term ***integrative taxonomy*** has become fashionable. These reflections and proposals are useful, but they fail to address the core problems of the taxonomic impediment which are the considerable manpower shortage, barriers to the collection of specimens in natural habitats in many parts of the world and problems faced by museums (see below). Their search for 'magical solutions' is insufficient to address taxonomic urgency. In particular, the 'triumvirate adjoining a unitary taxonomic cyberstructure + automated DNA barcoding + molecular phylogeny' may be aptly described as 'a threefold myopia' (de Carvalho et al., 2008).

It is doubtless that 'collecting new species in the field will remain the rate-limiting step' for the inventory of species (May, 2004), i.e., earth's species diversity ('speciodiversity'; Dubois, 2017a). New species are not found in computers and sequencers of cities of the 'North', but in the forests, savannahs, mountains, rivers, lakes, marshes and seas of the whole planet, and especially in tropical regions. No purely technical solution will bring species from the field to the laboratory, DNA barcoding analysis notwithstanding. In order to face taxonomic urgency, we need a dramatic acceleration of active field work by competent taxonomists worldwide. This requires an increase in the number of positions for professional taxonomists, especially in countries of the 'South', and in the funds allocated to fieldwork, institutional collections of specimens, and taxonomic revisions and publications.

Despite strong words heard or read (e.g., May, 1988; Wheeler, 2004, 2024; Wheeler et al., 2004), it is clear that the international community of professional biologists has not yet grasped the full extent of this problem. A clear indication that this community is not properly playing its role is the fact that a large proportion of the new species discovered and described each year are now made known by 'amateur' taxonomists (Löbl and Leschen, 2005), i.e., by people who are not paid and lack formal theoretical training in biology. Although this fact is well known, nothing is being done to address it. It has even been suggested that most fieldwork for collecting and studying organisms should be entrusted to amateurs, non-governmental organisations and 'parataxonomists'. The basic premise is to work 'given the taxonomic facilities available worldwide' (Basset et al., 2000, p. 899), i.e., without considering the possibility of increasing these facilities. The same approach is behind the development of 'citizen science' or 'participatory research': why pay professionals to carry out work that can be entrusted to passionate pensioners? Such proposals are examples of a widespread contemptuous attitude towards taxonomy. No one is proposing that research in oceanography, physics, chemistry, physiology, neurology, medicine, etc., be carried out at lower cost by 'parascientists', amateurs, students or other non-professionals. As for laboratory research, unlike many molecular studies that can be carried out by any well-trained technician, efficient and high-quality taxonomic work requires years of experience on the group under study. Amateur contributions to taxonomy should be supported, but in the context of a community of professional researchers, curators and their students.

Taxonomy is a complex, analytical and synthetic scientific discipline that relies on deep and diversified scholarship. It does not require 'miracle solutions' based on technology in the place of manpower, but 'theoretical training, more professionals, a lasting commitment to collections, and recognition as a robust science by peers and policy-makers, without which taxonomy itself may fall victim to extinction' (de Carvalho et al., 2005). It needs specialised pre-doctoral and doctoral teaching in taxonomy, post-doctoral contracts and stable positions for taxonomists (in museums, universities and other research institutions), proper funding for fieldwork and the permanent storage of specimens, tissues and information attached to them in specialised institutions, and major programs for taxonomic surveys of regions of the globe, revisions of large taxonomic groups and publications presenting these data in detail. Such publications are voluminous and need to be handled by editors aware of the idiosyncrasies of taxonomy, which requires the existence of periodicals and book publishers specialised in this discipline, to avoid errors made by many generalist publishers (Dubois, 2003).

Using amphibians as an example, Dubois et al. (2021) advocated the implementation of two strategies for the taxonomic survey of zoological groups of 'manageable' size that can be efficiently discovered and collected by competent field taxonomists: 'exhaustive taxonomy' and 'preventive taxonomy'.

Mankind is capable of realising, in a short period of time, gigantic projects which would have at first been believed impossible, if motivated by the military (e.g., the 'Manhattan Project' or the 'Global Positioning System'), medical (e.g., the 'human genome project'), political or financial priorities (e.g., the social networks or the A.I.). If the decision to finance it was taken, an almost complete inventory of species still present on earth would be possible for many zoological groups. Exhaustive taxonomy will never be possible for extinct species, most of which disappear without leaving fossils. This aim would probably be unrealistic for groups with gigantic numbers of species of small size and living in habitats difficult to access, but such a project is realistic for groups of relatively large-sized animals living in habitats readily accessible to humans. Today, there remains time for the community of taxonomists to commit to the exploration of all countries, ecosystems and habitats of the planet. International meetings could decide to promote the objective of achieving an almost exhaustive taxonomic account of life in the first half of this century. This would require a bold political decision, a shift in priorities and a modification of well-entrenched habits, particularly regarding the competition between research teams to be 'the first one' to describe new taxa and attach one's 'authorship' to it. Removing the names of authors after the nomina of taxa would help in this respect (Dubois, 2008b). Field collection of specimens of such groups would not require expensive techniques and equipment, but mostly manpower, brains and arms, i.e., salaries, plenty of working time and free access to natural habitats, with the legal possibility to collect specimens for permanent collections – which has become increasingly difficult in recent decades.

Even if such an international collaborative strategy sounds utopian today, it is possible to develop a new 'culture' in taxonomy. In many countries, the destruction of natural habitats is progressing at a very fast pace, leaving little time for long- or medium-term programs for the collection and study of biodiversity. In such cases, it would be useful to define priorities for urgent collecting campaigns, especially when habitat destruction can be foreseen. From a taxonomic point of view, the urgency of exploration and specimen collection is particularly high for habitats and ecosystems which are threatened by programmed deforestation, changes in agricultural practices, construction of roads or buildings, flooding by dam lakes or various other foreseeable habitat destructions. The community of taxonomists could consider developing special tools to deal with such situations, using the examples of preventive archaeology and ecological Rapid Assessment Programs. As soon as a threat to habitat is identified, special fieldwork could be organised, not to 'save' the species, which in most cases is unrealistic, but to collect specimens, tissues and information, to index and store them in safe conditions. This would require the implementation of *ad hoc* techniques and to solve the legal problems associated with the collection of specimens in such special circumstances. Implementation of preventive taxonomy would probably need the special training of teams of field taxonomists devoted to such 'last minute' collections. Even if time and manpower are currently lacking for properly studying these specimens, a material memory of their existence could be kept for future generations.

It is not enough to collect specimens and tissues. It is crucial to store them permanently in buildings devoted to their safe conservation (e.g., against fire), curation and study by researchers from the whole world. Natural history museums are highly specialised institutions for this purpose. The difference between museum collections and other natural history collections (e.g., those stored in universities or research institutions) is that museums are designed primarily for the function of long-term conservation and management of collections, with buildings, space, staff and budgets especially dedicated to this mission. The specimens kept for hundreds of years in museums carry information that can be useful for many purposes, in a virtually unlimited variety of domains (Alberch, 1993; Winker, 2004; Anonymous, 2006; Dubois, 2010a). The collections of major museums of the world will be essential to understand the history, evolution, structure, functioning and problems of biodiversity. As one example, specimens collected 50 or 100 years ago in precise localities provide baseline evidence of the impact of climate or other environmental changes on natural populations in different ecosystems, at different altitudes, etc.

Because of the high cost of their maintenance, natural history museums exist mostly in the richer countries of the 'North'. It would be crucial for the understanding and preservation of biodiversity that many countries of the 'South' progressively establish modern natural history museums and libraries and maintain them with permanent staff and funding. The maintenance and management of such establishments require an experienced and specialised staff with technical knowledge, buildings, space and equipment, and permanent funding. These needs are very specific, and different from those of organisations focused on research and teaching, like universities. Museums should not be subject to constraints imposed on the latter institutions, such as short-term objectives linked to targeted projects and based on contracts with governments or private companies. Rather, they should be considered 'a part of the world's scientific infrastructure', as stated by the 'Buffon Declaration'.

Collections do not consist only of specimens or other samples (tissues, biomolecules, seeds, living or deep-frozen strains), but also of databases associated with them (from hand-written to computerised catalogues), and libraries which are sometimes the only source for old books, periodicals and separates of importance in taxonomy. It is justified to speak of an 'organic link between taxonomy, collections and libraries' (Dubois, 2010b). Although digitalisation of such old works is progressing, the process is far from complete. Thus, quality research in taxonomy cannot rely only upon online bibliographic resources. All these collections, databases and resources should be permanently supported and not dependent on short-term trends in scientific research for their funding.

Reasons exist to be worried about the fate of several natural history museums, their collections, archives and libraries. Recent trends raise concern regarding priorities, which are changing even among the oldest and most famous museums (e.g., Naggs, 2022; Deuve, 2023). A very significant example concerns collections and libraries[5] being moved far away from big cities and their research laboratories in order to free space for exhibitions and other public-oriented activities. This separates taxonomic research from supposedly 'more noble' scientific disciplines remaining in the big cities. Another concern is security. Collections and/or libraries of several museums have been partially or entirely destroyed by fire in recent years.

## TRYING TO UNDERSTAND INERTIA

Many recent papers complain about the crisis of taxonomy, pointing to its decline and 'inertia' (Tancoigne and Dubois, 2013) and arguing, sometimes convincingly, for changes in current practices without clear justification. The motives behind these calls for change are at least in part sociological or political, rather than scientific. One explanation is provided by the ecological principle of competition for available resources (professional positions, salaries, facilities, budgets for equipment and functioning, etc.) applied to scientific disciplines, specifically between taxonomy and 'concurrent disciplines' devoted to biodiversity (evolutionary biology, ecology, conservation biology).

Taxonomy was one of the first disciplines of what became biology, formally dating from the middle of the 18th century. Like all scientific domains, before trying to address questions like 'when?', 'how?' and 'why?', biology had to start with the question 'what?' (Mayr, 1997). The purpose of taxonomy was to explore the world, collect specimens, describe, classify and name them. It progressively developed methods, techniques and concepts which transformed it into a scientific discipline distinguishing it from the traditional empirical, utilitarian, mythical and religious approaches to animals and plants. Since its creation, this discipline has constantly evolved incorporating new concepts and theories, some from other fields of biology.

Taxonomy first developed increasingly elaborate methods of analysis for morphology and anatomy, then incorporated methods and data from morphometrics, developmental biology, cytogenetics, ethology, bioacoustics, immunology, protein electrophoresis, etc., allowing fine qualitative and (univariate and multivariate) quantitative comparisons of the holomorphs of specimens. In the middle of the 20th century, phenetics achieved a high degree of technical sophistication, but, following the works of Willi Hennig, taxonomists realised that phenetic comparisons, resulting in elaborate measurements of 'similarities' and 'differences', were misleading, as they did not take evolution into account. New concepts, including plesiomorphy, apomorphy, paraphyly and cladograms, led to a revolution of taxonomic thought and practice and, by the end of the 20th century, had prompted numerous works on morphology, anatomy, development, and a refinement of the old concept of homology to produce outstanding revisionary publications for various taxonomic groups. This significant intellectual and scientific progress remained unnoticed by most biologists who continued to consider taxonomy an old-fashioned science. Aspects of taxonomic progress were slowed by the appearance of nucleic acid sequencing and 'molecular phylogeny' studies, some of which were clearly phenetic. We have now entered the era of genomics which is highly promising. Nevertheless, genomic data, which cannot be used for most fossils, says virtually nothing about phenotypes and adaptations, and adds to, but will never replace, taxonomy based on morphology.

Today, most professional taxonomists base taxa and classifications on a combination of morphological and molecular data, and on phylogenetic and phylogeographic hypotheses derived from them. Nonetheless, taxonomy is still perceived as an 'old' discipline by biologists who chase fads and novelties (so-called 'innovation'), and who are 'fascinated by the tool' (Dubois, 2008a), forgetting that techniques are not an aim in themselves but are at the service of science.

Taxonomy has benefited from the development of new techniques in many domains from computing to molecular biology. But in some cases, progress has had a negative impact on the discipline, such as when the search for automatisation and standardisation results in thought impoverishment or theoretical regression. A good example is DNA barcoding, a technique useful for identifying specimens, i.e., to allocate them to species already defined and described, or to match specimens of different sexes or of different life cycle stages. Its use to erect and delimit species or other taxa, however, based on *a priori* arbitrary 'distances' and 'threshold values' constitutes a return to the worst phenetic thinking of the middle of the 20th century.

The growing dependence of scientific domains on ever more sophisticated and costly techniques is encouraged by research institutions and the governance of science. Scientists know that it is much easier to obtain funds to buy costly equipment (computers, sequencers, chemicals, etc.) than to pay salaries for field or laboratory work. It is significant to note that funds invested in equipment are returned to the 'market' and benefit the firms that provide them, and their investors, while salaries go to the intellectual and technical workers who do the job, but do not generate added value.

The problems faced by taxonomy do not stop there. The discipline is confronted by important difficulties of a widely different nature, which can be summarised in a few words: reductionism, teleology, predictability and homogenisation. In biology, the reductionist approach consists of the belief that all biological phenomena can be explained by universal laws that apply to all organisms, just like the laws of physics and chemistry. While this is true for many phenomena that concern the biochemical, physiological, genetic, developmental, etc., functioning of organisms (e.g., the genetic code or the Krebs cycle), this does not apply to evolutionary phenomena.

As stressed repeatedly by Ernst Mayr, the concepts and criteria of the epistemology of the physicochemical sciences do not apply to evolutionary biology. Organic evolution is not teleological. It does not result from a program (contrary to the ontology of organisms), it is not directed towards any conscious or unconscious aim. It does not obey any unifying 'law' and cannot be predicted. It results from a variable combination of 'chance and necessity' (Monod, 1970), and as such is not deterministic or predictable. Since Darwin (1859), chance has often been believed to be confined to the domain of appearance of genetic variation (now called mutation), which provides the material on which natural selection acts, while natural selection was viewed by many evolutionary biologists as a universal 'law'. Natural selection is a statistical phenomenon, which describes the general ***tendencies*** leading to differential reproductive success among individuals of a population. Especially when resources are abundant or competition, predation and parasitism weak, it is far from being infallible, as has always been known by experienced field naturalists. If the 'film' of life was to be unfolded again, it would produce a completely different pattern of biodiversity (Gould, 1989), as the species and other taxa were not the result of any *a priori* plan or of deterministic laws. This means that the results of evolution cannot be predicted but only observed. The existence and characteristics of species are unpredictable, the only way to discover them is to explore natural ecosystems, collect organisms and study them; they are not knowable from 'models' derived from a superficial knowledge of small subsamples.

These simple ideas are not shared by the whole scientific community. For a long time, many biologists thought that it would be useless to collect and study all the species of the planet as this would be 'redundant'. It would be sufficient to study a few 'model organisms' (*Mus*, *Drosophila* or *Caenorhabditis*) to know the 'laws of life' or the 'laws of evolution'. These ideas have their roots in the Aristotelian conception of science as dealing only with general laws, not with particular cases. They are still shared by many biologists, including some studying biodiversity, as testified by the persistent use in the literature of terms like 'model' to designate an organism, like evolutionary or reproductive 'strategies', or 'predictive taxonomy'.

Molecular phylogenetic trees are particular models that provide hypotheses of the relationships between species. Most additions of species, specimens or genes, as well as changes in methodology of construction of trees, result in modifications of these trees. As time passes and the inventory of species progresses, most trees tend to stabilise, but they will always remain provisional because the vast majority of extinct species, which considerably outnumber the species currently living on earth, will never be known to us, and because in some cases the inclusion of a single species may have drastic consequences on the structure of a tree.

In fact, even in a well-known zoological group, including hundreds or thousands of very similar species, it is commonplace to find 'intruders', i.e. species that have completely unexpected morphologies, behaviours, reproductive modes or life histories, which could not have been guessed by any model and could only be discovered and studied. Amphibians provide an excellent example of this situation.

The ideas above may sound obvious to taxonomists and more largely to biologists with a naturalist background. They may be summarised as follows: the best way to learn something about living organisms is to study them; each particular case may provide unexpected and interesting information, so that increasing the sample is not redundant for comparative and evolutionary biology.

We are still very far from having given a satisfying reply to the 'what?' question concerning species diversity on earth. If we have discovered and described 20% of the living animal species of the planet or less, we are still very far from the usual scientific standards of quality of scientific work (5%, 1%, 0.1% of 'acceptable' errors or missing data, or even less). What scientific field other than taxonomy would feel satisfied with a rate of missing data of 80% or more? If our aim is to have a good picture of evolutionary phenomena and biodiversity, we need to increase drastically our coverage of the diversity of living species while time exists to do it. Many species are likely to disappear in coming decades without having told us their 'secrets', and even without having left at least a trace. At the same time, many new fossils discovered (especially of large vertebrates) raise considerable interest and prompt detailed studies. This inexplicable double standard is quite interesting but does not speak in favour of the rationality of scientists.

## SHORTCUTS AND PROXIES

Faced with the difficult situation of taxonomic research, some taxonomists have turned to shortcuts and proxies. The use of barcoding mentioned above, and more generally of 'genetic distances' to replace missing data on morphology, life history, behaviour, hybridisation and gene flow, in order to erect 'new species', are one of the plagues of recent taxonomy. These ignore the rich literature on the 'species problem' and do not rely on any particular species concept.[6]

Molecular data can serve as a proxy for the study of organisms. When they display discrepancies with conclusions of holomorphological characters, how do taxonomists solve them? In most cases, they discard the morphological data and retain only the molecular. Such an attitude does not allow progress in understanding the causes of these discrepancies. In many cases, the problem is largely artificial. Incompatibility of data may be resolved by more thorough morphological studies. In other cases, it persists, but then becomes scientifically interesting and deserves closer scrutiny, which discarding one set of data does not allow.

Another recent development, problematic from an epistemological point of view, is the hopeless search for a 'unified species concept', e.g., as 'independent lineages' (de Queiroz, 1998). It deliberately ignores some evolutionary situations such as hybridogenetic entities, which depend in each generation on backcrosses with one of the parental species from which they descended through hybridisation, as well as various other kinds of sexual, parasexual or asexual entities which have unusual systems of gametopoeisis (eumeiosis, metameiosis or ameiosis) and/or kinetogenesis (zygogenesis, gynogenesis, parthenogenesis or absence of gametopoeisis), and non-clonal, hemiclonal or meroclonal heredity, and some of which are not panmictic 'independent lineages' and have recourse to sexual parasitism for reproduction (see Dubois, 2011). In fact, the phenomenon of interspecific hybridisation has long been underestimated in zoology, and the recent discovery that it is widespread and plays a major evolutionary role will require a drastic re-examination of some of our basic beliefs. The plea for a unified species concept does not help our understanding of speciation and of the basic 'bricks' of biodiversity, which would be much easier to understand under a pluralistic conception of species.

Other domains currently poorly addressed in zootaxonomy are criteria for erection and delimitation of taxa referred to the taxonomic category and nomenclatural rank of genus (Dubois, 1988, 2004) and criteria for the construction of hierarchical taxonomies that account for the structure of phylogenetic trees (Dubois et al., 2021). Unfortunately, the current period of taxonomy, in which taxonomists 'chase new techniques', is less suitable for theoretical reflections on the discipline.

The main proxy of our epoch for genuine taxonomic research is the flourishing domain of online databases supposed to provide state-of-the-art taxonomic and nomenclatural knowledge. Most of these databases do not distinguish between nomenclatural information, which is fixed and does not allow 'interpretations', and taxonomic information, which is labile and susceptible to change in the light of new data, methods and concepts. These databases would be useful if they just provided bibliographic references to the taxonomic literature, allowing users to choose among conflicting opinions through the study of the publications themselves. Unfortunately, most exceed this function and present as 'universal truths' the opinions of their compilers, who in some cases go as far as taking new taxonomic decisions (e.g., subjective synonymisation or revalidation of nomina) or nomenclatural acts (e.g., new combinations or spelling changes) which are nomenclaturally unavailable, not having been published online in immutable format and following Zoobank pre-registration. This would not be a problem if users were aware that these databases, contrary to genuine scientific works, are not submitted to peer review, but are just working documents that save time in the search for information, but have no 'official status'. It would be of great help to understanding and clarification if these databases adopted the policy of the National Center for Biotechnology Information (https://www.ncbi.nlm.nih.gov/Taxonomy/Browser/wwwtax.cgi), which displays the following declaration on all its pages: 'Disclaimer: The NCBI taxonomy database is not an authoritative source for nomenclature or classification – please consult the relevant scientific literature for the most reliable information.'[7] However, in recent years a movement has been developing, not only in favour of databases, but also of declaring some of them to be 'official lists' for the taxonomic groups they cover. If pushed to its extreme, this technocratic demand for 'taxonomic stability,' at the expense of the quality and reliability of taxonomic and nomenclatural information, will freeze taxonomic lists and thereby contribute to the death of taxonomy as an active scientific field (Dubois, 1998).

## CONCLUSION

Several conclusions can be drawn from the ideas presented above.

1. We are now in the century of extinctions. Millions of animal species are in the process of disappearing forever from our planet for a number of reasons, the most important of which is, by far, the perturbation, spoiling or destruction of ecosystems caused by the growth of human populations and the increase of its impact on the environment.
2. Whatever our efforts, and contrary to what wishful thinking suggests, we will not impede most of these extinctions. Furthermore, our efforts in this respect are largely anthropocentric, being concentrated on large vertebrates, and having no impact on the bulk of biodiversity, which could be protected only by stopping aggressions on ecosystems and habitats.
3. The vast majority of species that still live on earth and are threatened with extinction are yet unknown to science, undescribed and unnamed, and reducing significantly this taxonomic gap would require drastic decisions.

However, we are engaged in a race against extinctions and have limited time for carrying out this work of exploration and survey.

4. Each living species carries unique, irreplaceable and unpredictable information on living organisms, their evolution and adaptations. In order to save as much of this irreplaceable information as possible, it is urgent to undertake an unprecedented effort for the taxonomic survey of all natural habitats, the collection of specimens, tissues and information about them, and their safe storage in permanent, well-curated and appropriately funded natural history collections, such as those of museums.
5. Altogether, these new problems and ideas constitute a new paradigm for biology, which may be summarised by the phrase 'taxonomic urgency'. The community of biologists, not to mention the scientific community and society at large, have not yet recognised this novelty, let alone afforded it the high priority it deserves.
6. These problems concern two major disciplines of biology: taxonomy and conservation biology. So far, these two disciplines have largely ignored each other, do not master the core questions, concepts, terms and aims of the other, and in many situations tend to act as 'adversaries'. For the time being, conservation biologists, who have not shown any convincing results of their endeavours for the protection of most threatened species on earth, often have an anti-collection attitude that acts as a strong barrier against better knowledge of species while there is still time to collect and study them (Dubois, 2009). So far, neither taxonomy nor conservation biology have proved to be able to stop, or even significantly reduce, species loss.[8] As long as conservation biology acts against the collection of specimens, it will act against taxonomic research, and thus against knowledge and, in the long run, against the conservation of biodiversity itself. It is urgent that this situation change and that the actors of these two disciplines learn to cooperate.
7. Collecting specimens and tissues is the main challenge of our time for the study of biodiversity, which should be given the highest priority. This is much more urgent than carrying out monographic works on a few selected species, and even the phylogenetic studies of the relationships among known species, which represent a subsample of current biodiversity, the representativeness of which is impossible to test. Material collections of vouchers and tissues will be inexhaustible sources of information on biodiversity, much more informative than simple photos, descriptions or even molecular sequences. Models and trees are just hypotheses. As such, their fate is to be challenged and replaced by better ones. It is clear that taxonomy, like all other sciences, will never reach 'final knowledge' – hence the inappropriateness of the search for 'taxonomic stability', for which Gaffney (1979) coined the phrase 'stability is ignorance'. In contrast, voucher specimens can be endless sources of information, and will remain useful and precious under any taxonomic arrangement or theory of evolution.
8. Developing such a program would require not only funds for exploration and collections but numerous fixed positions and decent salaries for field and laboratory work by adequately formed and passionate researchers and

technicians, all around the planet. In our society, some people earn millions of dollars while others who work hard are 'volunteers'. This suggests that their work is not indispensable, which is unfair. Furthermore, incorporating these people into the professional community would allow them to work more 'professionally', being in contact with people who have benefited from an academic background. In my personal experience, some 'amateur taxonomists' have only a phenetic approach to the study of specimens and many of their taxonomic conclusions are later shown to be wrong.

9. The difficulties confronting such a project should not be underestimated. It would be liable to meet strong resistance for the reasons evoked above. Its ambition being transnational, the absence of world governance would not facilitate it. Many papers by taxonomists, some very well-argued and convincing, have already presented these or similar ideas, but their effects on the politics of science worldwide have been very limited. The time has come to consider the possibility of an alternative strategy (Dubois, 2003). The community of taxonomists has so far failed to build a powerful international organisation to promote their discipline. This may result in part from the fact that taxonomists have long worked as isolated individuals engaged in strong competition to be the first to name the new taxa and 'attach their name' to them ('mihilism'). In recent decades, however, largely because of the development of molecular surveys of many specimens from many origins, taxonomic research has become much more collaborative – but this has not reduced significantly the mihilism, the competition being now mostly between teams, rather than individuals. However, united we stand, divided we fall. If taxonomists are not able to build a strong international organisation, acting as a lobby to advocate taxonomy, as biologists of many other disciplines have done, the decisions concerning these questions will remain in the hands of others.

Most people defend the need to study the species of our planet stress that this knowledge has an infinity of applications useful to man in various domains, that species diversity plays a most important role in the functioning of the biosphere, and that its study can bring answers to many questions about the way the biological world has evolved. More rarely, mankind's intellectual curiosity about the biodiversity of our planet, and an amazement and fascination for its beauty and inventiveness, are put forward, as if these motivations were shameful and outdated. This attitude can indeed be seen as a loss of curiosity among some biologists for 'what exists' in the world but may soon have vanished, as if science was only justified by utilitarian aims, including trying to 'explain' the world (in terms of proximal and distal causes) and to act on it, but not for our intellectual and aesthetic satisfaction. As stressed by the disappearance among academic disciplines of terms like zoology and botany, animals and plants, many biologists nowadays express no interest in organisms themselves, but only in the models supposed to account for their phylogeny and sometimes for the evolutionary mechanisms responsible for current biodiversity. If taxonomy survives and develops in coming decades, we will discover many novelties, some of which will prove useful to us, but many will 'simply' be wonderful and instructive about the endless ability of organic evolution to 'invent' characters, adaptations and solutions to the problems posed to living beings in natural ecosystems.

Having studied amphibians for more than 60 years, and spent hundreds of nights with them in their habitats, I often had the opportunity to be fascinated by new discoveries concerning these animals, such as the umbraculum on the cornea of some South African frogs, the use of their hands by some Indian frogs to cover their eggs with protective mud, or the extraordinary diversity of modes of reproduction and development of these animals, including that of the gastric brooding frogs of Australia discovered in 1973 and extinct in 1990, the use of the vocal sacs of males or of cavities in the backs of females for the early development of offspring, and many others. Most, if not all, of these discoveries, were made in the field by taxonomists whose purpose was to 'see what exists', and were not part of ambitious research programs aiming at solving difficult theoretical questions. I am grateful to them and many others, who have modestly contributed to discovering what Darwin (1859) called 'endless forms most beautiful and most wonderful'.

Of course, our plea for an awakening of taxonomy as a major science before it is too late may remain in vain. If so, Robert May's (1988) sentence will sound like a lugubrious but justified premonition: 'I believe future generations will find it blankly incomprehensible that we are devoting so little money and effort to the study of these questions'.

## NOTES

1 Needless to say, for centuries, looting and destruction of biodiversity have been massive in colonised countries on the whole planet, and they continue nowadays in a post-colonial context through many 'proxies' such as international industrial, agro-alimentary or mining companies, or even the Big Tech companies, whose activities result in deforestation, destruction of wetlands and chemical, physical (including nuclear) or biotic pollution. This should be fought without weakness whenever possible, but this does not mean that local populations should be exempted in this respect on the pretext that they have long been overexploited themselves. Anyway, this fight cannot be carried out by scientists alone, but by the citizens of the planet, even in many cases against their own governments.
2 A good example of this attitude is the reluctance of some authorities or even 'conservationists' to the eradication of invasive species in the name of their 'right to live'.
3 Note that the 'unknown' species, which considerably outnumber the 'known' ones, were not even mentioned!
4 This idea has long been prevalent even among taxonomists, as testified by the use of the term 'duplicata' to designate specimens referred to the same taxon in museum's collections, but no serious taxonomist would defend it today.
5 Among many examples of the 'institutional vandalism' (Naggs, 2022) currently raging, in 2023, the herpetological library of the Paris Museum, which existed since the beginning of the 19th century and was one of the richest in the world (particularly concerning separates of works currently unfindable), was destroyed, its holdings being in part distributed in various other libraries and in part pounded.
6 Species so 'defined' have received the ironical designation of 'boleons', from the 'Barcode of Life, BOL' (Dubois, 2017b).
7 Although I examined dozens of taxonomic and nomenclatural databases, I have not yet found a single one which did not show weaknesses or errors that could be found in a few minutes. The idea that databases could provide serious taxonomic and nomenclatural data is misguided: in fact, they just reduce energy, work and salaries dedicated to the latter.

8 In fact, if mankind wanted to stop species loss, and more generally fight the bio-geo-climatic crisis, it would have to wrest the political power from the states, companies (including those of the Big Tech), banks, and other actors of our societies, whose industrial, agricultural, transports, energy and other choices are responsible for the current disaster.

## REFERENCES

Anonymous [Systematics Agenda 2000] (1994) *Charting the Biosphere: A Global Initiative to Discover, Describe and Classify the World's Species.* Technical Report. American Museum of Natural History, American Society of Plant Taxonomy, Society of Systematic Biologists and the Willi Hennig Society, New York.

Anonymous [The International Council of Museums] (2006) Development of the museum definition according to ICOM statutes (1946–2001), https://icom.museum/en/resources/standards-guidelines/museum-definition/

Anonymous [The Buffon Symposium] (2008) The Buffon Declaration. Natural History Institutions and the Environmental Crisis. Concluding Message from the Buffon Symposium. October 18th and 19th, 2007. Muséum National d'Histoire Naturelle, Paris, https://www.cbd.int/doc/external/external-meetings/buffon-2007-10-final-en.pdf

Anonymous [IUCN] (2023) *The IUCN Red List of Threatened Species.* Version 2023–1, https://www.iucnredlist.org

Alberch, P. (1993) Museums, collections and biodiversity inventories. *Trends in Ecology and Evolution*, 8: 372–375, https://doi.org/10.1016/0169-5347(93)90222-B

Basset, Y., et al. (2000) Experience with parataxonomists and digital photography in Papua New Guinea and Guyana. *BioScience*, 50(10): 899–908, https://doi.org/10.1641/0006-3568(2000)050[0899:QBEWPA]2.0.CO;2

Carvalho de, M.R., Bockmann, F.A., Amorim, D.S., de Vivo, M., de Toledo-Piza, M., Menezes, N.A., de Figueiredo, J.L., McEachran, J.D., et al. (2005) Revisiting the taxonomic impediment. *Science*, 307: 353–353, https://doi.org/10.1126/science.307.5708.353b

Carvalho de, M.R., Bockmann, F.A., Amorim, D.S., et al. (2008) Systematics must embrace comparative biology and evolution, not speed and automation. *Evolutionary Biology*, 35: 150–157, https://doi.org/10.1007/s11692-008-9018-7

Cowie, R.H., Bouchet, P. and Fontaine, B. (2022) The sixth mass extinction: Fact, fiction or speculation? *Biological Reviews*, 97: 640–663, https://doi.org/10.1111/brv.12816

Darwin, C.R. (1859) *On the Origin of Species by Means of Natural selection, or the Preservation of Favoured Races in the Struggle for Life.* J. Murray, London.

Deuve, T. (2023) *À quoi songe donc ce printemps.* Conflans-Sainte-Honorine: Magellanes.

Diamond, J.M. (2005) *Collapse: How societies choose to fail or succeed.* Viking Press, New York.

Dubois, A. (1988) The genus in zoology: A contribution to the theory of evolutionary systematics. *Mémoires du Muséum National d'Histoire Naturelle, Série A Zoologie*, 140: 1–123.

Dubois, A. (1998) List of European species of amphibians and reptiles: Will we soon be reaching 'stability'? *Amphibia–Reptilia*, 19(1): 1–28.

Dubois, A. (2003) The relationships between taxonomy and conservation biology in the century of extinctions. *Comptes Rendus Biologies*, 326: S9–S21, https://doi.org/10.1016/s1631-0691(03)00022-2

Dubois, A. (2004) Developmental pathway, speciation and supraspecific taxonomy in amphibians. 2. Developmental pathway, hybridizability and generic taxonomy. *Alytes*, 22: 38–52.

Dubois, A. (2006) Species introductions and reintroductions, faunistic and genetic pollution: Some provocative thoughts. *Alytes*, 24: 147–164.
Dubois, A. (2008a) Identifying some major problems and their possible solutions. In: *Future trends of taxonomy*, EDIT Symposium, Carvoeiro (Portugal), 21–23 January 2008: 38–42.
Dubois, A. (2008b) A partial but radical solution to the problem of nomenclatural taxonomic inflation and synonymy load. *Biological Journal of the Linnean Society*, 93: 857–863.
Dubois, A. (2009) The special message from threatened amphibians. *Alytes*, 27 (1): 26–36.
Dubois, A. (2011) Species and 'strange species' in zoology: Do we need a 'unified concept of species'? *Comptes Rendus Palevol*, 10: 77–94, https://doi.org/10.1016/j.crpv.2011.01.002
Dubois, A. (2017a) The nomenclatural status of *Hysaplesia*, *Hylaplesia*, *Dendrobates* and related nomina (Amphibia, Anura), with general comments on zoological nomenclature and its governance, as well as on taxonomic databases and websites. *Bionomina*, 11: 1–48.
Dubois, A. (2017b) The need for reference specimens in zoological taxonomy and nomenclature. *Bionomina*, 12: 4–38, https://doi.org/10.11646/bionomina.12.1.2
Dubois, A. (2010a) Zoological nomenclature in the century of extinctions: Priority vs. 'usage'. *Organisms, Diversity and Evolution*, 10: 259–274, https://doi.org/10.1007/s13127-010-0021-3
Dubois, A. (2010b) La liaison organique taxinomie-collections-bibliothèques. *Biosystema*, 26: 21–44.
Dubois, A. (2010c) Postface. Un naturaliste face à l'extinction massive des espèces. In: J. Delord, *L'extinction d'espèce: histoire d'un concept et enjeux éthiques*, Paris, Muséum National d'Histoire Naturelle: 617–666.
Dubois, A. and Nemésio, A. (2007) Does nomenclatural availability of nomina of new species or subspecies require the deposition of vouchers in collections? *Zootaxa*, 1409: 1–22.
Dubois, A., Ohler, A. and Pyron, R.A. (2021) New concepts and methods for phylogenetic taxonomy and nomenclature in zoology, exemplified by a new ranked cladonomy of recent amphibians (Lissamphibia). *Megataxa*, 5: 1–738, https://doi.org/10.11646/megataxa.5.1.1
Dubois, A. and Testart, J. (2013) La résurrection d'espèce, bonne ou mauvaise idée? *Mediapart*, 21 Mars 2013. https://sciencescitoyennes.org/la-resurrection-despece-bonne-ou-mauvaise-idee/.
Dufresnes, C., Monod Broca, B., Bellati, A., Canestrelli, D., Ambu, J., Wielstra, B., Dubey, S., Crochet, P.-A., Denoël, M. and Jablonski, D. (2024) Piercing the barcoding puzzle of Palearctic water frogs (*Pelophylax*) sheds light on amphibian biogeography and global invasions. *Global Change Biology*, 30, e17180, https://doi.org/10.1111/gcb.17180
Engel, M.S., Ceríaco, L.M.P., Daniel, G.M., Dellapé, P.M., Löbl, I., Marinov, M., Reis, R.E., Young, M.T., Dubois, A., Agarwal, I., Lehmann, A.P., Alvarado, M., Alvarez, N., Andreone, F., Araujo-Vieira, K., Ascher, J.S., Baêta, D., Baldo, D., Bandeira, S.A., Barden, P., Barrasso, D.A., Bendifallah, L., Bockmann, F.A., Böhme, W., Borkent, A., Brandão, C.R.F., Busack, S.D., Bybee, S.M., Channing, A., Chatzimanolis, S., Christenhusz, M.J.M., Crisci, J.V., D'Elía, G., Costa, L.M. da, Davis, S.R., Lucena, C.A.S. de, Deuve, T., Fernandes Elizalde, S., Faivovich, J., Farooq, H., Ferguson, A.W., Gippoliti, S., Gonçalves, F.M.P., Gonzalez, V.H., Greenbaum, E., Hinojosa-Díaz, I.A., Ineich, I., Jiang, J., Kahono, S., Kury, A.B., Lucinda, P.H.F., Lynch, J.D., Malécot, V., Marques, M.P., Marris, J.W.M., Mckellar, R.C., Mendes, L.F., Nihei, S.S., Nishikawa, K., Ohler, A., Orrico, V.G.D., Ota,

H., Paiva, J., Diogo Parrinha, D., Pauwels, O.S.G., Pereyra, M.O., Pestana, L.B., Pinheiro, P.D.P., Prendini, L., Prokop, J., Rasmussen, C., Rödel, M.-O., Rodrigues, M.T., Rodríguez, S.M., Salatnaya, H., Sampaio, Í., Alba Sánchez-García, A., Shebl, M.A., Bruna S. Santos, B.S., Solórzano-Kraemer, M.M., Sousa, A.C.A., Stoev, P., Teta, P., Trape, J.-F., Van-Dúnem Dos Santos, C., Vasudevan, K., Vink, C.J., Vogel, G., Wagner, P., Wappler, T., Ware, J.L., Wedmann, S. and Kusamba, C.Z. (2021) The taxonomic impediment: A shortage of taxonomists, not the lack of technical approaches. *Zoological Journal of the Linnean Society*, 22: 381–387, https://doi.org/10.1093/zoolinnean/zlab072

Gaffney, E.S. (1979) An introduction to the logic of phylogeny reconstruction, pp. 79–111. In: *Phylogenetic Analysis and Paleontology* (eds J. Cracraft and N. Eldredge). Columbia University Press, New York.

González-Oreja, J.A. (2008) The Encyclopedia of Life *vs*. The Brochure of Life: Exploring the relationships between the extinction of species and the inventory of life on earth. *Zootaxa*, 1965: 61–68, https://doi.org/10.11646/zootaxa.1965.1.3

Gould, S.J. (1989) *Wonderful life. The Burgess Shale and the Nature of History.* W.W. Norton, New York.

Hoffmann, M., et al. (2010) The impact of conservation on the status of the world's vertebrates. *Science*, 330: 1503–1509, https://doi.org/10.1126/science.1194442

IPBES (2019) Global assessment report on biodiversity and ecosystem services of the Intergovernmental Science-Policy Platform on Biodiversity and Ecosystem Services. In *IPBES Secretariat* (eds E.S. Brondizio, J. Settele, S. Díaz and H.T. Ngo). Bonn, Germany. 1148 p, https://doi.org/10.5281/zenodo.3831673

Löbl, I. and Leschen, R.A.B. (2005) Demography of coleopterists and their thoughts on DNA barcoding and the Phylocode, with commentary. *The Coleopterists Bulletin*, 59: 284–292, https://doi.org/10.1649/850.1

May, R.M. (1988) How many species are there on earth? *Science*, 241: 1441–1449, https://doi.org/10.1126/science.241.4872.1441

May, R.M. (2004) Tomorrow's taxonomy: Collecting new species in the field will remain the rate-limiting step. *Philosophical Transactions of the Royal Society (B)*, 359: 733–734, https://doi.org/10.1098/rstb.2003.1455

Mayr, E. (1997) *This is Biology. The Science of the Living World.* Belknap Press, Cambridge, Mass.

Monod, J. (1970) *Le Hasard et la Nécessité. Essai sur la Philosophie Naturelle de la Biologie Moderne.* Seuil, Paris.

Naggs, R. (2022) The tragedy of the Natural History Museum, London. *Megataxa*, 7: 85–112, https://doi.org/10.11646/megataxa.7.1.2

Pereira, H.M., et al. (2010) Scenarios for global biodiversity in the 21st century. *Science*, 330: 1496–1501, https://doi.org/10.1126/science.1196624

Queiroz, K. de (1998) The general lineage concept of species, species criteria, and the process of speciation: A conceptual unification and terminological recommendations, pp. 57–75. In *Endless Forms: Species and Speciation* (eds D.J. Howard and S.H. Berlocher). Oxford University Press, New York.

Rostand, J. (1939) *Pensées d'un Biologiste.* Stock, Paris.

Tancoigne, E. and Dubois, A. (2013) Taxonomy: no decline, but inertia. *Cladistics*, 29 (5): 567–570. http://dx.doi.org/10.1111/cla.12019

Wheeler, Q. (2024) *Species, Science and Society. The Role of Systematic Biology.* Routledge, Oxon.

Wheeler, Q.D. (2004) Taxonomic triage and the poverty of phylogeny. *Philosophical Transactions of the Royal Society (B)*, 359: 571–583, https://doi.org/10.1098/rstb.2003.1452

Wheeler, Q.D., Raven, P.H. and Wilson, E.O. (2004) Taxonomy: Impediment or expedient? *Science*, 303: 285, https://doi.org/10.1126/science.303.5656.285

Wilson, E.O. (1985) The global biodiversity crisis: A challenge to science. *Issues in Science and Technology*, 2: 20–29, http://www.jstor.org/stable/43310356

Winker, K. (1996). The crumbling infrastructure of biodiversity: The avian example. *Conservation Biology*, 10: 703–707.

Winker, K. (2004) Natural history museums in a postbiodiversity era. *BioScience*, 54: 455–459, https://doi.org/10.1641/0006-3568(2004)054[0455:NHMIAP]2.0.CO;2

# 7 Bringing Taxonomy Back into the Spotlight

*Pablo Muñoz-Rodríguez*

## INTRODUCTION

In an article published in *Scientific Reports*, May 2021, Gabriel Damasco and colleagues presented an exhaustive revision of *Protium heptaphyllum* (Aubl.) Marchand, a hyperdominant Amazon tree species in the reasonably well-known genus *Protium* Burm.f. (family Burseraceae Kunth.). In spite of the apparently good knowledge existing about this species, the study by Damasco and collaborators revealed that, as currently delimited, the species is not homogeneous throughout its distribution: it includes at least 'eight distinct evolutionary lineages' with different morphologies, distribution areas and ecological requirements. The authors conclude that there is enough evidence, gathered from multiple sources, to recognize these independently evolving lineages as distinct species (Damasco et al., 2021).

The article by Damasco and colleagues is an exemplary exercise on integrative taxonomy, combining data from multiple sources—morphology, ecology, genomics—in the taxonomic process, and it aligns well with recent calls for integrative taxonomy (Scotland and Wood, 2012; Williams et al., 2014; Muñoz-Rodríguez et al., 2019; Orr et al., 2020; Favret, 2024). However, their article takes an unexpected turn: despite gathering and presenting substantial evidence and successfully identifying the distinct species, the authors do not formally describe the new taxa in their study. Instead, they briefly state that 'Updates regarding the taxonomic treatment [...] are in progress [...] and detailed descriptions of new species are in preparation as part of a taxonomic revision' (Damasco et al., 2021 p. 4). At the time of submission of this chapter—over three years since the article's publication — the anticipated taxonomic revision has yet to appear, and the new species remain unpublished. In the interim, more than eight hundred new references citing '*Protium heptaphyllum*' have been published (cf., Google Scholar), yet none of these studies have incorporated the taxonomic novelties, as they remain unpublished. Given the scarcity of data for most plant species globally (Feeley and Silman, 2011; Feeley, 2015; Goodwin et al., 2015, 2020; Wiens 2021), it is evident that the authors had sufficient evidence to include the taxonomic changes in the paper where they presented the supporting data. While their aim to publish a complete taxonomic treatment is laudable, the fact is that the scientific machinery continues to advance, and later works do not incorporate their important results. This may lead to conflicting, misleading, or simply outdated outcomes that do not take into account the existence of the different '*Protium heptaphyllum*' lineages. More worrisome is the fact that, as long as the different lineages are treated as the same species within the Amazon basin—i.e., not named as distinct


DOI: 10.1201/b22822-7

taxa—the chances of any specific conservation measures being implemented for any one of them are slim.

In a broader context, it is hard to imagine, for example, a biochemistry paper that identifies and characterizes a new protein but defers its description to another publication, or an astronomy paper reporting the discovery of a new planet without providing the necessary information for others to locate and study it. The study on *Protium heptaphyllum* is just one of many cases that exemplify a peculiarity affecting taxonomy: it is perhaps the only scientific discipline where the formal description of new discoveries—new species—can be so easily separated from the evidence that led to their discovery, often without reproach.

In this chapter, I briefly explore the reasons behind the persistent detachment of taxonomy from other biological sciences and how it has come to be. I then turn to what we, as taxonomists, can do to reverse this trend, focusing on two key aspects: how we conduct our work and how we communicate its value to the broader scientific community.

## THE ROLE OF TAXONOMY IN MODERN SCIENCE

Throughout history, taxonomy has played a vital role in deepening our understanding of nature. By studying and organizing the diversity of life on Earth, taxonomy illuminates the relationships between species and their environments and provides essential evidence to support scientific theories and concepts. A classic example is Charles Darwin's two-volume taxonomic monograph on barnacles (Darwin, 1851, 1854). Darwin's monograph advanced our understanding of barnacles by providing a much-needed revision of existing knowledge and new observations—including the confidence that they are crustaceans rather than molluscs (Love, 2002)—and by describing many new species. Additionally, Darwin's research helped establish his reputation as a zoologist, earning him the Royal Medal in 1853. More importantly, it laid the ground for his later theories on natural selection (Darwin, 1859; Deutsch, 2010) and gave him credibility to pronounce his ideas on evolution (Rainbow, 2011). Darwin's work illustrates how the implications of thorough taxonomic research can extend beyond the immediate subject, impacting the broader scientific discourse and enhancing our understanding of biological diversity.

However, 170 years later, taxonomy's foundational role is increasingly overlooked. Taxonomists now face extraordinary pressures to survive in academia, largely due to the lack of support from institutions and policymakers who fail to recognize the critical importance of taxonomy and the resources it depends upon (Escobar and Vogel, 2018; Naggs, 2022; Löbl et al., 2023; Pennisi, 2024; Wheeler, 2024). This situation is further complicated by the tension between traditional taxonomy and the rise of data-driven biodiversity science. The promise of big data has created an environment where the sheer volume of information can overshadow the detailed, species-level insights that taxonomists provide (Hortal et al., 2015; Devictor and Bensaude-Vincent, 2016).

Yet, the value of big data and global biodiversity repositories is inherently tied to the accuracy of the taxonomic information they contain—a point often overlooked by scientists presented with this vast amount of data (Yesson et al., 2007).

Much of these data are derived from specimens in natural history collections, which have expanded significantly in recent decades. For example, herbaria have doubled in number over the last sixty years (Goodwin et al., 2015), and collections worldwide now house an estimated 400 million plant specimens (Thiers, 2023). However, the usefulness of these specimens and the data derived from them is compromised when their identifications are incorrect, as is often the case (Goodwin et al., 2015), or when taxon boundaries remain inadequately defined due to incomplete studies (Bebber et al., 2010, Goodwin et al., 2020). Using poorly identified specimens or data can directly impact research outcomes, distorting our understanding of the natural world.

In response to these challenges, many researchers now advocate alternative approaches to expedite taxonomic work, or even to replace taxonomists altogether. Methods such as DNA barcoding (Tautz et al., 2002, 2003) and computer-driven species delimitation (Karbstein et al., 2024) have gained traction in this regard. While these approaches can be effective for certain groups or purposes, they often fail to address the core issue underlying many biodiversity studies: the lack of basic knowledge about most species. For example, DNA barcoding is ineffective without a reliable reference database—and we must be aware that most species have never been sequenced, even for a single DNA region (Sayers et al., 2021). Similarly, AI-based methods struggle to accurately delimit species without adequate data to train their models. As a consequence, despite the availability of new resources and tools, comprehensive taxonomic studies remain rare, and the pace of species discovery has not significantly increased (RBG Kew, 2016). For many species, knowledge is so limited that subsequent studies are not feasible, making it impossible to apply these technological approaches effectively.

This lack of fundamental knowledge is especially pronounced in tropical organisms. For instance, it is estimated that 90% of tropical plant species are so poorly known that they remain invisible to conservation efforts (Feeley and Silman, 2011; Feeley, 2015). Many groups of tropical plants have never been comprehensively studied, and our understanding of numerous species is so rudimentary that it remains in a 'discovery' phase—despite having been described decades or even centuries ago (Goodwin et al., 2020).

As this concerning trend continues, experts have warned about the potential consequences of detaching taxonomy from other disciplines, and numerous articles have called for reintegration (e.g., Wheeler et al., 2004; Wheeler, 2014; Wilson, 2004; Drew, 2011; Garnett and Christidis, 2017; Thomson et al., 2018). Unfortunately, despite the growing awareness of the ongoing 'biodiversity crisis' among the public, the detachment of taxonomy has intensified in the 21st century. A Google Books Ngram Viewer analysis (Figure 7.1) illustrates this, revealing a significant decline in the frequency of the terms 'taxonomy' and 'systematics' in books published since the 1980s, even as the use of the term 'biodiversity' has risen exponentially.

This is a striking paradox: while taxonomy is the very discipline that recognizes and describes biodiversity, its prominence in scientific literature appears to wane. At a time when taxonomy should be at the forefront of biodiversity science, it is relegated to the sidelines, often regarded merely as a complementary tool for more advanced research.

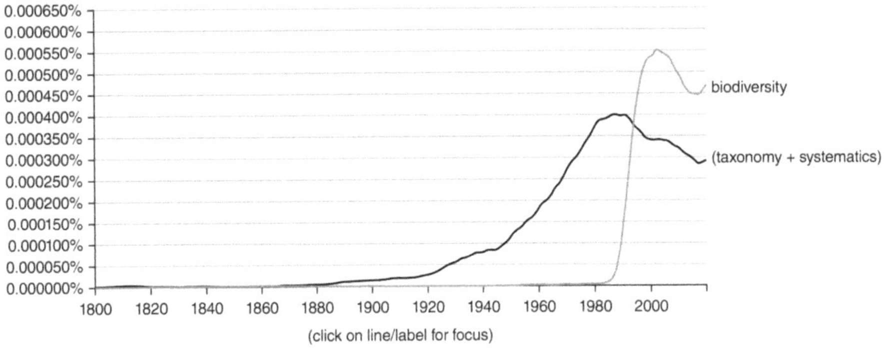


**FIGURE 7.1** Google Books Ngram Viewer graph showing the frequency of use of the terms 'taxonomy' and/or 'systematics' (black line) compared to the term 'biodiversity' (grey line) in books published between 1800 and 2019. The frequency of use of the term 'taxonomy/systematics' has declined steadily since the 1980s, while the frequency of use of 'biodiversity' has increased exponentially.

## HOW TO INCREASE VISIBILITY: SELL THE PRODUCT

At every gathering of taxonomists, there are recurrent discussions about the lack of incentives for conducting thorough taxonomic work, limited institutional support, inadequate funding, and the barriers to publishing in high-impact journals—all of which hinder career progression (Wheeler, 2020, 2024). Given these challenges, it is imperative to consider how taxonomists can reclaim their pivotal role in science, assuming that the system will not change for us. Specifically, we must ask how, as taxonomists, we can effectively navigate the current scientific landscape (i.e., keep our jobs) and restore the prominence of our discipline. In this section, I focus on three aspects: publication of results, leveraging available resources, and advocating for our work in broader fora.

First, one of the most pressing issues taxonomists face today is the tendency to leave taxonomy aside, sometimes publishing incomplete results, if that means accessing higher-impact journals. Be this our own decision or a requirement from journal editors, we must change this approach. A robust taxonomy is the backbone of biodiversity research, and it should be included in the main publication, not relegated to another article or, worse, omitted altogether. Too often, journals prioritize flashy results or large-scale data analyses at the expense of the solid taxonomic foundation that underpins the work. This trend hinders scientific progress, as it overlooks the critical step of defining and describing the very organisms we are studying. If the journal we submit our work to refrains from publishing taxonomic results, we should actively seek alternative journals that value taxonomy as a core part of research. Comprehensive, well-rounded taxonomic treatments need to be given the spotlight they deserve, ensuring that our findings are both complete and meaningful. This, I acknowledge, may be easier said than done for early career researchers, who often feel pressure to publish in high-impact journals. However, there are fortunately high-quality alternatives that value and support the publication of comprehensive taxonomic research.

In addition to ensuring thoroughness in our publications, taxonomists must embrace the broad range of resources available to us in the digital age to make our results accessible. While traditional publications remain essential for career progression, they can be slow to produce, rarely updated, and sometimes difficult to access for colleagues at other institutions. To overcome these limitations, we must use all resources at our disposal, including websites, digital platforms, and global repositories, where users can easily access curated data. For example, my collaborators and I have dedicated two decades to the study of *Acalypha*, a megadiverse plant genus. The project website, formally the *Acalypha Taxonomic Information System* (www.acalypha.es), includes comprehensive taxonomic information, including identification keys, photos, illustrations, maps, and information on over 40,000 specimen records we have consulted during this time. We update it regularly to include the latest information and revise existing content, and it is linked to the Global Biodiversity Information Facility for publishing our taxonomically curated datasets. Our work is just one example, with other initiatives worth exploring, such as the Legume Data Portal (https://www.legumedata.org/), the Solanaceae Source (https://solanaceaesource.myspecies.info/), and the Begonia Resource Centre (https://padme.rbge.org.uk/Begonia/). These platforms enable researchers to keep our taxonomic research dynamic and current. By embracing this multiplicity of channels, we ensure that taxonomic data remain relevant and easily accessible. Our work should not only exist in static formats but should live and breathe within the larger scientific community, evolving with new discoveries.

Finally, it is crucial that we take the discussion beyond the taxonomic community. Far too often, taxonomic research remains siloed within niche conferences and journals, self-limiting its reach and impact. We need to speak up and present our findings in interdisciplinary forums where our work can reach a broader audience. The relevance of taxonomy extends beyond its own discipline—into ecology, conservation, climate science, and more. By making our voices heard in these broader arenas, we can elevate the importance of taxonomy and ensure that our contribution is fully recognized. When taxonomists and taxonomy advocates take part in cross-disciplinary dialogues, we not only raise the visibility of our work but also demonstrate its integral role in addressing global challenges. See, for example, three events recently organized by the Systematics Association and The Linnean Society on 'Biodiversity and finance: making space for a common planet', with Elree Winnett Seelig, Mike Maunder and Lucy Fitzgeorge-Parker (https://www.youtube.com/watch?v=58rnM-BuqCo); 'The role of science in government', with Sir Ian Boyd (https://www.youtube.com/watch?v=DpElEfnbzT8); or 'A future for fungi: lichens, the environment, and current conservation', with Dr Rebecca Yahr (https://www.youtube.com/watch?v=X5xLEveQQPk).

By raising the profile of taxonomy within the scientific community and using the range of resources readily available, we can not only improve our understanding of biodiversity but also tackle the global challenges related to biodiversity discovery and conservation. Taxonomy is not just a foundational discipline; it is essential for informing effective policies and strategies that protect our planet's biological heritage for future generations.

## A 21ST CENTURY APPROACH: COMPREHENSIVE, INTEGRATIVE, FAST

As taxonomists, we should advocate for the visibility and importance of our work within the scientific community. Yet, we must also recognize that we need to be more ambitious in our approach. Amid the ongoing climate and biodiversity crisis, it is not enough to accelerate the description of new species; we also need to increase and improve the information we have about them. In this section, I will argue that one of the most effective ways to achieve this is through the production of taxonomic monographs. While this may evoke in the reader images of 19th-century works, I contend that taxonomic monographs—when conducted at scale and truly integrating different disciplines—can significantly enhance our understanding of biodiversity and its implications for conservation and policy (Scotland and Wood, 2012; Muñoz-Rodríguez et al., 2019).

Taxonomic monographs represent the gold standard of systematics research, providing comprehensive treatments that serve as reference for entire groups of organisms (order, family, genus). They are often valuable for more than a century, as exemplified by Darwin's and other authors' work (e.g., Morison, 1672; Grant, 1924; Rosendahl et al., 1936; Hughes, 1998). Historically, these monographs have been thought of as lengthy undertakings with limited immediate value, leading many in the field to prioritize shorter publications that align more closely with career advancement. However, in the 21st century, it is essential that we recognize their transformative potential. To meet the demands of a rapidly evolving scientific landscape, authors must go beyond traditional taxonomy. Modern monographs need to be comprehensive and integrative, rooted in evolutionary biology, genomics, and the vast amounts of biodiversity data now available. They must also be global in scope, conducted through international collaboration and teamwork, and crucially, they must embrace new technologies that allow for more rapid and precise species identification, description, and analysis (Muñoz-Rodríguez et al., 2024). When done right, taxonomic monographs are not merely static records of the diversity of life; they become living, evolving resources that propel science forward.

In 2020, my colleagues and I published a monograph on *Ipomoea* L., a plant genus in the family Convolvulaceae Juss., in the American continent (Wood et al., 2020). This project started as a plan to accelerate the pace of taxonomy by the two lead authors (Scotland and Wood, 2012), who previously demonstrated the potential of their approach by producing a monograph of the genus *Convolvulus* L. (Wood et al., 2015). These works, I think, exemplify the modern approach monographs should take. In the case of *Ipomoea*, with about 810 species distributed across tropical and subtropical regions worldwide, it presented a significant challenge, not only because of its sheer diversity but also due to the gaps in existing knowledge (Muñoz-Rodríguez et al., 2023). Addressing the genus required a team effort, integrating multiple disciplines, working on a continental scale, and employing advanced DNA sequencing technologies. A detailed explanation of our approach has been published recently (Muñoz-Rodríguez et al., 2024). We conducted herbarium research and fieldwork simultaneously with lab-based molecular work, ensuring our species delimitation hypotheses were as robust as possible and integrated both morphological and genetic evidence. By combining long-term established taxonomic practices with modern genomics, we achieved a more nuanced understanding of *Ipomoea.*

The results were both extensive and impactful. We identified and described dozens of new species, dramatically improving our understanding of the genus (Wood et al., 2016, 2017a, b, 2018). We reduced synonymy by 69% (e.g., Muñoz-Rodríguez et al., 2019), organizing an enormous amount of information published over 250 years of research and ensuring that future research would be built on a solid foundation of correctly identified species. Alongside this, we produced detailed identification keys, illustrations, photographs, and distribution maps that make the monograph a practical tool for scientists, conservationists, and policymakers. The phylogenies we developed provided deeper insights into the evolutionary history of *Ipomoea* and the sweet potato (*Ipomoea batatas* (L.) Lam.), adding a temporal dimension to the study of the group and enabling us to study it within a broader evolutionary context (Muñoz-Rodríguez et al., 2018, 2019, 2022a). Additionally, we compiled a database of over 30,000 specimens, creating a digital repository that will continue to serve as a valuable resource for future research.

In the 21st century, the role of taxonomists should extend beyond species descriptions. We must act as integrators of knowledge, bridging the gap between traditional taxonomy and the new frontiers of biodiversity research. To do this effectively, we need to work more efficiently, collaborate across disciplines, operate on a global scale, and embrace technology to ensure our work is continuously updated with new data. I am convinced taxonomic monographs provide an unparalleled strategy to achieve this. By following these steps, we will ensure that taxonomy remains at the forefront of scientific discovery and continues to play a crucial role in understanding and preserving Earth's biodiversity.

## LEAD PROJECTS: EXPLORE GLOBAL QUESTIONS

Taxonomists invest significant effort—often a lifetime—detailing the taxonomy of a group of organisms. When such a work, maybe in a monograph format, is produced, it is often viewed as the culmination of the research, leading the taxonomist to shift their focus to another group in need of taxonomic study. While it is understandable for taxonomists to remain within their areas of expertise, it is now crucial that we take the initiative to lead projects that build upon our taxonomic findings and address specific questions in ecology, agriculture, conservation, and other fields. Doing so is the only way to ensure that our work has a broader impact and, equally important, highlight the importance of taxonomy and secure funding for subsequent research.

Beyond the immediate taxonomic advancements, our *Ipomoea* monograph has had a wide-reaching impact across various fields. In the years since its publication, it has spurred the description of seven new species in the Americas (Jara et al., 2020; Santos et al., 2021; Wood et al., 2024), demonstrating that taxonomic monographs are not static; they foster ongoing scientific inquiry and discovery. The monograph has also supported theoretical research in evolutionary biology and systematics, where the large and well-documented dataset has been used to test new methods and models (Carruthers, 2019; Carruthers et al., 2020; Wells et al., 2021; Muñoz-Rodríguez et al., 2022a, b; Muñoz-Rodríguez et al., 2023). It has provided insights into pressing issues such as climate change, food security and crop wild relatives (Muñoz-Rodríguez et al., 2018, 2022a; Deloya Brito et al., 2023; Onen et al., 2023) and has played a consultative role in biodiversity management, helping guide conservation strategies in regions as diverse as India and New Guinea (Cámara-Leret et al., 2020; Wood et al., 2022). Our work even prompted a re-examination of *Ipomoea* remains in the archaeological record of Easter Island, linking taxonomy with archaeological and palaeobotanical research (Muñoz-Rodríguez et al., 2022c). The *Ipomoea* monograph, along with related works, has accumulated over 500 citations across diverse disciplines (cf., Google Scholar), further illustrating its broad impact beyond the field of taxonomy.

In summary, by combining taxonomy, genomics, and herbarium collections, we were able to produce a monograph that not only advanced our understanding of *Ipomoea*, but also created a framework supporting ongoing research across biological, environmental, and even cultural fields. This example illustrates how a well-conceived, comprehensive monograph is much more than a species catalogue; it is a foundational work that influences a wide range of disciplines. When developed with an evolutionary framework in mind, at a global scale, and supported by technology, a monograph becomes a dynamic tool that generates knowledge long after its publication.

In terms of funding, our work on *Ipomoea* has obtained continuous support for over a decade, leading to the results summarized above. Recently, it has provided the necessary evidence to extend beyond *Ipomoea* and apply a similar approach to the study of another megadiverse plant genus, *Acalypha*. The project I lead titled 'Origin and evolution of tropical plant lineages of economic interest: integrated studies on the megadiverse genus *Acalypha*', has received funding from the Spanish Ministry of Science, Innovation and University (ref. PID2022-139634NA-I00). We aim to produce a comprehensive taxonomic monograph of *Acalypha* while investigating the origin and diversity of its species of medicinal interest and relevant compounds. Our objectives include producing identification tools, comprehensive genomic phylogenies, conservation assessments, and analyzing how climate change might affect their future distribution. This proposal underscores the potential of taxonomic research to inform broader scientific research and contribute to conservation and economic sustainability.

## CONCLUSIONS

As we navigate the complexities of the 21st century, it is imperative that taxonomists embrace an ambitious vision for our discipline. By increasing the visibility and relevance of our work, we can secure a prominent place for taxonomy within the broader scientific landscape. We must actively engage in collaborative projects that transcend traditional boundaries, integrating taxonomy with other disciplines such as genomics, ecology, and conservation biology. This multidisciplinary approach not only enriches our understanding of biodiversity but also enhances our ability to tackle global challenges.

We should take the lead in driving innovative projects that showcase the transformative potential of taxonomy, ensuring that our contributions extend beyond the confines of our field. In doing so, we can redefine the narrative surrounding taxonomy and firmly establish its critical role in shaping the future of biological research and conservation.

## REFERENCES

Bebber, D.P., Carine, M.A., Wood, J.R.I., Wortley, A.H., Harris, D.J., Prance, G.T., Davidse, G. et al. (2010) Herbaria are a major frontier for species discovery. *Proceedings of the National Academy of Sciences*, 107: 22169–22171, https://doi.org/10.1073/pnas.1011841108

Cámara-Leret, R., Frodin, D.G., Adema, F., Anderson, C., Appelhans, M.S., Argent, G., Arias Guerrero, S. et al. (2020) New Guinea has the world's richest island flora. *Nature*, 584: 579–583, https://doi.org/10.1038/s41586-020-2549-5

Carruthers, T. (2019) What can we learn about plant evolution from a robust phylogenetic framework? PhD thesis. University of Oxford, Oxford.

Carruthers, T., Muñoz-Rodríguez, P., Wood, J.R.I. and Scotland, R.W. (2020) The temporal dynamics of evolutionary diversification in *Ipomoea*. *Molecular Phylogenetics and Evolution*, 146: 106768, https://doi.org/10.1016/j.ympev.2020.106768

Damasco, G., Baraloto, C., Vicentini, A., Daly, D.C., Baldwin, B.G. and Fine, P.V.A. (2021) Revisiting the hyperdominance of neotropical tree species under a taxonomic, functional and evolutionary perspective. *Scientific Reports*, 11: 9585, https://doi.org/10.1038/s41598-021-88417-y

Darwin, C. (1851) *Living Cirripedia. A monograph on the sub-class Cirripedia, with figures of all the species. The Lepadidae; or, pedunculated cirripedes.* The Ray Society, London.

Darwin, C. (1854) *Living Cirripedia. The Balanidae, or sessile cirripedes.* The Verrucidae. The Ray Society, London.

Darwin, C. (1859) *On the origin of species* [...]. John Murray, London.

Deloya Brito, E.C., Clay, C., McIntyre, P.J., Piña-de la Rosa, I.A., Scotland, R.W., Muñoz-Rodríguez, P. and Cacho, N.I. (2023) Climatic amplitude is a predictor of geographic range size in Mexican morning glories (*Ipomoea* L., Convolvulaceae). *Botanical Sciences*, 101(4): 1016–1033, https://doi.org/10.17129/botsci.3322

Deutsch, J. (2010) Darwin and barnacles. *Comptes Rendus Biologies*, 333: 99–106, https://doi.org/10.1016/j.crvi.2009.11.009

Devictor, V. and Bensaude-Vincent, B. (2016) From ecological records to big data: The invention of global biodiversity. *History and Philosophy of the Life Sciences*, 38(4): 1–23, http://www.jstor.org/stable/44744739

Drew, L.W. (2011) Are we losing the science of taxonomy? As need grows, numbers and training are failing to keep up. *BioScience*, 61: 942–946, https://doi.org/10.1525/bio.2011.61.12.4

Escobar, H. and Vogel, G. (2018) 'It was a foretold tragedy'—fire destroys Brazil's National Museum and its prized science collections. *Science*, 2018, https://doi.org/10.1126/science.aav3064

Favret, C. (2024) The 5 'D's of Taxonomy: A User's Guide. *The Quarterly Review of Biology*, 99: 131–156, https://doi.org/10.1086/732044

Feeley, K. (2015) Are we filling the data void? An assessment of the amount and extent of plant collection records and census data available for tropical South America. *PLoS ONE*, 10: e0125629, https://doi.org/10.1371/journal.pone.0125629

Feeley, K.J. and Silman, M.R. (2011) The data void in modeling current and future distributions of tropical species. *Global Change Biology*, 17: 626–630, https://doi.org/10.1111/j.1365-2486.2010.02239.x

Garnett, S.T. and Christidis, L. (2017) Taxonomy anarchy hampers conservation. *Nature*, 546: 25–27, https://doi.org/10.1038/546025a

Goodwin, Z.A., Harris, D.J., Filer, D. Wood, J.R.I. and Scotland, R.W. (2015) Widespread mistaken identity in tropical plant collections. *Current Biology*, 25: R1066–R1067, https://doi.org/10.1016/j.cub.2015.10.002

Goodwin, Z.A., Muñoz-Rodríguez, P., Harris, D.J., Wells, T., Wood, J.R.I., Filer, D. and Scotland, R.W. (2020) How long does it take to discover a species? *Systematics and Biodiversity*, 18: 784–793, https://doi.org/10.1080/14772000.2020.1751339

Grant, A.L. (1924) A monograph of the genus *Mimulus*. *Annals of the Missouri Botanical Garden*, 11(2/3): 99–388.

Hortal, J., De Bello, F., Diniz-Filho, J.A.F., Lewinsohn, T.M., Lobo, J.M. and Ladle, R.J. (2015) Seven shortfalls that beset large-scale knowledge of biodiversity. *Annual Review of Ecology, Evolution, and Systematics*, 46: 523–549, https://doi.org/10.1146/annurev-ecolsys-112414-054400

Hughes, C. 1998 Monograph of *Leucaena* (Leguminosae-Mimosoideae). *Systematic Botany Monographs*, 55: 1–244, https://doi.org/10.2307/25027876

Jara, E., Muñoz-Rodríguez, P., Wood, J.R.I. and Beltrán, H. (2020) *Ipomoea noemana* (Convolvulaceae) a new species from Ancash eastern slope in Peru. *Phytotaxa*, 461: 286–294, https://doi.org/10.11646/phytotaxa.461.4.5

Karbstein, K., Kösters, L., Hodač, L., Hofmann, M., Hörandl, E., Tomasello, S., Wagner, N.D. et al. (2024) Species delimitation 4.0: Integrative taxonomy meets artificial intelligence. *Trends in Ecology and Evolution*, 39: 771–784, https://doi.org/10.1016/j.tree.2023.11.002

Löbl, I., Klausnitzer, B., Hartmann, M. and Krell, F.-T. (2023) The silent extinction of species and taxonomists—An appeal to science policymakers and legislators. *Diversity*, 15: 1053, https://doi.org/10.3390/d15101053

Love, A.C. (2002) Darwin and *Cirripedia* prior to 1846: Exploring the origins of the barnacle research. *Journal of the History of Biology*, 35: 251–289, http://www.jstor.org/stable/4331734

Morison, R. (1672) *Plantarum Umbelliferarum Distributio Nova, per Tabulas Cognationis et Affinitatis ex Libro Naturae Observata et Detecta*. Oxford University Press, Oxford.

Muñoz-Rodríguez, P., Carruthers, T., Wells, T., Sumadijaya, A., Wood, J.R.I. and Scotland, R.W. (2024) The research behind a taxonomic monograph: A case study from *Ipomoea* (Convolvulaceae). Kew Bulletin, 79, 897–914, https://doi.org/10.1007/s12225-024-10184-6

Muñoz-Rodríguez, P., Carruthers, T., Wood, J.R.I., Williams, R.R.M., Weitemier, K., Kronmiller, B., Ellis, D. et al. (2018) Reconciling conflicting phylogenies in the origin of sweet potato and dispersal to Polynesia. *Current Biology*, 28: 1246–1256.e12, https://doi.org/10.1016/j.cub.2018.03.020

Muñoz-Rodríguez, P., Carruthers, T., Wood, J.R.I., Williams, B.R.M., Weitemier, K., Kronmiller, B., Goodwin, Z. et al. (2019) A taxonomic monograph of *Ipomoea* integrated across phylogenetic scales. *Nature Plants*, 5: 1136–1144, https://doi.org/10.1038/s41477-019-0535-4

Muñoz-Rodríguez, P., Wells, T., Wood, J.R.I., Carruthers, T., Anglin, N.L., Jarret, R.L. and Scotland, R.W. (2022a) Discovery and characterisation of sweet potato's closest tetraploid relative. *New Phytologist*, 234: 1185–1194, https://doi.org/10.1111/nph.17991

Muñoz-Rodríguez, P., Wood, J.R.I. and Scotland, R.W. (2022b) Diagnosability and cryptic nodes in Angiosperms: A case study from *Ipomoea*, pp. 131–142. In *Cryptic Species: Morphological Stasis, Circumscription and Hidden Diversity* (eds A. Monro and S. Mayo). Systematics Association and Cambridge University Press, London.

Muñoz-Rodríguez, P., Wood, J.R.I. and Scotland, R.W. (2022c) Sweet potato on Rapa Nui: Insights from a monographic study of the genus *Ipomoea*, pp. 63–83. In *The prehistory of Easter Island (Rapa Nui): Towards an integrative interdisciplinary framework* (eds V. Rull and C.M. Stevenson). Springer, Cham, Switzerland.

Muñoz-Rodríguez, P., Wood, J.R.I., Wells, T., Carruthers, T., Sumadijaya, A. and Scotland, R.W. (2023) The challenges of classifying big genera such as *Ipomoea*. *Taxon*, 72: 1201–1215, https://doi.org/10.1002/tax.12887

Naggs, F. (2022) The tragedy of the Natural History Museum, London. *Megataxa*, 7: 86–112, https://doi.org/10.11646/megataxa.7.1.2

Onen, H., Farooq, S., Muñoz-Rodríguez, P., Alharbi, S.A. and Alfarraj, S. (2023) *Ipomoea tricolor* (Convolvulaceae) in Turkey: New occurrence record and potential spread areas under current climatic conditions. *Journal of King Saud University – Science*, 35: 102543, https://doi.org/10.1016/j.jksus.2023.102543

Orr, M.C., Ferrari, R.R., Hughes, A.C., Chen, J., Ascher, J.S., Yan,Y.-H., Williams, P.H. et al. (2020) Taxonomy must engage with new technologies and evolve to face future challenges. *Nature Ecology and Evolution*, 5: 3–4, https://doi.org/10.1038/s41559-020-01360-5

Pennisi, E. (2024) 'A tragic mistake': Decision to close Duke University's herbarium triggers furor. *Science*, 2024, https://doi.org/10.1126/science.zopwqlr

Rainbow, P.S. (2011) Charles Darwin and marine biology. *Marine Ecology*, 32: 130–134, https://doi.org/10.1111/j.1439-0485.2010.00421.x

RBG Kew (2016) *The State of the World's Plants Report 2016*. The Board of Trustees of the Royal Botanic Gardens, Kew, Richmond.

Rosendahl, C.O., Butters, F.K. and Lakela, O. (1936) *A Monograph on the Genus Heuchera*. University of Minnesota Press, Saint Louis.

Santos, D., Souza, E.B. and Buril, M.T. (2021) *Ipomoea lanifolia* sp. nov. (Convolvulaceae), a new species endemic to the Ibiapaba plateau in northeastern Brazil. *Rodriguésia*, 72: e00782020, https://doi.org/10.1590/2175-7860202172076

Sayers, E., Cavanaugh, M., Clark, K., Pruitt, K.D., Schoch, C.L., Sherry, S.T. and Karsch-Mizrachi, I. (2021) GenBank. *Nucleic Acids Research*, 49: D92–D96, https://doi.org/10.1093/nar/gkaa1023

Scotland, R.W. and Wood, J.R.I. (2012) Accelerating the pace of taxonomy. *Trends in Ecology and Evolution*, 27: 415–416, https://doi.org/10.1016/j.tree.2012.04.013

Tautz, D., Arctander, P., Minelli, A., Thomas, R.H. and Vogler, A.P. (2002) DNA points the way ahead in taxonomy. *Nature*, 418: 479–479, https://doi.org/10.1038/418479a

Tautz, D., Arctander, P., Minelli, A., Thomas, R.H. and Vogler, A.P. (2003) A plea for DNA taxonomy. *Trends in Ecology and Evolution*, 18: 70–74, https://doi.org/10.1016/S0169-5347(02)00041-1

Thiers, B. (2023) *Index Herbariorum: A Global Directory of Public Herbaria and Associated Staff.* New York Botanical Garden's Virtual Herbarium, https://sweetgum.nybg.org/science/ih/

Thomson, S.A., Pyle, R.L., Ahyong, S.T., Alonso-Zarazaga, M., Ammirati, J., Araya, J.F., Ascher, J.S. et al. (2018) Taxonomy based on science is necessary for global conservation. *PLoS Biology*, 16: e2005075, https://doi.org/10.1371/journal.pbio.2005075

Wells, T., Carruthers, T., Muñoz-Rodríguez, P., Sumadijaya, A., Wood, J.R.I. and Scotland, R.W. (2021) Species as a heuristic: Reconciling theory and practice. *Systematic Biology*, 71: 1233–1243, https://doi.org/10.1093/sysbio/syab087

Wheeler, Q.D. (2014) Are reports of the death of taxonomy an exaggeration? *New Phytologist*, 201: 370–371, https://doi.org/10.1111/nph.12612

Wheeler, Q.D. (2020) A taxonomic renaissance in three acts. *Megataxa*, 1: 4–8, http://dx.doi.org/10.11646/megataxa.1.1.2

Wheeler, Q.D. (2024) *Species, Science and Society: The Role of Systematic Biology.* Routledge, Abdingdon.

Wheeler, Q.D., Raven, P.D. and Wilson, E.O. (2004) Taxonomy: Impediment or expedient? *Science*, 303: 285, https://doi.org/10.1126/science.303.5656.285

Wiens, J.J. (2021) Vast (but avoidable) underestimation of global biodiversity. *PLoS Biology*, 19: e3001192, https://doi.org/10.1371/journal.pbio.3001192

Williams, B.R.M., Mitchell, T.C., Wood, J.R.I., Harris, D.J., Scotland, R.W. and Carine, M.A. (2014) Integrating DNA barcode data in a monographic study of *Convolvulus. Taxon*, 63: 1287–1306, https://doi.org/10.12705/636.9

Wilson, E.O. (2004) Taxonomy as a fundamental discipline. *Philosophical Transactions of the Royal Society of London. Series B: Biological Sciences*, 359: 739, https://doi.org/10.1098/rstb.2003.1440

Wood, J.R.I., Martinez Ugarteche, M.T., Muñoz-Rodríguez, P. and Scotland, R.W. (2018) Additional notes on *Ipomoea* (Convolvulaceae) in Bolivia. *Kew Bulletin*, 73: 57, https://doi.org/10.1007/s12225-018-9784-z

Wood, J.R.I., Muñoz-Rodríguez, P., Degen, R. and Scotland, R.W. (2017a) New species of *Ipomoea* (Convolvulaceae) from South America. *PhytoKeys*, 88: 1–38, https://doi.org/10.3897/phytokeys.88.12891

Wood, J.R.I., Muñoz-Rodríguez P. and Scotland, R.W. (2022) Priorities for taxonomic studies of Indian Convolvulaceae. *Journal of Economic and Taxonomic Botany*, 46: 5–14.

Wood, J.R.I., Muñoz-Rodríguez, P., Wells, T., Espinel, D., Romoleroux, K., Cerón Martínez, C.E., Cornejo, X. and Scotland, R.W. (2024) Notes on *Ipomoea* from Ecuador. Kew Bulletin, 79, 915–939, https://doi.org/10.1007/s12225-024-10186-4

Wood, J.R.I., Muñoz-Rodríguez, P., Williams, B.R.M. and Scotland, R.W. (2020) A foundation monograph of *Ipomoea* (Convolvulaceae) in the New World. *PhytoKeys*, 143: 1–823, https://doi.org/10.3897/phytokeys.143.32821

Wood, J.R.I., Urbanetz, C. and Scotland, R.W. (2016) *Ipomoea pantanalensis*, a new species of *Ipomoea* L. (Convolvulaceae) from the Pantanal, Brazil. *Kew Bulletin*, 71: 6, https://doi.org/10.1007/s12225-016-9619-8

Wood, J.R.I., Vasconcelos, L.V., Simão-Bianchini, R. and Scotland, R.W. (2017b) New species of *Ipomoea* (Convolvulaceae) from Bahia. *Kew Bulletin*, 72, 8, https://doi.org/10.1007/s12225-017-9678-5

Wood, J.R.I., Williams, B.R.M., Mitchell, T.C., Carine, M., Harris, D. and Scotland, R.W. (2015) A foundation monograph of *Convolvulus* L. (Convolvulaceae). *PhytoKeys*, 51: 1–282, https://doi.org/10.3897/phytokeys.51.7104

Yesson, C., Brewer, P.W., Sutton, T., Caithness, N., Pahwa, J.S., Burgess, M., Gray, W.A. et al. (2007) How global is the Global Biodiversity Information Facility? J. Beach [ed.]. *PLoS ONE*, 2: e1124, https://doi.org/10.1371/journal.pone.0001124.t003

# 8 Systematics and Biogeography, Ontology, and Vicariance

*Visotheary Ung and Anaïs Grand*

> 'In this definition it seems to me particularly essential that the task of 'ordering' (and what means the same thing, of systematics), lies in considering the unit as a member of an ordered whole [...] In terms of quantum theory, or class theory, each single element may occur in differently composed partial quantities, or partial classes, within a whole quantity, depending on which of the many relationships existing among the elements of the total quantity are chosen to form the partial quantities'
>
> *Hennig, 1966, p. 4*

> 'In this volume we experiment with a perspective that allows past work to be related to questions still open'
>
> *Nelson and Platnick, 1981, pp. ix–x*

## INTRODUCTION

In 1981, Gareth Nelson and Norman Platnick published *Systematics and Biogeography, Cladistics and Vicariance*, also known as 'the blue book' (Nelson and Platnick 1981, available at https://content.ucpress.edu/chapters/nelson_systematics_and_biogeography.pdf).

In our perspective, 'the blue book' is more than just a volume of knowledge; it represents a corner stone in the field of Evolutionary Biology. We would qualify it as the last revolutionary work since the publication of Willi Hennig's book *Phylogenetic systematics* (Hennig, 1966). The historical echo between these events emphasizes the long-lasting impact of *Systematics and Biogeography*, elevating it to a status that recalls these transformative moments in the history of evolutionary science.

*Systematics and Biogeography* is organized into three parts (Space, Time and Form), each considered a distinct aspect of Comparative Biology. Biogeography focuses on *space*, Ontogeny and Paleontology at the dimensions of *time*, and Systematics investigates the complexities of *form*. Each section demonstrates the authors' in-depth exploration of the subject.

> Comparative biology, if it is to allow us to understand the living world, must thus, deal with three distinguishable elements: (1) similarities and differences in the


DOI: 10.1201/b22822-8

> attributes of organisms, (2) the history of organisms in time, and (3) the history of organism in space
>
> *Nelson and Platnick, 1981, p. 5*

The present chapter is not meant to be a review; instead, it aims at being a kind of tribute to the authors of *Systematics and Biogeography*. Their work stands as a foundational piece, a pioneering journey into the realms of Comparative Biology. It becomes obvious that the limitations they faced were not in their theories or ideas but rather in the technological tools available at that time. In this chapter, we would like to acknowledge their perspicacity, recognizing that everything they conceptualized in their book can now, in principle, be implemented, thanks to the progress of technology. What might have been perceived as analytical challenges can now be addressed and overcome. This chapter aims to honor not just the content of their work but also the vision they possessed, anticipating a future where the tools would catch up with the originality of their ideas. They have meticulously created a comprehensive outline of comparative biology, and it becomes evident that their initial thoughts and theories find confirmation in the ongoing evolution and advancement of technologies and techniques. In this chapter, our aim is to emphasize that evolutionary theory should be considered an integrative science and elucidate its ontology through a semantic approach. This exploration aligns with Nelson and Platnick's perspective on Comparative Biology, as presented in their book. Evolutionary theory stands out as the unifying theory in biology through precise formalization and modelling. We will anchor our discourse in a particular branch of Philosophy of Sciences, i.e., the semantics conception of scientific theories and we will illustrate how the graphic representation of a rooted tree, i.e., a hierarchy, plays a pivotal role in this formalization. This mathematical depiction is essential in enabling the characterization of various facets of the theory and fostering interdisciplinary connections. Notably, it integrates methods, resources, and technologies within a unified model, providing a holistic perspective of evolutionary theory, including systematics and biogeography.

## THE ART OF CLASSIFICATION: TAXONOMY, SYSTEMATICS, AND BIOGEOGRAPHY

As human beings, our natural tendency is to categorize and classify the world around us. This tendency is closely linked to the act of naming those objects. These names are intellectual constructions, facilitating our understanding of the world. If we are saying 'this is our house', everyone will understand that we are referring to the place we live, i.e., our home. This collective understanding is the result of cultural, linguistic, and societal references, which enables clear communication and mutual comprehension. Extending this example, if we are saying 'this is our house with our two cats' additional levels of meaning are conveyed. The word 'cats' brings up images of furry, independent, and perhaps even spoiled, mammals. Again, this understanding is not individual but collective, rooted in shared experiences and cultural contexts.

In the field of biology, the discipline dealing with the classification of organisms is known as *Taxonomy*. This science was practiced by Carl Linnaeus (Carl von Linné), who introduced a hierarchical system now known as Linnaean taxonomy for organizing and categorizing living organisms. The word 'taxonomy' was introduced later by de Candolle, who provided the first extended essay on the subject (Candolle, 1813). Linnaeus also introduced 'binomial nomenclature', a system for assigning two-part scientific names to biological entities: the genus name, and the species name, as in *Homo sapiens*. Linnaeus significantly transformed the landscape of taxonomy through his work, including the first edition of *Systema Naturae* in 1735, *Species Plantarum* in 1753, and the 10th edition of *Systema Naturae* in 1758–1759, the most significant edition of that publication as it is the starting point of zoological nomenclature. The classification of organisms relies on a hierarchical representation of the objects under study, i.e., taxa. This system provides an organized structure for categorizing living entities into groups based on their shared characteristics. However, it's important to note that these classification schemes do not inherently include information about the evolutionary history of the taxa – systematics is the discipline that addresses the evolutionary aspect.

Systematics reflects the diversification of organisms, both past and present, exploring their relationships through time. This is often visualized using branching diagrams, which depict the evolutionary connections between different taxa. Nelson and Platnick's Chapter 2, 'Systematic History: Kinds of branching diagrams' explores the various ways in which diagrams can be constructed and interpreted to represent evolutionary relationships. In mathematics and logic, diagrams are designed to convey a mathematical relationship between entities to facilitate valid reasoning (Allwein and Barwise, 1996). When applied to the study of evolution, these diagrams often take the form of a rooted tree, aligning with graph theory's definition as 'an indirect graph in which any two vertices are connected by exactly one path, or equivalent to a connected acyclic undirected graph' (Bender and Williamson 2010, p. 171). The rooted tree structure, whether in biological taxonomy, systematics or mathematical graph theory, reflects a hierarchy. In the context of evolution, a rooted tree diagram illustrates not just the classification of organisms but also their historical relationships, capturing the branching and divergence of lineages over time. Systematics is also very much interconnected with biogeography, another sub-discipline of biology. The term biogeography was coined both in German and English in the late nineteenth century (Ebach, 2015).

> 'If present-day features of the environment are not sufficient to explain distributional patterns, what could the explanation be? The only alternative seems to be history: we can conclude that present-day distributions are the result of the history of taxa and of the areas in which they have lived. Historical biogeography, then, has two components: the history of life on earth and the history of earth itself'.
>
> *Nelson and Platnick, 1981, p. 43*

Biogeography is the study of plant and animal geography over time, or the study of the distribution of plants and animals over time (Nelson and Platnick, 1978, 1981; Morrone, 2009).

While scientists may not explicitly identify themselves as 'biogeographers', biogeography has been widely studied. A simple request using Google Scholar with the keyword 'biogeography' yields ca. 760,000 results[1] which emphasize the extensive interest and research in this field. And a simple examination of the results shows how diverse biogeography is: It can be a subject, a method, a concept, or even a tool. Several comprehensive books, including those by Morrone (2009) and Parenti and Ebach (2009), have explored various aspects of biogeography. These works highlight the division of the field into two sub-categories: one that looks at the *geographic history of taxa*, exploring distribution patterns within specific groups, and another that concentrates on the *geographic history of areas*, investigating the spatial dynamics of ecosystems and regions which has been called *comparative biogeography*. Once again, the foundational work of Nelson and Platnick is essential, particularly in exploring the relationships between regions. Their book, as highlighted in the 7th chapter, focuses on the subject of biogeography. This section of their work examines the connections and dependencies among different geographical areas, presenting these relationships in the familiar form of branching diagrams, aligning with their earlier emphasis on using such diagrams in the context of systematics and evolutionary studies.

A shared element among the subjects of Taxonomy, Systematics and Biogeography is the use of branching diagrams, particularly in the form of rooted trees. These diagrams, representing a hierarchy, serve as a visual tool to illustrate the relationships between the objects under study, whether they are taxa or geographic areas. Here, our argumentation will focus on historical biogeography, the study of the evolutionary history of biogeographic areas. As we investigate this common ground, it becomes apparent that another level of abstraction is necessary. This calls for a clear conceptualization, i.e., a model that can formalize Evolutionary Biology (including Systematics and Biogeography), into a cohesive and coherent theoretical framework. Formalizing should be understood as characterizing the content of a theory using logical and mathematical tools. This conceptualization can also be applied to the temporal dimension, which corresponds to Nelson and Platnick's 3rd chapter. To achieve this, Philosophy of Science becomes imperative (De Haro, 2019). We will focus specifically on the semantic conception of scientific theories, one subfield which interprets models semantically and shows how we can relate it to an implementation in Computer science. The connection between the semantic conception of scientific theories in the Philosophy of Science and semantic modelling in informatics becomes apparent as we dig into the notions of semantics, models, ontology and integration. Through this model, the semantics of taxic and biogeographic evolution are elucidated, accounting for the ontology of these objects which can then be implemented in computer sciences.

## PHILOSOPHY OF SCIENCE AND COMPUTER SCIENCE: ONTOLOGY AND ONTOLOGIES

### The Semantic Conception of Scientific Theories

As we expand into the discussion on semantics, it's essential to clarify the definitions for the terms we will be using. What is an ontology? Guarino and Giaretta made a clear distinction between Ontology with an uppercase 'O' and ontology without an uppercase 'o' (Guarino and Giaretta, 1995). In Philosophy of Science, an Ontology (upper case O) is 'the branch of philosophy which deals with the nature and the organization of reality', or in other words 'what is being?' Although an ontology (lowercase o) is 'a logical theory which gives an explicit, partial account of a conceptualization' (both definitions from Guarino and Giaretta, 1995, p. 32), in this case, it is a data model that handles concepts and relationships enabling the modelling of knowledge within a specific domain. We will demonstrate that the same rooted tree representation, a hierarchy, can be applied across various disciplines, thereby confirming its universality.

The semantic conception of scientific theory states that a model is a **structure composed of objects and relationships** to which we add a **constraint** (Figure 8.1).

It emerges at a time when the 'syntactic conception of scientific theories' was predominant (Coffa, 1991; Friedman, 1999; Creath, 2014; Uebel, 2014) which was settled by the Circle of Vienna in the 1920s. This subfield, dedicated to interpreting models semantically, emerged in the 1960s, with contributions from Patrick Suppes, Bas van Fraasen, Frederic Suppe and Ronald Giere, building upon Tarski's model-theoretic approach (Vaught, 1974, 1986). More than a pictorial representation of the world, a model is a set of objects and relationships satisfying a set of axioms (i.e., constraints). Models allow mathematical interpretation of axioms. Under a semantic framework, a scientific theory has to be formalized as a model. The main purpose of this subfield is to relate different theories or different parts of the same theory, which can be also viewed as a 'reductionist' approach. In the realm of biology, Elisabeth Lloyd, John Beatty, and Paul Thompson have applied semantic models to population genetics and mendelian theory (Beatty, 1987; Lloyd, 1986, 1987, 1988; Thompson, 1983, 1986, 1987). This approach, initially presented by Beatty (1987) and Lloyd (1986, 1987, 1988) has found continued relevance, as seen in the work by Gildenhuys (2013). Thompson (1986) extended this semantic conception to view evolutionary theory as a network of interacting theories, highlighting the interconnectedness of the theory of heredity with the theory of natural selection. He stated that '[...] the semantic conception of theories provides a framework within which a formalization of evolutionary theory understood as a family of interacting theories can be given [...]' (Thompson, 1986, p. 82). Similarly, it is crucial to highlight that comparative biogeography also engages with phylogenetic systematics. Consequently, a semantic model designed to incorporate comparative biogeography is domain-independent and such a model inherently includes phylogenetic systematics as well.

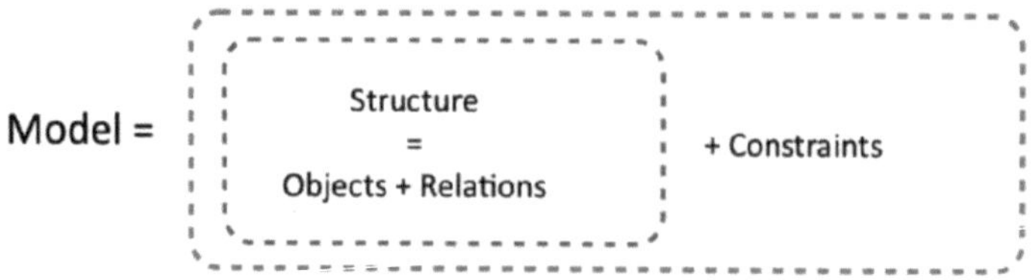


**FIGURE 8.1** Representation of a model under the semantic conception of scientific theories. A model is a structure composed of objects and relations to which constraints are added.

## Application to Systematics and Biogeography

In phylogenetic systematics, a cladogram is viewed as a model representing nested groups, elucidating relationships among monophyletic taxa following Hennig's *phylogenetic systematics* (Hennig, 1966). With Hennig's contributions, evolutionary trees (patterns) highlight monophyletic groups i.e., taxa sharing homologous characters. A monophyletic taxon can be defined as a class grouping an ancestor and all of its descendants. Related to that, the notion of homology was defined as the relationship of similarity among parts of organisms inherited from a common ancestor (Hennig, 1966). There is an extensive literature concerning the concepts of homologues and homology that is not discussed further here (Nelson and Platnick, 1981; Patterson, 1988; Rieppel, 1988; de Pinna, 1991; Kitching et al., 1998; Williams, 2004; Williams and Ebach, 2008, amongst others). This present work is an attempt to analyze cladistics using the semantic conception of scientific theories, achieved through a review of current evolutionary representations in systematics and biogeography within a semantic framework. This analysis allows us to draw conclusions about the ontology of evolutionary theory, its alignment with reality, and its interdisciplinary nature. The monophyly constraint (Hennig, 1966) offers effective ontological constraints. Through a formal model of embedded classes representing monophyly, hierarchical knowledge and data models can be derived (Ung, 2018, figure 1).

Taxa are formal objects linked by hierarchical relationships of inclusion which is the structure of the model, adhering to the monophyly condition (the ontological constraint). Similarly, within comparative biogeography, the objects of study are areas of endemism inferred from taxa distributions. These areas are pragmatically defined before the application of semantic modelling. Taxa and areas are the Tips or Terminals as illustrated in Figure 8.2(A–I). Figure 8.2A illustrates an evolutionary tree that can be either a phylogenetic tree (a phylogeny) or an areagram (a cladogram of biogeographic areas or a biogeographic tree), the terminals are taxa or areas of endemism, respectively. The same tree can be represented as a Venn diagram (Figure 8.2B and C) that will show only the sets of objects, i.e., clades (or classes) of taxa or biogeographic areas. Nelson and Platnick (1981) stated that monophyletic taxa 'have unique sets of characters' which is the definition of a clade. We can extend this definition to the domain of biogeography and say that a monophyletic biogeographic area has 'unique sets of taxa'. In this context, we can go further and add that a terminal is simply a particular class of an evolutionary tree with a single element. Figure 8.2B obviously shows that both representations are equivalent: there are six nodes (or classes) in the tree (Figure 8.2A) and Figure 8.2C has six sets. A node is equivalent to a set. This does not contradict the necessity for defining terminals as objects of the model, whether they represent taxa or areas of endemism, prior to their modelling. However, semantic modelling incorporates and elucidates their ontology within the framework.

This formalization of both systematics and biogeography within the same ontological (semantic) model has been discussed in correspondence by Ung (2018) and Ung and Buttigieg (2023), key terms are summarized in Table 8.1.

Maintaining fidelity to a hierarchical model across the stages that link the conceptual level (i.e., knowledge level) with the empirical or data level (i.e., observations,

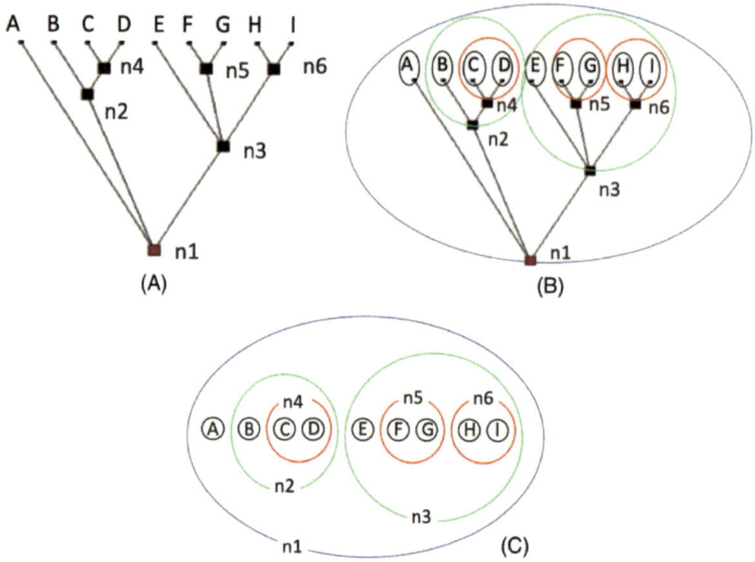


**FIGURE 8.2** A model of an evolutionary tree (A to I are OTUs, n1 to n6 are nodes of the tree) represented as a hierarchy (A), the same hierarchy and a Venn diagram superimposed (B), Venn diagram representation (C). Modified from Ung (2018).

measures) is crucial for ensuring internal consistency. Philosophers that formalized the semantic conception of theories have stressed the importance of implementing isomorphic models (e.g., Van Fraassen, 1994; Frigg, 2002; French, 2003). The hierarchical model can indeed be implemented at the character level. Furthermore, given that this model recognizes the interdisciplinary nature of theories, it can be applied consistently across both phylogenetics and biogeography. Once more, the hierarchical model of inclusion can be directly linked to the innovative work of Nelson and

**TABLE 8.1**
**List of terms used in Systematics and Biogeography**

| | Systematics | Biogeography |
|---|---|---|
| Terminal = Tip | Taxa | Areas of endemism |
| Node of a tree | Character state | Biogeographic area |
| Homology hypothesis | Character tree (terminals are states of characters, nodes are states of characters | Phylogeny (terminals are taxa, nodes are clades) |
| Character (mixed structure) | Taxa bearing the character state (terminals are taxa, nodes are state of characters) | Taxon-Area cladogram (terminals are areas of endemism, nodes are clades) |
| Results of a congruence analysis | Phylogeny (terminals are taxa, nodes are clades) | Area cladogram (terminals are areas of endemism, nodes are biogeographic areas) |

Platnick, who have already highlighted the parallel between systematics and biogeography (e.g., Nelson and Platnick, 1981, see also Nelson and Ladiges, 1991a, 1991b, 1992).

The hierarchical model of embedded classes provides an ontology for taxa and areas of endemism. Some years ago, Crother and Murray (2011, 2013) introduced an ontology for areas of endemism, which they termed the 'areas of endemism as individuals' theory' (Crother and Murray, 2011, 2013). This theory was built upon the framework of the 'species as individuals theory' as proposed by Ghiselin (1974) and Hull (1978). Crother and Murray wrote: 'The critical first step for the argument of areas of endemism as individuals is the recognition of species as individuals' (Crother and Murray, 2011, p. 1010).

They offer five conditions to establish their ontology:

1. An area must be a particular thing;
2. There cannot be any instance of it;
3. An individual is defined ostensively and not intentionally;
4. Individuals have boundaries in time and space; and
5. Individuals experience cohesiveness in their response to change.

However, it becomes evident that the 'individual thesis' lacks the monophyly constraint since an individual is bounded in time (having a birth and death), it cannot encompass all its descendants. Consequently, areas viewed as individuals do not encompass all their descendants (e.g., areas that emerged after a vicariance event) and are not monophyletic. A theory that accommodates objects seen as individuals cannot be integrated into a model. Instead, this theory is formalized as an unrestricted structure. We will not discuss each of the five points proposed by Crother and Murray (2011, 2013) describing such a structure, as our focus here is on models. Frigg (2002) has demonstrated that structures alone are insufficient for drawing fruitful conclusions. Referring to an ontology outside of a semantic framework is irrelevant because ontologies cannot be fully understood through a structure alone, but rather through a semantic model. Therefore, Crother and Murray's Ontology, despite its name, does not accurately represent the Ontology of areas of endemism, especially because there is no constraint (i.e., monophyly) as developed in the semantic conception of scientific theories.

## What About Time?

The second section (chapter 5) in Nelson and Platnick (1981) deals with *Time*: 'Ontogeny, Phylogeny, Paleontology, and the Biogenetic law'. It is a small chapter compared to the other two on *Form* and *Space*, with only 22 pages of a 500+ page book. The temporal dimension is a key element in evolutionary biology and temporal intervals can also be analyzed under the semantic conception of scientific theories. They are traditionally represented by a geological scale. However, Hennig had already suggested a different representation. We have reproduced his figure (Figure 8.3, after Hennig, 1966, p. 161, fig. 48) with additional terms to illustrate his point of view. This figure is part of the paragraph 'Measurement of Absolute ages of higher taxa'. In this paragraph, Hennig explains what he has called 'the age of differentiation' and the 'age of origin' of a group (Hennig, 1954, figs 268–272, for more details).

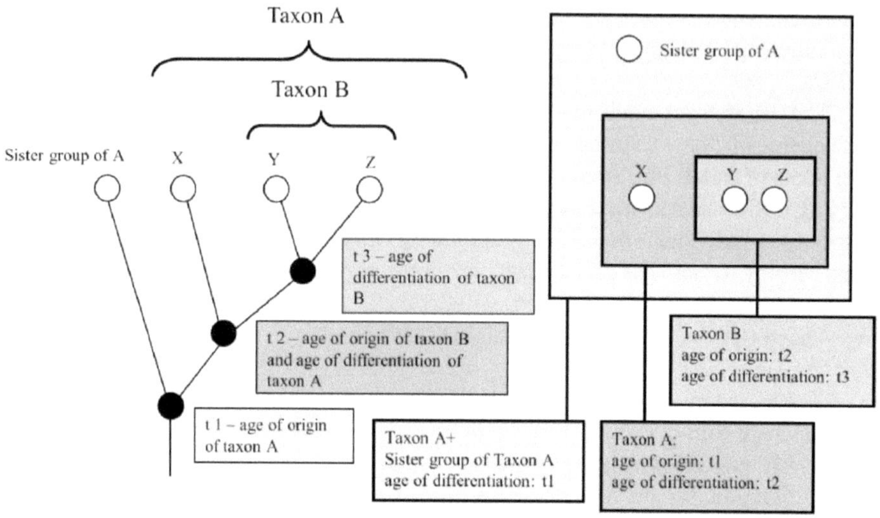


**FIGURE 8.3** Hennig's concept of 'age of origin' and 'age of differentiation' modified after Hennig 1966 (fig. 48). T1 is the age of origin of Taxon A which is composed of X, Y and Z. T2 is the age of origin of Taxon B (composed of Y and Z) and the age of differentiation of Taxon A. T3 is the age of differentiation of Taxon B into Y and Z.

The inferred age of origin of a monophyletic group is the same as that of its respective sister group; therefore, the age of clades is an attribute of the node, not of the branches. Consequently, ages in a cladogram are distributed hierarchically:

> 'From the point of view that naturally the age of differentiation of a group is identical with the age of origin of its oldest recent subgroups'.
>
> *Hennig, 1966, p. 161*

This concept has been reformulated by Zaragüeta-Bagils et al. (2004). They represented the temporal dimension as a hierarchy: a rooted tree of temporal intervals (Figure 8.4B). This representation means that a time period does not end with the start of the following one, but *includes* the next period of time. Zaragüeta-Bagils et al. did not question the relevance of the geological scale (Figure 8.4A) but proposed another representation of durations without involving modifications of the geological information. This new representation is relevant within a semantic framework. It illustrates the hierarchy of temporal periods as a semantic model. Here, the node units represent sets of periods, while the relationship between them is still a relation of inclusion. The constraint is inherent to the compatibility of the hierarchy's structure and the underlying stratigraphy.

Earlier, we introduced the term 'reductionist approach' to describe how various facets of a theory can be modelled within the same modelling framework. We have characterized both systematics and biogeography adopting the same hierarchical tree representation. Moreover, we have shown that the temporal dimension can also be effectively modelled using this same hierarchical tree representation. This specific

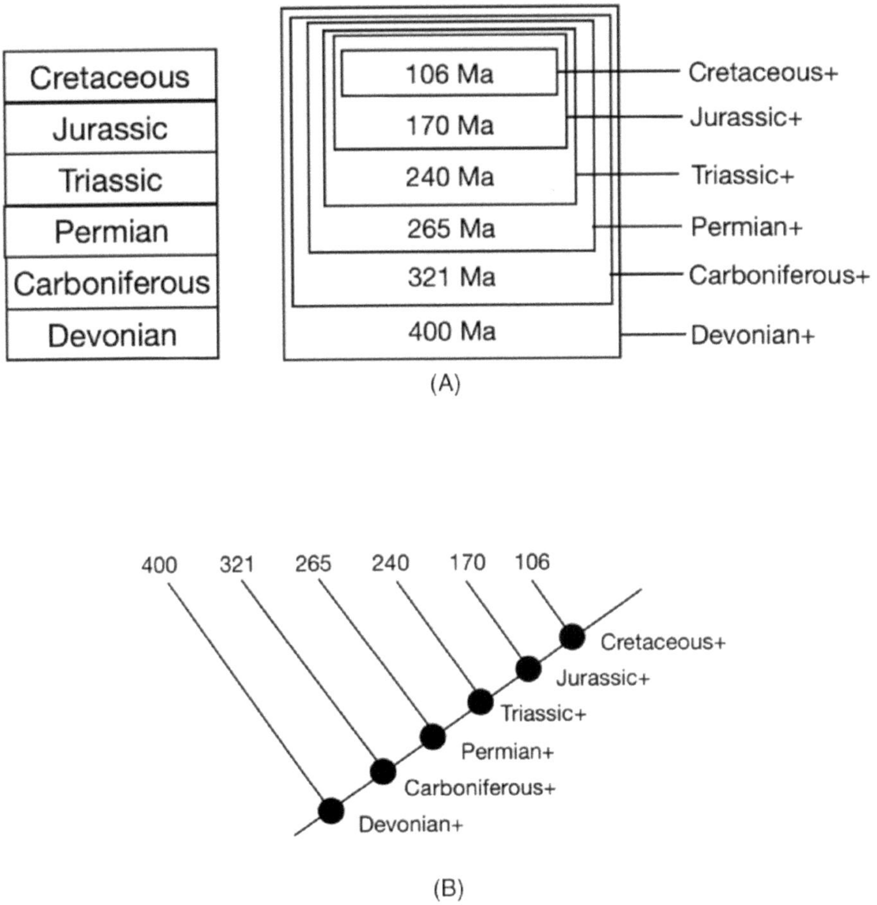


**FIGURE 8.4** The hierarchical representation of geological time (modified from Zaragüeta-Bagils et al., 2004). A: traditional representation of the geological scale: a temporal period ends when the next one starts. B: with a hierarchical representation of the temporal dimension, one temporal interval does not end when the next one starts but includes the following periods. A symbol '+' is added to express that those intervals do not replace the geological scale and names of corresponding periods of time.

aspect of interdisciplinarity can be extended to computer science where ontologies integrate the representation of a semantic model and combine technologies which can lead to novel insights and discoveries. This interdisciplinary exchange not only enriches each field individually but also fosters collaboration and cross-pollination of ideas, ultimately advancing our understanding and capabilities in each domain.

## Implementing an Informatic Ontology

An ontology is 'a logical theory which gives an explicit, partial account of a *conceptualization* [i.e., an intentional semantic structure which encodes the explicit rules constraining the structure of a piece of reality] (Guarino and Giaretta, 1995, p. 32).

This definition is rooted in a semantic framework that makes explicit the object and relationships (i.e., the structure) and the constraints that a model accommodates. Since the philosophical interpretation in logico-mathematic terms aligns well with the requirements of computer science developments, it is not surprising that a semantic model serves as an ontological model. Semantics facilitates the integration of the term 'ontology' across philosophy and computer science.

As a direct application, we have developed *BioRealm*, the first informatic ontology for comparative biogeography (Ung and Buttigieg, 2023). *BioRealm* is an immediate response to the current problem of biodiversity loss and climate change since it encodes a model-theoretic view of endemism using semantic web approaches, offering new avenues to express and analyze biogeographic units. This approach offers a means to identify monophyletic biogeographic areas for conservation, based on specific combinations of monophyletic endemic taxa. *BioRealm* is an ontology in the sense of being a computable and logically coherent representation of entities and offers a formalization of knowledge for comparative biogeography. We use semantic web standards, Resource Description Framework (RDF) and the Web Ontology Language (OWL) published by the World Wide Web Consortium (www.w3c.org) following the technical guidelines of the Open Biological and Biomedical Ontology (OBBO) foundry[2] and Library.

*BioRealm* was developed as an extension of the Environment ontology (ENVO, http://purl.obolibrary.org/obo/envo.owl; Buttigieg et al., 2013, 2016) contending with often confounded biogeographic entities including biogeographic areas, e.g., areas of endemism and endemic areas as well as their relationships. Generic ontological classes are used to define categories and individuals which belong to these classes are understood as instances of the class. All entities (classes, individuals, and relations) are associated with a unique identifier (a Uniform Resource Identifier, URI) which provides a rigorous, extensible and machine-actionable framework that can improve biogeographic analyses and interoperability between systems. Such an ontology provides knowledge representation solutions which support interoperability along the FAIR ('Findable, Accessible, Interoperable, Reusable', see below) principles (Wilkinson et al., 2016), thus fostering more consistent ecological informatics. In *BioRealm* we have created a FAIR and computationally actionable ontology as an implementation of discussions on the metaphysical nature of areas of endemism for comparative biogeography (Crother and Murray, 2011, 2013). Using a model of embedded (i.e., hierarchical) classes structured by the monophyly constraint, we reconcile elements of Crother and Murray's metaphysical Ontology on areas of endemism through formalization via logics and mathematics (Guarino and Giaretta, 1995).

*BioRealm* provides support to semantically describe and align data and databases in an open data and open model following the FAIR principles. Further developments will focus on linking descriptions of biomes and other ecological units related to those biogeographic areas to extend and enhance both *BioRealm* and ENVO. Moreover, we seek to apply these ontologies to enhance software such as the Global Biotic Interactions platform (https://www.globalbioticinteractions.org/) and more generic use of semantic web resources to annotate, mobilize, and analyze data. The use of ontologies *sensu largo* is part of a global solution – not a solution in itself but helps reach the goal: facilitating clear data and information flow for enhanced decision-making processes to better conserve the biodiversity and the planet.

## ONTOLOGIES, OPEN SCIENCE, AND STANDARDS

As a part of ENVO, *BioRealm* is available on the OBO portal. The OBO foundry has been established in 2003 as an 'international collaborative effort dedicated to developing and promoting high-quality, interoperable ontologies for biology, biomedicine, and related domains'. The organization is led by a volunteer team consisting of ontology maintainers and stakeholders, i.e., the 'OBO operations committee' which is divided into four working groups (editorial, technical, outreach and academy) to best perform its activities. As an international organization, they have created and published a set of principles that capture ontology best practices, in order to develop a family of interoperable ontologies. Those best practices are based on openness, collaborative development and interoperability through common syntax and relations. The portal lists 183 different ontologies[3] covering 15 domains of biology and biomedicine including 'metadomains' like 'upper' which provides more general syntax and vocabularies, e.g., RO, the Relation Ontology.

Indeed, open science and open research represent a transformative movement in the realm of scientific research, driven by the principles of accessibility, transparency, and collaboration. At its core, open science advocates for making scientific knowledge freely available to everyone, fostering an environment where information exchange is fundamental to scientific progress. One significant aspect of open science is the promotion of open-access publishing. In this model, scientific findings are made accessible to the public without financial barriers, allowing a broader audience to benefit from and build upon the research. This gap from traditional subscription-based models enhances the dissemination of knowledge and democratizes access to scientific information.

Moreover, open science encourages researchers to share not only their research findings but also the underlying data and source codes associated with their work. By promoting the open sharing of research data, scientists enable the verification and reproduction of results by their peers, contributing to the robustness and reliability of scientific outcomes. This collaborative approach enhances the quality of research, as findings can be investigated, validated, and expanded upon by the broader scientific community. In essence, open science aims to create a more inclusive, transparent, and accountable research ecosystem. By embracing these principles, the scientific community as a whole can accelerate the pace of discovery, foster innovation, and address societal challenges more effectively.

In order to comply with open science requirements, standards are key. What is a standard? In the context of standardization and specifications, a standard is *a set of criteria, guidelines, or specifications established to ensure consistency, interoperability, and quality within a particular field*. Standards serve as a reference point, offering a common language and framework that facilitates communication, understanding and compatibility among entities within a given industry or discipline.

There are two main types of standards, each with distinct characteristics. Official standards are established by standardization bodies, developed and endorsed by official standardization bodies officially approved by a state or, in the case of international standards, by international organizations, for example AFNOR (Association Française de Normalisation) in France, ISO (International Organization for Standardization) at the global level, ANSI (American National Standards Institute) in the United States, and many others. They are established through a formalized process involving expert committees, public consultation, and approval by the relevant standardization organization. Other standards established by user groups or specialists such as the Biodiversity Information Standards (TDWG, www.tdwg.org) emerge organically within specific user communities, often driven by a group of users who are experts in a particular field of research or practice. While not officially mandated, these standards are widely adopted based on their practical utility and acceptance within the relevant community. In the field of Biodiversity, the Darwin Core (developed by TDWG) is one of the most famous and is endorsed by the Global Biodiversity Information Facility (GBIF, www.gbif.org). The process of establishing a standard follows the review process of a scientific paper. The standard is reviewed by specialists and goes through a public review by the community before being endorsed by the organization. It may be more flexible and adaptable than official standards, allowing for quicker responses to evolving needs and technologies.

In summary, while official standards are developed and endorsed by recognized standardization bodies, other standards arise from the consensus and recommendations of user groups and specialists within a particular field and a particular organization. Both types play crucial roles in ensuring uniformity, interoperability, and quality across science.

The FAIR principles, introduced by Wilkinson et al. represent an essential moment in academia, marking a significant shift towards making science more open, accessible, and collaborative (Wilkinson et al., 2016). These principles emphasize four key elements: Findability, Accessibility, Interoperability, and Reusability, and they have played a transformative role in shaping the landscape of scientific data management. Additional information can be added to make data more findable like metadata which provides essential context and information about the data, making it easier to locate and identify. Each dataset should be associated with a global, persistent, and unique identifier, ensuring its distinctiveness and preventing ambiguity. To enhance *accessibility*, data should be openly accessible to the scientific community and, where possible, to the public. The FAIR principles emphasize the importance of clear policies and guidelines for data access, ensuring that everyone and particularly researchers can retrieve and use data with minimal barriers. *Interoperability* is achieved through the implementation of existing standards specific to different scientific fields. This guarantees that data can be easily integrated and exchanged between different systems and platforms. To promote *reusability*, metadata should include information about the conditions and context under which the data was generated. Establishing clear licensing and usage terms for datasets encourages responsible and ethical reuse by other researchers. The FAIR principles emphasize the practical implementation of these concepts, providing a framework for researchers to follow in order to enhance the openness, effectiveness, and interconnection of scientific data.

## CONCLUSION

### Simplicity as a Fact, Hierarchy as a Model

In their epilogue, Nelson and Platnick wrote:

> This book has explored the question of order in nature at two levels. Even the most elementary observations of the attributes of living organisms reveal a bewildering array of characters (similarities and differences). These elementary observations might seem to reveal only chaotic or random association of organisms (taxa), each indicated, and supported, by different characters. One might believe, therefore, that the world, in that sense, is complex. But the questions arise**: might the world really be simple, rather than complex? Might there be a single pattern of relationships (a general cladogram of taxa) for all organisms, such that any character whatsoever would define a component of that pattern?** [...]. Similarly, even the most elementary observations of the ranges of living organisms reveal a bewildering array of distributions (showing similarities and differences). These elementary observations might seem to reveal only chaotic or random associations of areas (regions), each indicated, and supported, by the distributions of different groups of organisms. One might believe, therefore, that the world, in that sense, is complex. But the questions arise again: **might the world really be simple, rather than complex? Might there be a single pattern of relationships (a general cladogram of areas) for all groups of organisms, such that the distribution of any group whatsoever would define a component of that pattern?**
>
> *Nelson and Platnick, 1981, pp. 539–540*

We could not agree more. *Might the world really be simple rather than complex? Might there be a single pattern of relationships (a general cladogram of taxa or areas) for all organisms or groups of organisms such that any character or any distribution of any group whatsoever would define a component of that pattern?* The hierarchical model provides a versatile framework applicable across various disciplines, from data modelling to knowledge representation, and remains a fundamental tool in the fields of systematics and biogeography (Figure 8.5). Figure 8.5 represents the classical pyramid model, illustrating the progression from data to knowledge and finally wisdom. The real world is the first level from which data are gathered. Information is obtained as data are organized and structured, then transformed into knowledge through analysis, producing generalized results such as cladograms of taxa in systematics or areas in comparative biogeography. Wisdom is reached when research provides actionable solutions to real-world problems assuring decision-making process. On the left side of the figure, the abstraction process, or formalization, is emphasized, while the right side illustrates the sequential steps of the research application. Throughout this process, a hierarchical representation remains consistent across all levels of abstraction. However, the actions undertaken within specific fields or domains of research may vary accordingly.

Nelson and Platnick's work was truly visionary and pioneering, laying the groundwork for this formalization process. In their time, technological advancements were insufficient to fully support their ideas, computers just started to enter homes becoming 'personal', thus they had to run their calculation by hand, which was quite hard and limited. One illustration is their '3-item methodology', which was outlined in

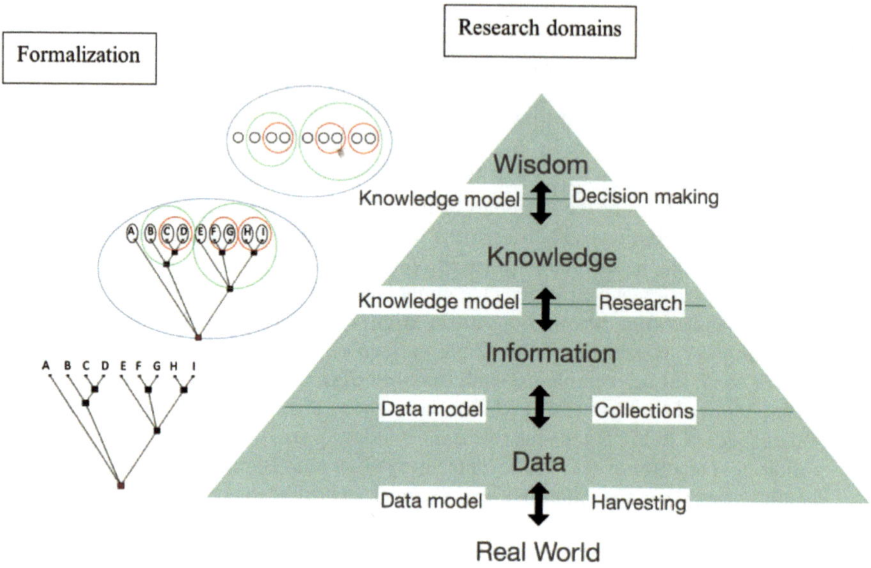


**FIGURE 8.5** From real-world data to wisdom.

Nelson and Platnick and developed a decade later (Nelson and Platnick, 1991). This approach gained popularity only upon the release of the Lisbeth software in 2012 (Zaragueta-Bagils et al., 2012; Grand et al., 2014; Hoagstrom et al., 2014, 2022; Ung et al., 2016a, 2016b; Murphy et al., 2019; Michaux and Ung, 2021). In this approach, the elementary information is represented as a 'three-item statement' (or 3is, a three-item hierarchy statement). Their main argument is that among three items (taxa or areas of endemism) two are always closer to each other than to the third (Nelson and Platnick, 1991 for further details). Lisbeth is the only software that implements a hierarchical representation of data amongst available programs for phylogenetic systematics and biogeography. Lisbeth is also the only software that implements 3-items analysis (Nelson and Platnick, 1991), paralogy-free subtree analysis (Nelson and Ladiges, 1996) and the transparent method (Ebach et al., 2005). This is quite unique in the history of biology to see that the transition from data to knowledge is facilitated by the application of a single and unaltered model: a hierarchy constrained by the monophyly condition. This model captures and represents data but also maintains a fidelity to reality which ensures its internal consistency. In the same way, the evolution of computers with augmented calculation capacities and computer science (i.e., algorithmic, semantic and ontologies) have offered new possibilities, along with the development of standards. All those improvements and progress align closely with Nelson and Platnick's original objective: to provide a concise and simple representation and explanation for the complexity of life and forms on Earth.

## On the Need to Change Our Socio-Cultural Paradigm ...

The globalization of the world profoundly impacts the academic and scientific domain. Technological advances have enabled an immeasurable progress in many scientific fields, for example, structural or functional genomics, bioinformatics or even the development of artificial intelligence (AI), which is becoming a catalyst for what could be qualified as 'more outdated' disciplines such as taxonomy. Indeed, photographic identification systems associated with learning algorithms will quickly come to support scientists (Shirai et al., 2022 and others), like barcoding (Hebert et al., 2003) has recently done. However, we are witnessing an unprecedented paradox. We have entered a period described as the sixth mass extinction (Kolbert, 2014; Régnier et al., 2015; Ceballos et al., 2015, 2020; Cowie et al., 2022 among others), biodiversity has been in decline for many years because of anthropogenic activity (Myers et al., 2000; Sachs et al., 2009; Butchart et al., 2010) and yet, techniques have never been so advanced. They produce a deluge of results and data that must be better managed and exploited for benefits on a global scale. Those data need to be accurately standardized to maximize their access, potential use and re-use: they need to be FAIR. In that regard, the implementation of ontologies provides an efficient solution.

This is about making a change that we would describe as a socio-cultural paradigm. We need to make a change in the culture of data science and research. We must be part of the construction of the research community and open science, create bridges between different disciplines and promote interdisciplinary science. Open science is a new way of conducting research, combining the production and sharing of research so that as many people as possible can access it freely (and for free), reuse, build and distribute these resources without barriers.

### ... to More Ethical Research

This involves integrating an ethical dimension into research, which is what open science promotes. Making research data accessible to everyone for everyone promotes inclusion. In 2010, the Convention on Biological Diversity (https://www.cbd.int/cop10/) adopted the 'Nagoya Protocol on Access to Genetic Resources and the Fair and Equitable Sharing of Benefits Arising from Their Utilization' – it came into force in October 2014. This protocol marks a first step towards more ethical research. More recently, in 2020, under the sponsorship of the RDA (Research Data Alliance, www.rda.org), the 'Global Indigenous Data Alliance' (GIDA, https://www.rd-alliance.org/gida-global-indigenous-data-alliance) was formed whose aim is to promote indigenous control of indigenous data. This alliance is made up of researchers, users, and indigenous activists who defend indigenous sovereignty and data governance within their nation-states and internationally. GIDA members published in 2020 the CARE principles (Collective Benefit, Authority to control, Responsibility, Ethics) which go along with the FAIR principles (Carroll et al., 2020; Carroll et al., 2022a, 2022b) and highlight the emphasis on considering the rights and sovereignty of indigenous peoples as the world engages in open science and data sharing.

## EPILOGUE

> 'One might ask, for example, whether prior to the commerce and traffic of recorded historical times, there existed natural populations of humans which were effectively restricted, by geography or ecology, to certain areas. [...] One might wonder, of course,

whether the similar distributions of organisms and languages could have the same cause (Whitten 1979; Meggers, 1979). If the distributions of the various groups of organisms are the result of the same events, one would expect to find repeated in each group the same pattern of relationships among the taxa restricted to the various areas of endemism. [...] What we have, then, are unanswered questions: about organisms in general, about their interrelationships, about areas of endeism, about their interrelationships, about human populations, and about their interrelationships. We hope that this volume may at least focus attention on the questions, and perhaps lead to some answers as well'.

*Nelson and Platnick, 1981, pp. 541–543*

These are the last sentences of Nelson and Platnick (1981). Forty-three years after the publication of their book, their visionary ideas are now implemented in software and an ontology which provides a rigorous model for systematics and biogeography underpinning evolutionary biology as a theory. The rooted tree representation, i.e., a hierarchy advocated by Nelson and Platnick, is an interdisciplinary model allowing the integration of subfields, methods, resources, and technologies as shown in this paper. Extended to the domain of Philosophy of Science, the semantic model and the addition of the monophyly constraint bridge the gap between the two disciplines and integrate the term 'ontology' across various areas.

## ACKNOWLEDGMENTS

In 1981, Visotheary Ung was 7 years old and Anaïs Grand was not yet born. On the 6th of October 2023, Visotheary Ung was invited into Gareth Nelson and Pauline Ladiges's home. There, he gave her what might be the last sealed copy of Systematics and Biogeography, now with their signatures. Pauline wasn't sure if it was right she signed, but finally agreed as Visotheary insisted – her signature now stands with Gary's. This event left her with a strong feeling, solidifying her connection with them. Those shared moments convinced her to pursue her research in a domain often qualified as 'old-fashioned' or 'out-of-date'. Despite the skepticism sometimes encountered, the sincere support and friendship offered by Gareth and Pauline have become a source of inspiration and gratitude. The 'Blue Book' has provided us with invaluable answers and has guided our attention to the questions raised by Systematics and Biogeography. It prompts us to explore how we can leverage emerging technologies such as ontologies and artificial intelligence, alongside advancements in computing power, to conduct precise investigations and address novel inquiries in an era of open science. We appreciate your support and friendship, which inspires us to collaborate across generations and disciplines, fostering a more inclusive and equitable world for future generations. Thank you for validating the title of this article Systematics and Biogeography, Ontology and Vicariance as a tribute to the blue book since we are convinced that the ontological representation is the missing link that bridges the gap.

We would like to sincerely thank Gareth Nelson and Pauline Ladiges for their advices and support, David Williams and Quentin Wheeler for their invitation to participate in this special volume and for fruitful discussions. We are grateful to Malte Ebach for providing comments that have greatly enhance this paper.

## NOTES

1 Access research date: 10 April 2024.
2 https://obofoundry.org/
3 Access date: 10 April 2024.

## REFERENCES

Allwein, G. and Barwise, J. (1996) *Logical Reasoning with Diagrams.* Oxford University Press, Oxford.

Beatty, J. (1987) On behalf of the semantic view. *Biology and Philosophy*, 2: 17–23, https://doi.org/10.1007/BF00127560

Bender, E.A. and Williamson, S.G. (2010) *Lists, Decisions and Graphs. With an Introduction to Probability*, https://cseweb.ucsd.edu/~gill/BWLectSite/Resources/C2U4GT.pdf

Butchart, S.H., Walpole, M., Collen, B., van Strien, A., Scharlemann, J.P., Almond, R.E., Baillie, J.E., Bomhard, B., Brown, C., Bruno, J., Carpenter, K.E., Carr, G.M., Chanson, J., Chenery, A.M., Csirke, J., Davidson, N.C., Dentener, F., Foster, M., Galli, A., Galloway, J.N., Genovesi, P., Gregory, R.D., Hockings, M., Kapos, V., Lamarque, J.F., Leverington, F., Loh, J., McGeoch, M.A., McRae, L., Minasyan, A., Hernández Morcillo, M., Oldfield, T.E., Pauly, D., Quader, S., Revenga, C., Sauer, J.R., Skolnik, B., Spear, D., Stanwell-Smith, D., Stuart, S.N., Symes, A., Tierney, M., Tyrrell, T.D., Vié, J.C. and Watson, R. (2010) Global biodiversity: Indicators of recent declines. *Science*, 328(5982): 1164–1168, https://doi.org/10.1126/science.1187512

Buttigieg, P.L., Morrison, N., Smith, B., Mungall, C.J. and Lewis, S.E. (2013) The environment ontology: Contextualising biological and biomedical entities. *Journal of Biomedical Semantics*, 4(1): 43, https://doi.org/10.1186/2041-1480-4-43

Buttigieg, P.L., Pafilis, E., Lewis, S.E., Schildhauer, M.P., Walls, R.L. and Mungall, C.J. (2016) The environment ontology in 2016: Bridging domains with increased scope, semantic density, and interoperation. *Journal of Biomedical Semantics*, 7(57), https://doi.org/10.1186/s13326-016-0097-6

Candolle, A.P. de (1813) *Théorie élémentaire de la botanique; ou, Exposition des prinicpes de la classification naturelle et de l'art de décrire et d'étudier les végétaux.* Déterville, Paris.

Carroll, S.R., Garba, I., Figueroa-Rodríguez, O.L., Holbrook, J., Lovett, R., Materechera, S., Parsons, M., Raseroka, K., Rodriguez-Lonebear, D., Rowe, R., Sara, R., Walker, J.D., Anderson, J. and Hudson, M. (2020) The CARE principles for indigenous data governance. *Data Science Journal*, 19(1): 43, https://doi.org/10.5334/dsj-2020-043.

Carroll, S.R., Garba, I., Plevel, R., Small-Rodriguez, D., Hiratsuka, V.Y., Hudson, M. and Garrison, N.A. (2022a) Using indigenous standards to implement the CARE principles: Setting expectations through tribal research codes. *Frontiers in Genetics*, 13: 823309, https://doi.org/10.3389/fgene.2022.823309.

Carroll, S.R., Plevel, R., Jennings, L.L., Garba, I., Sterling, R., Cordova-Marks, F.M., Hiratsuka, V., Hudson, M. and Garrison, N.A. (2022b) Extending the CARE principles from tribal research policies to benefit sharing in genomic research. *Frontiers in Genetics*, 13: 1052620, https://doi.org/10.3389/fgene.2022.1052620

Ceballos, G., Ehrlich, P., Barnosky, A., García, A., Pringle, R. and Palmer, T. (2015) Accelerated modern human–induced species losses: Entering the sixth mass extinction. *Science Advances*, 1(5), 1, e1400253 https://doi.org/10.1126/sciadv.1400253

Ceballos, G., Ehrlich, P.R. and Raven, P.H. (2020) Vertebrates on the brink as indicators of biological annihilation and the sixth mass extinction. *Proceedings of the National Academy of Sciences U.S.A.*, 117: 13596–13602, https://doi.org/10.1073/pnas.1922686117

Coffa, A.J. (1991) *The Semantic Tradition from Kant to Carnap: To the Vienna Station.* Cambridge University Press, Cambridge.

Cowie, R.H., Bouchet, P. and Fontaine, B. (2022) The sixth mass extinction: Fact, fiction or speculation?. *Biological Review*, 97: 640–663, https://doi.org/10.1111/brv.12816

Creath, R. (2014) Logical empiricism. In *The Stanford Encyclopedia of Philosophy* (Spring 2014 Edition) (eds E.N. Zalta et al.), https://plato.stanford.edu/archives/spr2014/entries/logical-empiricism.

Crother, B. and Murray, C. (2011) Ontology of areas of endemism. *Journal of Biogeography*, 38: 1009–1015, https://doi.org/10.1111/j.1365-2699.2011.02483.x

Crother, B. and Murray, C. (2013) Parsimony analysis of endemism under the "areas of endemism as individuals" thesis. *Cladistics*, 29(6): 571–573, https://doi.org/10.1111/cla.12023

De Haro, S. (2019) Towards a theory of emergence for the physical sciences. *European Journal for the Philosophy of Science*, 9: 38, https://doi.org/10.1007/s13194-019-0261-9

de Pinna M.G.G. (1991) Concepts and tests of homology in the cladistic paradigm. *Cladistics*, 7(4): 367–394, https://doi.org/10.1111/j.1096-0031.1991.tb00045.x

Ebach, M.C. (2015) *Origins of Biogeography: The Role of Biological Classification in Early Plant and Animal Geography.* Springer, Dordrecht. 173 pp.

Ebach, M.C., Humphries C.J., Newman, R.A., Williams, D.M. and Walsh S.A. (2005) Assumption 2: opaque to intuition? *Journal of Biogeography*, 32: 781–787, http://www.jstor.org/stable/3566268

French, S.R.D. (2003) A model-theoretic account of representation (or, I don't know much about art…but I know It involves isomorphism). *Philosophy of Science*, 70(5): 1472–1483, https://doi.org/10.1086/377423

Friedman, M. (1999) *Reconsidering Logical Positivism.* Cambridge University Press, Cambridge.

Frigg, R. (2002) *Models and representation: Why structures are not enough.* Measurement in Physics and Economics Project Discussion Paper Series DP [Dietsch, Peter (ed)] MEAS 25/02, London School of Economics, London, https://romanfrigg.org/wp-content/uploads/writings/Models_and_Representation.pdf

Ghiselin, M.T. (1974) A radical solution to the species problem. *Systematic Zoology*, 23: 536–544. https://doi.org/10.2307/2412471

Gildenhuys, P. (2013) Classical population genetics and the semantic approach to scientific theories. *Synthese*, 190: 273–291, https://doi.org/10.1007/s11229-012-0146-3

Grand, A., Zaragüeta Bagils, R., Vélez, L.M. and Ung, V. (2014) A cladistic re-analysis of the Gadiformes (Teleostei, Paracanthopterygii) using three-item analysis. *Zootaxa*, 3889: 525–552, https://dx.doi.org/10.11646/zootaxa.3889.4.3

Guarino, N. and Giaretta, P. (1995) Ontologies and knowledge bases: Towards terminological clarifications, pp. 25–32. In *Towards Very Large Knowledge Bases: Knowledge Building and Knowledge Sharing* (ed N. Mars). IOS Press, Amsterdam.

Hebert, P.D.N., Cywinska, A., Ball, S.L. and Waard, J.R. de (2003) Biological identifications through DNA barcodes. *Proceedings of the Royal Society, B: Biological Sciences*, 270(1512): 313–321, https://doi.org/10.1098/rspb.2002.2218

Hennig, W. (1954) Flügelgeäder und System der Dipteren unter Berücksichtigung der aus dem Mesozoikum beschriebenen Fossilen. *Beiträge zur Entomologie*, 4: 245–388.

Hennig, W. (1966) *Phylogenetic Systematics.* University of Illinois Press, Urbana. 264 pp.

Hoagstrom, C.W., Ung, V., Sweat, S., Matamoros, W. and Ennen, J. (2022) Comparative biogeography of North American turtle faunas: Neogene regionalization. *Frontiers of Biogeography*, 14(4), e57618, https://dx.doi.org/10.21425/F5FBG57618

Hoagstrom, C.W., Ung, V. and Taylor, K. (2014) Miocene rivers and taxon cycles clarify the comparative biogeography of North American highland fishes. *Journal of Biogeography*, 41: 644–658, https://dx.doi.org/10.1111/jbi.12244

Hull, D.L. (1978) A matter of individuality. *Philosophy of Science*, 45: 335–360, https://www.jstor.org/stable/187022

Kitching I.J., Forey, P.L., Humphries, C.J. and Williams, D.M. (1998) *Cladistics: The Theory and Practice of Parsimony Analysis*. 2nd ed. Oxford University Press, Oxford.

Kolbert E. (2014) *The Sixth Extinction: An Unnatural History*. Henry Holt and Company, New York, 336 pp.

Linnaeus, C. (1735) *Systema naturae, sive regna tria naturae systematice proposita per classes, ordines, genera, and species*, Leyde, Hilden.

Linnaeus, C. (1753) *Species plantarum: exhibentes plantas rite cognitas, ad genera relatas, cum differentiis specificis, nominibus trivialibus, synonymis selectis, locis natalibus, secundum systema sexuale digestas…*, Holmiæ, Impensis Laurentii Salvii.

Linnaeus, C. (1758) *Systema naturæ per regna tria naturæ, secundum classes, ordines, genera, species, cum characteribus, differentiis, synonymis, locis*. Holmiæ, Impensis direct. Laurentii Salvii (Salvius publ.), Zeist.

Lloyd, E.A. (1986) Thinking about models in evolutionary theory. *Philosophica*, 37, 87–100.

Lloyd, E.A. (1987) Confirmation of ecological and evolutionay models. *Biology and Philosophy*, 2: 277–293, https://doi.org/10.1007/BF00128834

Lloyd, E.A. (1988) *The Structure and Confirmation of Evolutionary Theory*. Princeton University Press, Princeton.

Michaux, B. and Ung, V. (2021) Biotectonics of Sulawesi: Principles, methodology, and area relationships. *Zootaxa*, 5068 (4): 451–484, https://dx.doi.org/10.11646/zootaxa.5068.4.1

Morrone, J.J. (2009) *Evolutionary Biogeography. An Integrative Approach with Case Studies*. Columbia University Press, New York.

Murphy, D.J., Ebach, M.C., Miller, J.T., Laffan, S.W., Cassis, G., Ung, V., Thornhill, A.H., Kerr, N. and Tursky, M.L. (2019) Do phytogeographic patterns reveal biomes or biotic regions? *Cladistics*, 35: 654–670, https://dx.doi.org/10.1111/cla.12381

Myers, N., Mittermeier, R.A., Mittermeier, C.G., da Fonseca, G.A. and Kent, J. (2000) Biodiversity hotspots for conservation priorities. *Nature*, 403(6772): 853–858, https://doi.org/10.1038/35002501

Nelson, G.J. and Ladiges, P.Y. (1992) Information content and fractional weight of three-item statements. *Systematic Biology*, 41(4): 490–494, https://doi.org/10.2307/2992590

Nelson, G.J. and Ladiges, P.Y. (1991a) Standard assumptions for biogeographic analysis. *Australian Systematic Botany*, 4: 41–58, https://doi.org/10.1071/SB9910041 [58 [Amendment. *Australian Systematic Botany,* 5: 247, https://doi.org/10.1071/SB9910041c].

Nelson, G.J. and Ladiges, P.Y. (1991b) Three-area statements: Standard assumptions for biogeographic analysis, *Systematic Zoology*, 40: 470–485, https://doi.org/10.2307/2992241

Nelson, G.J. and Ladiges, P.Y. (1996) Paralogy in cladistic biogeography and analysis of paralogy-free subtrees. *American Museum Novitates*, 3167: 1–58, http://hdl.handle.net/2246/3627

Nelson, G.J. and Platnick, N.I. (1978) The perils of plesiomorphy: Widespread taxa, dispersal, and phenetic biogeography. *Systematic Zoology*, 27 (4): 474–477, https://doi.org/10.2307/2412931

Nelson, G.J. and Platnick, N.I. (1981) *Systematics and Biogeography: Cladistics and Vicariance*. Columbia University Press, New York.

Nelson, G. and Platnick, N.I. (1991), Three-taxon statements: A more precise use of parsimony? *Cladistics*, 7: 351–366, https://doi.org/10.1111/j.1096-0031.1991.tb00044.x

Parenti, L.R. and Ebach, M.C. (2009) *Comparative Biogeography: Discovering and Classifying Biogeographical Patterns of a Dynamic Earth.* University of California Press, Berkeley.

Patterson, C. (1988) Homology in classical and molecular biology. *Molecular Biology and Evolution*, 5(6): 603–625, https://doi.org/10.1093/oxfordjournals.molbev.a040523

Régnier, C., Achaz, G., Lambert, A., Cowie, R., Bouchet, P. and Fontaine, B. (2015) Mass extinction in poorly known taxa. *Proceedings of the National Academy of Sciences*, 112(25): 7761–7766, https://doi.org/10.1073/pnas.1502350112

Rieppel, O. (1988) *Fundamentals of Comparative Biology.* Birkhäuser Verlag, Basel. 202 pp.

Sachs, J.D., Baillie, J.E.M., Sutherland, W.M., Armsworth, P.R., Ash, N., Beddington, J., Blackburn T.M. et al. (2009) Biodiversity conservation and the millennium development goals. *Science*, 325: 1502–1503, https://doi.org/10.1126/science.1175035

Shirai, M., Takano, A., Kurosawa, T. et al. (2022) Development of a system for the automated identification of herbarium specimens with high accuracy. *Scientific Reports*, 12, 8066, https://doi.org/10.1038/s41598-022-11450-y

Thompson, P. (1983) The structure of evolutionary theory: A semantic approach. *Studies in History and Philosophy of Science*, 14: 215–229, https://doi.org/10.1016/0039-3681(83)90013-4

Thompson, P. (1986) The interaction of theories and the semantic conception of evolutionary theory. *Philosophica*, 37: 73–86, 10.21825/philosophica.82524

Thompson, P. (1987) A defense of the semantic conception of evolutionary theory. *Biology and Philosophy*, 2: 26–32, https://doi.org/10.1007/BF00127562

Uebel, T. (2014) Vienna Circle. *The Stanford Encyclopedia of Philosophy* (Spring 2014 Edition) (eds E.N. Zalta et al.), https://plato.stanford.edu/archives/spr2014/entries/vienna-circle/

Ung V. (2018) The need for standard protocols in bioregionalisation: Comments on "The spectre of biogeographical regionalization" by Morrone (2018). *Zootaxa*, 4532(2): 296–300, https://doi.org/10.11646/zootaxa.4532.2.10

Ung, V. and Buttigieg, P.L. (2023) BIOREALM — An ontology of comparative biogeography: New insights into the semantics of biodiversity conservation. *Journal of Biogeography*, 50: 1576–1586, https://doi.org/10.1111/jbi.14631

Ung V., Michaux B. and Leschen R.A.B. (2016a). A comprehensive vicariant model for Southwest Pacific biotas. *Australian Systematic Botany*, 29: 424–439, https://doi.org/10.1071/SB16032

Ung V., Zaragueta-Bagils R. and Williams D.M. (2016b) Comparative biogeography of Southeast Asia and the West Pacific region. *Biological Journal of the Linnean Society*, 117(2): 372–385, https://dx.doi.org/10.1111/bij.12670

Van Fraassen, B. (1994) Interpretation of science: Science as interpretation, pp. 169–187. In *Physics and Our View of the World* (ed J. Hilgevoord). Cambridge University Press, Cambridge.

Vaught, R.L. (1974) Model theory before 1945, pp. 153–172. In *Proceedings of the Tarski Symposium, Proceedings of Symposia in Pure Mathematics*, vol. XXV (eds L. Henkin et al.). American Mathematical Society, Providence, RI.

Vaught, R.L. (1986) Tarski's work in model theory. *Journal of Symbolic Logic*, 51: 869–882, https://doi.org/10.2307/2273900

Wilkinson, M., Dumontier, M., Aalbersberg, I., Appleton G., Axton,M., Baak, A., Blomberg, N. et al. (2016) The FAIR Guiding Principles for scientific data management and stewardship. *Scientific Data*, **3**: 160018, https://doi.org/10.1038/sdata.2016.18

Williams D.M. (2004) Homology and homologues, cladistics and phenetics: 150 years of progress, pp. 191–224. In *Milestones in Systematics* (eds D.M. Williams and P.L. Forey). Taylor and Francis, London.

Williams, D.M. and Ebach, M.C. (2008) *Foundations of Systematics and Biogeography*. Springer, New York. 309 pp, https://doi.org/10.1017/S1477200009003053.

Zaragüeta-Bagils, R., Lelièvre, H. and Tassy, P. (2004) Temporal paralogy, cladograms and the quality of the fossil record. *Geodiversitas*, 26(3): 381–389.

Zaragüeta-Bagils, R., Ung, V., Grand, A., Vignes-Lebbe, R., Cao, N. and Ducasse, J. (2012) LisBeth: New cladistics for phylogenetics and biogeography. *Comptes Rendus Palevol*, 11: 563–566, https://doi.org/10.1016/j.crpv.2012.07.002

# 9 Nomenclatural Problems in Zoological Taxonomy

*Alain Dubois*

## INTRODUCTION

In our epoch of 'taxonomic urgency,' the need for clear rules for naming taxa recognised by phylogenetic and taxonomic research could not be greater. The role of the *International Code of Zoological Nomenclature* is to establish the unique valid scientific name (nomen) of any taxon in any classification. The *Code* is confronted today with several difficulties. Its complexity is necessary to address a wide range of concepts and questions in order to ensure automatic, unambiguous and stable universal functioning of its rules. This domain is currently in turmoil, submitted to contradictory forces which could have dramatic consequences for communication about taxa, and thus biology as a whole. In part because of imprecise, misleading and incomplete terminology, some basic concepts and rules of the *Code* are misunderstood, misapplied or ignored. These problems are multiform, concerning: the availability of nomina and nomenclatural acts and publication dates of papers and online publications; distinction between scientific nomina and non-scientific names; nomenclatural authorship; concepts of nomenclatural rank and taxonomic category; role of diagnoses in nomenclature; the ostensional system of taxonomic allocation of nomina through 'types'; conflicts between priority and usage; problems posed by databases; deliberate ignorance of the *Code* and of the Commission's decisions; ethical issues, including so-called taxonomic vandalism; and the absence of rules for the nomenclature of higher taxa. Reduction of these problems would require efforts from authors and editors, and improvement and updating of the *Code*.

## THE NEED FOR A *CODE*, PARTICULARLY IN THE CENTURY OF EXTINCTIONS

Our planet is inhabited by a vast diversity of organisms, including millions of animal species, fewer than 2 million (probably less than 10%, possibly much less) of which have been described so far. In order to deal with them, i.e., to study and use them, protect ourselves from them, and to predict their properties, we need a scientific discipline to discriminate among and classify them (***taxonomy****[1]), and another that designates them in an unambiguous and international way (***nomenclature****).

Biological nomenclature is a domain at the interface between science and society (Wheeler, 2024). It allows unambiguous communication about the biological classification of organisms, both among scientists and between them and society as a whole. In order to work efficiently and universally for this purpose, it must rely on elaborate and precise rules, formalised in international codes.

 DOI: 10.1201/b22822-9

In zoology, the *International Code of Zoological Nomenclature* (Anonymous, 1999; 'the *Code*' below) has been in force, more or less unchanged, for about a century, until a major change with the '2012 Amendment' concerning electronic publication (Anonymous, 2012). Its management and updating are currently in the charge of the International Commission of Zoological Nomenclature ('the Commission' below). In recent decades, various proposals to change its basic principles and rules were published, in various and often contradictory directions. Some of them can be viewed as desirable improvements, but others, if adopted, would introduce major confusion in a domain which requires strong stability of rules. The 'nomenclatural chaos' that could result from their implementation would be particularly inappropriate in our 'century of extinctions' (Dubois, 2003). Our epoch is a crucial one for increasing scientific knowledge of the organisms that live on our planet (Wheeler, 2024). The combination of the ***biodiversity crisis*** (resulting in mass extinctions) and the ***taxonomic gap*** (incompleteness of the taxonomic inventory of our planet) has resulted in ***taxonomic urgency*** (the need to accelerate considerably this work of taxonomic inventory) (Raposo et al., 2021). Priority should be given to ***collecting***, ***studying*** and ***describing*** the millions of still-unknown organisms on our planet before they vanish forever (Dubois, 2024). Great attention should be paid to the nomenclatural rules helping taxonomists in this urgent task, rather than diverting their time and energy to secondary questions or debates. The present *Code* has many properties that make it appropriate for this work, but it could be improved in several important respects, as suggested by the Linz Zoocode Committee (Dubois et al., 2019) or some of its members. This paper presents such improvements which are discussed in detail in works cited below.

## SOME BASIC CHARACTERISTICS OF THE *CODE*

The purpose of the *Code* is to provide a system of principles, rules and recommendations governing the nomenclature of ***taxa****. A very important point mentioned in the *Code*'s Preamble is its ***taxonomic neutrality***: the objects of the *Code* 'are to ensure that the name of each taxon is unique and distinct. All its provisions and recommendations are subservient to those ends and none restricts the freedom of taxonomic thought or actions'. Thus, the *Code* is not supposed to interfere with the erection and recognition of taxa, which are the domain of taxonomy.

The *Code* deals with three kinds of items: taxonomic publications, scientific names of taxa or ***nomina****, and ***nomenclatural acts****, i.e., published statements (such as first-reviser actions) that affect the status of publications and nomina. Nomina are referred to three 'name groups' or ***nominal-series****: the species-, genus- and family-series. Within each, according to the ***Principle of Coordination****, any nomen established for a taxon at any rank is deemed to have been simultaneously established for all taxa of other ranks, and to include its 'type.' The status of any nomen is settled through a ***Nomenclatural Process**** that goes through three subsequent stages (Dubois, 2005a): the nominal-series ***assignment**** and nomenclatural ***availability**** of nomina; their taxonomic ***allocation****; and their nomenclatural ***validity**** and ***correctness****. Each stage is largely independent from the preceding one, and these stages must be explored successively to identify the valid nomen of any taxon (Dubois, 2011).

The *Code* is an ***objective**** and ***ostensional**** nomenclatural system (Dubois, 2011). It is objective because the status of nomina depends on 'objects', i.e., specimens, in two respects: for the availability of nomina through ***diagnoses**** based on characters, and for the allocation of nomina to taxa. This allocation is not made through definitions of the taxa, but through special devices traditionally called 'types', or ***onomatophores**** (Simpson, 1940). This system is ***theory-free**** regarding taxonomy: it is not tied to a taxonomic paradigm, which has allowed its stability, and avoided the multiplication of 'nomenclatural revolutions' across the highly unstable history of taxonomy. Projects investing enormous amounts of working time, energy and funds to 'redefining' millions of nomina, justified by various reasons (either 'scientific' or 'ethical'; see 'Nomenclatural Availability' below), are in conflict with the need to accelerate the study of biodiversity and may be considered 'criminal operation[s] against the study of biodiversity' (Dubois, 2005b).

Numerous problems currently face zoological nomenclature, in the three stages of the Nomenclatural Process (Dubois and Ohler, 1997a; Dubois, 2000, 2011, 2017d; Dubois et al., 2013, 2019, 2021, 2022b). Some are due to the frequent confusion between taxonomy and nomenclature, but most are truly nomenclatural. They belong to three main categories: problems of interpretation of the *Code*, often due to imprecision of its text; genuine deficiencies of the *Code*; and precise problems in some articles. Below is a selection of five sources of problems important in the context of the biodiversity crisis, as they may interfere with the efficiency and quality of taxonomic research: [1] General problems; terminology; principles; ranks. [2] Problems of availability: the distinction between scientific nomina and non-scientific names; the role of characters in the availability of nomina; authorship of nomina; electronic publication. [3] Problems of taxonomic allocation: the bidirectional ostensional system of taxonomic allocation of nomina; 'type specimens'; 'type species' and 'type genera'. [4] Problems of validity: zygoidy and reversal of precedence; taxonomic and nomenclatural databases; deliberate ignorance of the *Code* and of the Commission's decisions. [5] Problems related to higher level nomenclature.

## GENERAL PROBLEMS

### TERMINOLOGY

The *Code* is a highly technical text providing a set of rules for creating nomina for zoological taxa and for identifying the valid nomen for a taxon as recognised under any ***ergotaxonomy**** (classification adopted as valid by taxonomists). As such, it makes use of a large number of special concepts and technical tools. Most domains of human activity relying on specialised concepts or techniques use precise designations for them, but this is not the case for the *Code* which employs very few 'technical terms,' mostly using 'common language terms' based on the unfounded assumption that this will make understanding concepts easier. The use of an ambiguous term (like 'name' for a spelling, or 'type' for an onomatophore[2]) is imprecise, and results in miscommunication. This leads to confusion and misinterpretations in the implementation of the *Code* (Dubois, 2011).

An important function of non-ambiguous terminology is that it provides an efficient tool for the computerisation of ***taxonominal**** (taxonomic and nomenclatural) databases, and more generally of zoological nomenclature and taxonomy. Such databases exist, but many of them use ambiguous categories of nomina and data that do not convey precise and accurate information. For instance, most databases include in the category of 'synonyms' very different kinds of nomina and spellings, such as ***isonyms**** (objective synonyms, having the same onomatophore), ***doxisonyms**** (subjective synonyms, having different onomatophores but designating the same taxon in a given ergotaxonomy), ***paronyms**** (various forms or avatars of the same nomen, differing in spelling, rank and if relevant combination) and ***chresonyms**** (subsequent usages of a nomen, under its original or a subsequent paronym, or for the same taxon or different taxa; for details, see Dubois, 2000). The use of the simple term 'synonym,' without distinguishing isonyms and doxisonyms, may lead users to consider doxisonyms as permanently invalid, to introduce new nomina for taxa for which nomina are already available, but 'hidden' in synonymic lists, increasing ***nomenclatural inflation*** and the ***synonymy load**** (Dubois, 2008a). Using a precise terminology for these categories would compel inquiry about the genuine status of the nomina and data before entering them into a database. This would enhance the efficiency of these databases for information storage and retrieval in taxonomy and improve communication by zootaxonomists among themselves and with the scientific community (see 'Taxonomic and Nomenclatural Databases' below).

Besides being confusing, current terminology of the *Code* is incomplete and should be expanded. For example, it lacks a term to designate a nomen replaced by a ***neonym**** (*nomen novum*); the term ***archaeonym*** was coined for this purpose (Dubois, 2006).

## Principles

The rules of the *Code* are expressions of underlying principles which have a more general value. The *Code* recognises six such principles, but in fact, its rules are supported by at least eighteen principles that were analysed in detail by Dubois (2011) and Dubois et al. (2019, 2022a). Twelve principles not explicitly stated in the *Code* are nonetheless followed 'surreptitiously' throughout. Their mention in a special chapter of future editions of the *Code* would help taxonomists to better understand its functioning.

## Ranks

The concept of ***nomenclatural rank**** is often considered equivalent to ***taxonomic category****, but they are widely different: the former is a nomenclatural concept pointing to a place in a hierarchy of taxa, the latter refers to a biological (taxonomic) concept (Dubois, 2005a; Dubois et al., 2021). The discussion below is restricted to nomenclatural ranks.

The *Code* uses nomenclatural ranks, such as genus, tribe or family, which are distributed in the three nominal-series mentioned above. Various authors (e.g., Smith, 1988; Sundberg and Pleijel, 1994; Minelli, 2000; Ereshefsky 2001; Pleijel and Rouse,

2003; Kluge, 2005) have claimed that ranks should be abandoned as they are subjective and arbitrary, and do not warrant comparisons between taxa of the same rank in different groups. This is true only if ranks are considered to have an absolute and fixed meaning, but not if they are understood as relative and labile. Taxonomic hierarchies, as reflected in nomenclatural ranks, are 'organisational models of relationships' (Knox, 1998) that are useful to express relationships among entities in a hierarchical system meant to reflect a phylogenetic model. Whether a given higher taxon is treated as a superfamily, an order or a class is largely a matter of tradition and general consensus among specialists for the group concerned. There would be no point in discussing whether a taxon 'is' a class or a family, as there is no theoretical background for defining ranks: ranks only point to a place in a hierarchy, and they are highly labile according to available information and changes in taxonomic arrangements. It is not necessary to abandon ranks but to realise that they have only relative meaning, informing us of the hierarchical structure of the classification, i.e., of hypothesised phylogenetic relationships. They are useful for storage and retrieval of taxonomic information but carry no further information. This is a matter of pedagogy, not of taxonomic theory.

Ranks are useful tools for taxonomy and for biology as a whole, provided they are used in a consistent way, the same rank being given, ***in a given clade***, to two taxa that are considered ***parordinate**** (sister-taxa) under a given phylogenetic hypothesis: thus, they provide information on the structure of the phylogenetic tree on which the classification is based (Dubois et al., 2021). Although advocated long ago (e.g., Hennig, 1966, 1974; Raikow, 1985), this approach is unfortunately rarely implemented in taxonomic works, presumably because taxonomists are worried by the fact that the rank of any taxon is liable to change easily when the cladogram changes, even if the definition and content of the taxon remain the same. This may be viewed as a problem if the rank is considered to be absolute and permanently attached to a nomen, but this is not required by the zoological *Code* (contrary to the botanical code), thanks to the existence of nominal-series. In many recent classifications, the same rank is given to two taxa one of which is subordinate to, or considered sister to, a taxon afforded the same rank. This widespread approach is arbitrary and a misunderstanding of the function of ranks (Dubois, 2008b).

## PROBLEMS RELATED TO THE AVAILABILITY OF NOMINA

### Nomenclatural Availability and the Distinction between Scientific Nomina and Non-Scientific Names

A central, but often ignored, idea that underlies all discussions about nomina of taxa is that they are neutral and meaningless labels, the function of which is simply to designate unambiguously the taxa as recognised within the frame of reference of a given ergotaxonomy, not to provide information about the characters of these taxa (diagnoses or definitions), about their origin, phylogeny, distribution, significance to mankind, or to praise their authors or namesakes (eponymous nomina). Nomina may be inappropriate (e.g., a species may be called *viridis* even if brown) without having to be abandoned or changed. They may even be meaningless 'arbitrary combinations of letters'. Once attached to a taxon, a nomen loses its original 'meaning' (if it had one) and becomes an inert 'tag' whose function is to point unambiguously to a specimen or group of specimens (its 'type') which makes a permanent and objective connection between the world of organisms and that of language. All the general reasons[3] allowing taxonomists to reject and replace previously established nomina are inadmissible; they represent a misunderstanding of the nature of scientific names (which are neutral labels, not words carrying extra-taxonomic 'messages') that poses a threat to zoological nomenclature and taxonomy and, ultimately, to biology.

Zoological nomenclature is not a peculiar or specialised field of 'common language'. It is a distinct domain, with its own terminology and conventions that have more to do with legal systems than linguistics, grammar and vocabulary. It is not properly a science but a technical domain tightly connected with the scientific discipline of taxonomy. The function of the latter is to furnish biology with a scientific system of classification of organisms and relevant information about their characters and evolution. The nomina of taxa are therefore part of the scientific realm and as such are entirely different from so-called 'vernacular names' or other kinds of names that can be given in common languages to organisms for various purposes.

Scientific nomenclature rests on conventions that must be understood and followed in order to have an international system of technical terms and scientific nomina to communicate about biodiversity, and that put fixed and clear barriers between this system and all other human communication systems. These limitations are necessary in order to have a functional international zoological nomenclature. If they did not exist, any name used in any language, at any time during mankind's history, and under any form (including oral language, writings on stone or wood, single-copy paper documents, on museum specimen labels, or in publications failing to comply with criteria of Chapter 3 of the *Code*) could be potentially considered the valid nomen of a taxon. This would apply not only to names used for animals by Aristotle, Pliny the Elder or Albertus Magnus, but also those used in the past or now locally in all human languages, including not only well-known ones like English, Mandarin or Swahili but also Irantxe, Tiwi or Newar among hundreds of others.

This is the justification for the concept of nomenclatural availability, which allows for the universality and unambiguity required by scientific communication about taxa. In order to 'exist' in zoological nomenclature and to be a potential valid nomen of a biological taxon, a name must first have been established fulfilling the conditions of Articles 1–20 of the *Code* (see 'Nomenclatural authorship' below).

The first one is that it must have been 'published' after 1757, i.e., distributed in several identical copies with a fixed format and content, and respecting a few other conditions. The *Code* uses the term ***publication**** in three distinct senses: a published document; the act of public distribution of a document; and the act of public distribution of a document that complies with the conditions of the *Code* for nomenclatural availability, better called ***promulgation**** (Pavlinov, 2014; Dubois, 2020b). To be promulgated and available, a new nomen must then comply with a variety of requirements (not all mentioned here), including the presence, in the original publication, of a scientific characterisation of the taxon for which it was proposed. This may be a diagnosis, a definition, a description or, for nomina published before 1931, a mere illustration of a specimen. These two criteria, promulgation in a document distributed after 1757 and characterisation of the taxon aiming at its scientific delimitation and recognition, are crucial for the distinction between scientific nomina and all other kinds of names.

In recent years, several proposals to abandon the concept of nomenclatural availability for so-called 'ethical' reasons were obligingly published by some periodicals, including 'famous' ones with no previous interest or competence in taxonomy and nomenclature. Their authors proposed to abandon or replace some categories of nomina in order to promote an 'inclusive terminology' (Cheng et al., 2023). The two main proposals were to [1] replace nomina considered by them to be offensive or inappropriate (e.g., Hammer and Thiele, 2021; Smith and Figueiredo, 2021a, b; Smith et al., 2022; Thiele et al., 2022) and [2] to replace nomina with names given to animals in the languages of human populations living in the area of their occurrence (e.g., Gillman and Wright, 2020; Knapp et al., 2020; Rivas et al., 2024).

Several authors (e.g., Mosyakin, 2021, 2022a, b; Palma and Heath, 2021; Ceríaco et al., 2023; Pethiyagoda, 2023; Dubois et al., 2024; Jiménez-Mejías et al., 2024) criticised these proposals on various grounds and argued that their implementation, which could concern thousands of nomina, would have devastating consequences in zoological nomenclature and should be rejected. All of these proposals are based on the characterisation of the nomina to be replaced as 'inappropriate,' although, as stressed above, appropriateness is irrelevant to the function of nomina which are inert and meaningless labels. Many arguments, not all discussed here, were provided against these proposals, but three are particularly relevant. This would entail considerable nomenclatural instability at a time of biodiversity crisis and taxonomic urgency which require clear, unambiguous and universally accepted nomenclatural rules. Different ethical values and criteria exist among taxonomists, human groups and individuals, so that none can claim universality. And, regarding so-called 'indigenous names', proposals are based on a confusion between scientific nomina and vernacular ('common' or 'local') names for 'kinds' of animals or plants, not for scientific concepts.

This third point deserves attention. Vernacular names are not based on a study of the characters and distribution of organisms, or on attempts to understand their relationships and evolution. In considering that 'indigenous peoples' had already 'named' entities that are now recognised by science, and suggesting that their names be 'restored', the science of taxonomy is underrated and the rigour of its theories and methods denied. Rivas et al. (2024) stated that a long-recognised species of snake actually consisted of two species and used, to name the purported new species, a local name that had been 'indigenously used to designate this species for at least hundreds (and perhaps even thousands) of years before the use of any of the other synonyms' and 'was certainly in use in 1758 when the Code[4] started counting names as valid' even though this species had never been distinguished from the other one by the local population[5]. Such statements that traditional cultures had already provided an acceptable taxonomy and nomenclature for organisms through empirical, mythical or religious approaches amounts to saying that there was no need to develop a whole body of concepts and methods to classify and name the units of biodiversity. This approach is not limited to a criticism of the current codes of nomenclature but is a frontal attack on the scientific discipline of biological taxonomy.

Differences between 'vernacular names' and scientific nomina include the absence of publication, date of promulgation, diagnosis based on characters, nomen-bearing specimens, criteria of validity in cases of homonymy and synonymy, the correctness of spelling, etc. Ignoring these aspects of the problem points to a misunderstanding of the particularities of the whole nomenclatural system.

The comments above are not a criticism of the use of vernacular names in local communications, in public media or in non-scientific literature, but on the need to distinguish both domains. Many vernacular names have been borrowed by taxonomists to name species or genera in the frame of scientific nomenclature, and this should be encouraged. But in no way is a nomen using the same spelling (or pronunciation) as a vernacular name identical with it, as they belong in different realms. The 'nature' of terms so 'borrowed' is changed once they have been proposed in agreement with the rules of the *Code*. They acquire a publishing date and diagnosis, an onomatophore and other characteristics that make them 'new entities'. Terms of both kinds are not synonyms or homonyms, and priority does not apply across them, as the nomenclatural concepts of homonymy, synonymy and priority, as regulated by the *Code*, apply only within the context of scientific taxonomy.

## Emendations vs. Incorrect Spellings

The *Code* distinguishes several different kinds of spellings of nomina: original and subsequent correct and incorrect spellings, justified and unjustified emendations. Only some of them are available, which has important nomenclatural consequences. The distinction between emendations and incorrect spellings depends on the intentionality of the change of spelling. The rules of the *Code* regarding these distinctions should be drastically improved. In particular, the wording of some Articles in the 1999 edition that remove the distinction between well-defined categories of subsequent spellings, in order to allow the validation of spellings considered to be 'in prevailing usage,' is deficient and confusing (see Dubois, 2017b; Dubois and Aescht, 2019, p. 137).

## The Role of Characters in the Availability of Nomina

In zoological nomenclature, after 1930, to be duly promulgated, a new nomen 'must be accompanied by a description or a definition that states in words characters that are purported to differentiate the taxon' (Article 13.1.1). This means that a nomen first published without the explicit mention in words of diagnostic characters is unavailable, and qualifies as a *nomen nudum* (Dubois, 1999, 2017d; Bauer et al., 2011; Dubois et al., 2018). This rule has an important consequence: in order to be available, nomina must have been proposed on the basis of real specimens of animals, not of 'hypothetical concepts' (Article 1.3.1) such as mythical animals, or as 'clades' proposed as results of phylogenetic analyses. No definition of the term ***character**** is given in the body of the *Code* itself. The list of 'exclusions' given in Article 12.3 is incomplete: other exclusions could be listed, such as a geographical distribution or a position in a phylogenetic tree. All these exclusions are based on statements based on ***extrinsic information*** or ***hypotheses*** regarding ***taxa***. In the *Code*'s glossary, the term 'character' is defined as 'any attribute of organisms used for recognizing, differentiating, or classifying taxa'. This definition is incomplete as it does not mention the need for these attributes to be based on ***observable intrinsic*** features (such as size, colour, shape, etc.), that are considered taxonomically meaningful (i.e., excluding injuries or other consequences of accidents during life), of the ***organism(s)*** for which the taxon is erected. The following definition of this term is more complete: 'Any intrinsic feature of organisms used for recognizing, comparing, differentiating or classifying taxa. In a given taxon, the same character may occur under several distinct alternative character states' (Dubois, 2017d). It avoids the frequent confusion between characters and ***character states****, and excludes ***extrinsic*** features or attributes (such as vernacular names, or geographical or stratigraphic origin or distribution), as well as ***relacters**** (real interactions between organisms, such as 'crossability', not comparisons between character states of different organisms).[6] This also excludes phylogenetic relationships, as these are not observed facts but models or hypotheses. Thus, the nomen of a new taxon erected on the sole basis of its place in a tree, without any explicit mention of diagnostic characters, is a *nomen nudum*. Unfortunately, this is not yet understood or accepted by all taxonomists, with many such cases having been published in recent decades (for some examples, see Dubois, 2017d).

Three further common misunderstandings concerning this question are worth mentioning.

First, the requirement for the mention in words of characters does not mean that only morphological characters may be used. Other kinds of intrinsic features of organisms may be used for this purpose, including molecular characters, but these must be presented as such, e.g., as sequences that are considered ***diagnostic**** or ***apognostic**** of the taxon (see Dubois, 2017d). In contrast, a position in a tree is not a character: it is a hypothesis, based on a model of relationships between taxa (which can be based on molecular, morphological, both, or other data).

Second, the mention of diagnostic characters is indispensable for the nomenclatural availability of a new nomen but does not imply that this diagnosis is accurate or complete. A wrong diagnosis may be refuted by a subsequent taxonomic study and result in the taxonomic invalidation of the nomen, but not in its unavailability.

Finally, a common error is the belief that any expansion of the diagnosis or content of a taxon may result in the erection of a new taxon and in the creation of a new, junior homonym, nomen. After its erection, a new taxon continues to be studied which often results in better knowledge and in modifications of its diagnosis. Thus, in subsequent works, its nomen may be applied to more or less comprehensive taxa. If some of them do not include its onomatophore, this qualifies as misidentification and, as such, does not amount to the establishment of new nomina. In order for a new homonymous nomen to be established in a new work, this should be expressly stated in the work. Although this was mentioned as a rule only in Article 16.1 of the 1999 edition of the *Code*, it is true also for older works – otherwise the taxonomic literature would be filled up with such invalid junior homonyms.

## Nomenclatural Authorship

In zoological nomenclature, the term ***authorship**** designates the name(s) of the author(s) of works, nomina or nomenclatural acts. Authorship is viewed by many taxonomists as a very important issue in nomenclature, and many papers mention problems related to it. According to the *Code*, to be available, all zoological nomina published after 1950 must have a stated authorship, but this is not indispensable before 1951. However, authorship does not play any role in the Nomenclatural Process that leads from availability to validity. It is not part of the nomen and its mention in the ***nominal-complex****, along with its effective year of promulgation, after the nomen of a taxon – in the form 'Nomen + Author(s) + Date' –, is 'optional, although customary and often advisable' (Article 51.1). Its function then is not to glorify its author, but to provide a 'shortened bibliographic reference' to the original publication of the nomen (Ng, 1994). This function has lost most of its utility nowadays with the multiplication of online databases. In cases of homonymy, the date alone, without the authorship, is usually sufficient for disambiguation, except in very rare cases (e.g., the case of the nomen *Hylomantis* in Dubois and Frétey, 2016, p. 19).

Today, the ***mention*** of authorship of nomina in their nominal-complex raises at least two major problems, which lead to questioning its relevance and utility.

For more than two centuries, this was meaningful and did not cause any problem, as the authors of nomina were real taxonomists, who had contributed to the recognition, description and naming of the new taxon. They consisted mostly of a single person, more rarely two or more, so their mention, once in a paper, did not occupy much space and was informative. Today, many works proposing new nomina have long and cumbersome lists of 'authors', only a few of which were involved in the taxonomic work itself, while the others have contributed to some of the phylogenetic and other analyses supporting the new taxon, or merely to the collection of specimens and tissues or other ancillary information, such as being responsible for the research team conducting the work (Dubois and Ohler, 2021). Although these persons are formally 'authors' of these nomina, their mention after the Latin nomen is not justified scientifically and is only a source of long and unpalatable lists of names. This may be even worse in the case of a new nomen proposed with a list of 'authors' which is distinct from the list of authors of the work wherein it is included, thus leading to nominal complexes in the form 'Nomen + Author(s) of nomen + *in* + Author(s) of the work + Date' (Recommendation 51E). This possibility offered by the *Code* is confusing, as the first 'authorship' (of the nomen) is irrelevant when tracing a bibliographic reference. This practice results from an overemphasis put on authorship discussed below and could be abandoned without nomenclatural harm. Replacing mention of the 'author(s)' of nomina with that of the dates of the publications where they were promulgated would be simpler and sufficient to trace the reference.

Much more problematic is that the mention of the names of authors of nomina after the latter is a reason for the publication of many unwarranted nomina based on insufficient data and/or poor taxonomic methodology, nomina which will be synonymised and contribute to nomenclatural inflation and increase of the synonymy load. The reason for this flourishing is ***mihilism**** (Bruun, 1950), i.e., the tendency of some taxonomists to publish unjustified new nomina to increase the list of nomina for which they can claim authorship. The ambiguous term 'taxonomic vandalism' has been used (e.g., Kaiser et al., 2013) to refer to this phenomenon. Reducing mihilism significantly could not be obtained through a modification of the *Code* removing its taxonomic neutrality. Opening this Pandora's box could be considerably harmful to the urgent inventory of the earth's biodiversity in focusing the work of taxonomists on endless discussions of so-called 'ethical' matters. The important point here is not to have a 'unitary' taxonomy, which is currently unrealistic for various reasons, but that taxonomists collect and provide to the community specimens (preferably allowing molecular investigation), tissues and data and be very clear on the taxonomic concepts and methods they are using, in order to allow for a revision of their conclusions by taxonomists of the future. In contrast, an efficient way to reduce the impact of mihilism on taxonomy would be to stop mentioning the authorship after nomina (Dubois, 2008a). The *Code* (Article 51.3) considers it more important to cite the author's name than the date, which is misguided. The reverse would be more judicious, as the date is enough to trace a nomen, even among homonyms, and it plays a role in the validity of nomina, which is not the case of authorship.

In fact, the mention of authorship is not only useless, it is misleading. The mention of the author of a nomen just means that this person has promulgated the nomen, not that it had established the concept of the taxon as recognised nowadays in any ergotaxonomy, or justified it scientifically. Most new nomina established by Linnaeus are still considered valid, but very few of them for taxa having the same definitions and contents as in his works. In order to refer to a taxon as currently understood, some authors add '*sensu*' after its nomen, followed by a taxonomic reference, but this is unnecessarily burdensome as it would be the case for almost all nomina. Scientific nomina are not definitions, they have no phylogenetic or other meanings, they are just labels designating 'boxes', each bound to include only one item, its onomatophore. According to the classification adopted at any given time, the content of the box may change widely, except for the 'anchor' provided by the onomatophore. Mentioning the authorship of a nomen merely refers to the fact that this nomen was the first one promulgated for the taxon of this nominal-series including its onomatophore. It is not justified to see this as a 'glorification' of its author because in many cases the reasons for which this author had established it are not the same as those for which it now considered valid, and may even be completely incompatible with them! The only important pieces of information are the onomatophore, that governs the allocation of the nomen,[7] and the publication date, that governs its validity against potential senior homonyms and synonyms. Starting from these two elements, the *Code* provides almost automatically (except when nomenclatural ambiguities have not yet been resolved) the valid nomen of a taxon, and this removes this task from the shoulders of taxonomists who have more important and urgent scientific matters to address in delimiting and defining taxa rather than merely knowing which are their valid nomina (see 'The Need of a *Code*' above). Enough nomenclatural problems can appear when only matters of the *Code* are involved, so taxonomists should refrain from adding problems through their opinions, tastes or value judgments regarding the works where nomina were established or their authors.

Instead of the term 'authorship,' present in the *Code*, some taxonomists use the term 'authority'. This is unwarranted, first because the latter term is used in the *Code* in two widely distinct meanings (referring to the *Code* itself and to the Commission, but not to nomina), and mostly because authors of nomina do not exert the slightest 'authority' on them. The only act they can do is to coin nomina or borrow them from some external source, and promulgate them in a *Code*-compliant manner. Once this is done, they have no special power to intervene in their fate, be it to change their spelling or their onomatophore (except through first-reviser actions, just like any other authors[8]), and even less regarding their validity, which depends either on automatic rules (in cases of homonymy or objective synonymy) or on the taxonomic actions of the international community. Ignorance or misuse of the terminology of the *Code* causes problems of communication. The term authority should not be used to designate authorship or authors in zoological nomenclature.

### Electronic Publication

In 2012, the Commission adopted an Amendment to the *Code* which allowed the electronic promulgation of taxonomic works, new nomina and new nomenclatural acts. This 'revolution' was welcomed by many taxonomists. However, this Amendment has led to many problems, including the publication of high numbers (probably thousands) of unavailable works, nomina and nomenclatural acts, and uncertainties regarding promulgation dates. The most prominent problems derive from failure to pre-register works on *Zoobank*, from the incompleteness of the coverage of the latter (that does not allow the registration of nomenclatural acts other than the promulgation of works and nomina), from absence of information on the dates of distribution of final versions of works first published as preliminary versions, and from relegation of nomenclatural information indispensable for availability to unregistered online supplementary material. A discussion of this complex matter here would be much too long. Detailed analyses of these problems and proposals to solve them were provided by Dubois *et al.* (2013, 2022b, 2024), Dubois and Frétey (2023), and works cited by them.

## TAXONOMIC ALLOCATION OF NOMINA

### The Bidirectional Ostensional System of Taxonomic Allocation of Nomina

The concept of 'nomenclatural type', as distinct from that of 'taxonomic type' or other type concepts, did not exist in the early days of taxonomy. In zootaxonomy, it appeared at the beginning of the 19th century (Dubois, 2022a). Nowadays, it is probably one of the most misunderstood concepts of taxonomy for two distinct reasons.

First, because it is based on the term 'type', which leads some biologists to believe that the function of type specimens is to act as ***semaphoronts****, i.e., to provide characters for taxonomic, not nomenclatural purposes. Of course, these specimens do bear characters, and often these characters are helpful in allocating these specimens to taxa, particularly when several species occur in the region where they had been collected, but both functions (definition of taxa and allocation of nomina to taxa) are distinct, the first one being taxonomic and the second one nomenclatural.

The second reason is the imprecision of the *Code* regarding the way so-called 'typification' works, as shown by the definition of 'name-bearing type' in the Glossary of the *Code*, stating that it 'provides the objective standard of reference whereby the application of the name of a nominal taxon can be determined'. This definition puts a justified emphasis on the 'objective' nature of the onomatophores of nominal taxa: whatever their nominal-series, they are all ultimately based on objects (specimens). A nominal taxon (***taxomen****) consists of the permanent and indissoluble connection between a nomen and a specimen or series of specimens (in the species-series) or a subordinate taxon or set of taxa (in the genus- and family-series). Nominal taxa have no defined content (***extension****) or limits, and they do not change over time. This distinguishes them from so-called 'taxonomic taxa', whose definitions and contents are provided by the provisional ergotaxonomy adopted at a given time in the history of taxonomy, and which in most taxonomic groups change regularly. But what does 'can be determined' mean in this definition? How does this work?

The supporters of *Phylocode*, an alternative nomenclatural system (Cantino and de Queiroz, 2020) have repeatedly pretended that the *Code*'s system is subjective and inoperative because onomatophores do not provide the contents and limits of the taxa to which they apply so that it would be necessary to provide 'phylogenetic definitions' of the nomina. Such an idea is based on a misunderstanding of the system – which admittedly is not clearly explained in the *Code* or in any taxonomic textbook (see Dubois, 2020a). If nomina were attached to taxa only through their own onomatophores, any nomen established for an animal species or subspecies would potentially apply to all the taxa that include its onomatophore, from the base to the top of the hierarchical taxonomy. But this is not the way the system works, as all practising taxonomists 'intuitively' know. In fact, in any given ergotaxonomy, a nomen applies potentially to any taxon ***recognised as valid*** that ***includes*** its onomatophore but also that ***excludes*** the onomatophores of all other related taxa ***recognised as valid***. This system of ***bidirectional ostension**** is therefore based on two devices, not one: a fixed internal landmark, the onomatophore of the nomen, that provides its anchor or nucleus, and an external landmark, its ***onomatostasis****, that provides its limit, and hence its extension (content).[9] The onomatophore is fixed, but the onomatostasis may change according to the ergotaxonomy adopted. The long failure to recognise this fact has given birth to the 'myth', present in many works (including taxonomic textbooks), that the allocation of nomina to taxa is made only by their own onomatophores.

Therefore, the same nomen may designate different taxa in different classifications, as long as they include its onomatophore. This is not a weakness of the *Code*'s system, but one of its strengths, as it allows ***nomenclatural parsimony****: it does not require creation of a new nomen every time the ergotaxonomy changes to account for changes in phylogenetic hypotheses, and, in cases of recognition of new subordinate taxa referred to the same nominal-series, one of them will bear the same nomen as its superordinate taxon.

In the publication where a new nomen is established, it may be based on a single onomatophore (***monophory****), several of them (***symphory****) or none (***aphory****). The first situation is clear, but the other two may be ambiguous and require clarification through appropriate nomenclatural acts, but then this implies some potential problems, as several different acts may have been taken by different authors, and only the first published one will be valid. This applies both to the designation of lectotypes among syntypes and to the subsequent designation of type species among originally included nominal species of nominal genera or subgenera.[10] It will then be the responsibility of subsequent authors to trace the first appropriate nomenclatural act, which may be much more difficult to find than the original publication where a new nomen was established, and to ensure that the memory of and reference to this action is kept. In this respect, too many taxonomic publications or nomenclatural databases mention lectotypes or subsequently designated type species 'naked', as if they were 'well known', without citing the reference where this was done. They are then incomplete and potentially misleading, and cannot be used as reliable sources of nomenclatural information.

## 'Type Specimens'

So-called 'type specimens' (***onymophoronts****), i.e., holotypes, syntypes, lectotypes or neotypes, play a crucial role in zoological nomenclature, not only for the taxonomic allocation of species-series nomina but also for nomina of higher nominal-series which are ultimately based on them.

Ideally, in order to efficiently play their role as anchors for the allocation of nomina to taxa, they should [1] have been collected from a well-identified natural population (present in its 'type locality' or ***onymotope****), [2] be preserved in a well-curated collection accessible to taxonomists and in good condition of conservation (allowing for the study of their characters),[11] and [3] include or be accompanied, when possible, by tissues allowing molecular sequencing. These three criteria should be taken into account in the choice of onymophoronts.

For many ancient specimens, these conditions are unmet and, as long as this is the case, the nomina based on them may be *nomina dubia*, a situation viewed by some (e.g., Dayrat, 2005) as a major, often inextricable, problem, although in many cases it may be resolved rather easily. Many old onymophoronts have been damaged, destroyed or lost, or lack a precise or reliable locality, but are known by their original illustrations or descriptions. In many cases, at least some of the original syntypes are extant in identified collections. The fact that some syntypes are missing does not negate their function as onymophoronts. In such cases, the *Code* does not prescribe the designation of an extant specimen as lectotype as necessary: a missing specimen may be designated as 'virtual lectotype' if this best serves nomenclatural clarity and stability. This is particularly useful if a missing syntype is the only one known to have a precise origin. Such designations may be useful in avoiding reliance on syntypes of unknown or imprecise origins, or in bad condition of preservation, which then become paralectotypes, allowing the onymotope to have a precise locality. When the lectotype is missing, it is legitimate to designate a well-preserved neotype from the same locality thus allowing morphological and molecular study, and 'rejuvenation' of the onymophoront of the taxon (Dubois and Ohler, 1997b; Dubois et al., 2024). This method may alleviate the 'archaeological' work of taxonomists tracing the origin and status of old specimens, which may be motivated more by a 'fetishisation' of ancient specimens than by real scientific needs. However, this elegant procedure cannot be used if all the original types are extant and poorly informative: in such cases, the only way to discard them may be to ask the Commission to do so through the use of its Plenary Power, as currently, unlike the botanical code, the zoological *Code* does not recognise ***isotypes***, a possibility which would be worth exploring. Regardless, the widespread belief that taxonomists are 'prisoners' of defective old specimens is unjustified. Nomenclature is not history of science. In some cases, we will never know which species was indeed in the hands of an ancient author, but this should not paralyse us. Appropriate actions respecting the *Code* should allow taxonomists to remove the doubts and go forward through an 'objectivisation' of nomenclature.

This question can be put quite differently for recent taxonomic works. A growing number of papers propose to erect new taxa on the sole basis of molecular data, without solid morphological information, a trend that doubtless denotes a worrying loss of taxonomic expertise among the taxonomic community (Wheeler, 2024). Although this is acceptable for nomenclatural availability, this is very poorly informative from a biological point of view. This defect can be obviated when the new taxon is supported by a ***hypodigm**** (Simpson, 1940) conserved in a permanent scientific collection, comprising several specimens or even a single one (the onymophoront), but this becomes insufficient when no specimen has been preserved, the molecular data having been obtained, for example, through a drop of blood from a specimen photographed and released.

Strangely, for almost a century, since its foundation under the title *Règles* (Blanchard, 1905), the *Code* did not require the explicit designation of an onymophoront for the availability of a new species-series nomen. Article 16.4 of the 1999 edition introduced the requirement of the explicit fixation of such a specimen but not of its conservation in a collection. A large international movement involving an overwhelming majority of practising zootaxonomists asked for a change in the *Code*, requiring the deposition of at least one specimen in a collection for nomenclatural availability of a new nomen or a special procedure allowing exceptions when this is impossible (Dubois and Nemésio, 2007; Ceríaco et al., 2016), but the Commission (Anonymous, 2017) rejected this request and maintained the possibility for any taxonomist to name new species today based on missing specimens. Such 'typeless species' may become more and more common owing to the multiplication of laws and regulations forbidding the collection of specimens or their international transport, which are a considerable nuisance for taxonomy and, by way of consequence, for biological knowledge and conservation. This rule should be modified and the possibility of such descriptions in exceptional cases should be first submitted for validation by a qualified body (Dubois, 2017c).

Because of their profound importance in taxonomy and nomenclature, permanent curated collections conserving nomen-bearing and other specimens, and their accessibility to researchers, should be considered major international infrastructures for research, protected against the many threats they face, and benefit from considerable increases in staff, budgets, and facilities. This question has much wider implications not developed here.

## 'Type Species' and 'Type Genera'

The onomatophores of taxa of the genus- and family-series are not specimens but nominal taxa, respectively so-called 'type species' (***nucleospecies****) and 'type genera' (***nucleogenera****), which in turn are based on the onymophoronts of the former. Nucleospecies allow the taxonomic allocation of genus-series nomina, but this remains an internal process of nomenclature: they are nominal, not biological or taxonomic species, and it is therefore wrong and misleading to replace them, in taxonomic works or databases, by the valid nomina of the taxa as adopted in an ergotaxonomy, as this can lead to gross nomenclatural mistakes (e.g., Dubois et al., 2024).

Until 1930, new genus-series nomina could be established without any mention of included nominal species or with several of them, none being fixed as 'type'. After 1930, the explicit designation or implicit fixation (through 'monotypy' or tautonymy) of a nucleospecies is compulsory for the availability of new genus-series nomina (Article 13.3). A nomen proposed after 1930 without such a designation is permanently unavailable, even if several included species were mentioned: no subsequent selection of one of them will render such a nomen available, except if expressly presented as a new nomen with new authorship.

No such problems exist for the fixation of the nucleogenera of family-series nomina, as the latter are based on available genus-series nomina then considered valid, which are therefore *ipso facto* designated as onomatophores by ***implicit etymological designation**** – except in the rare cases of genus-series nomina having ***homographic**** stems, hence the addition in Article 16.2 of the 1999 *Code* of the need to mention explicitly the nucleogenus for the availability of new family-series nomina.

## VALIDITY OF NOMINA

### Zygoidy and Reversal of Precedence

Not all available nomina and nomenclatural acts are valid. They may be invalid in situations of ***zygoidy**** (conflict of precedence). Most rules concerning such situations are simple and straightforward as long as taxonomists agree to follow them, while a few raise genuine problems. Attention must be given to the fact that situations of conflict of ***zygonymy**** (synonymy or homonymy) between nomina are not always settled by ***priority**** but are sometimes, in cases of synchronic publication, solved by ***airesy**** ('first-reviser action'[12]) or ***proedry**** (rank precedence). As explained by Dubois (2010a), the rules concerning airesy have been made unnecessarily complex in Article 24.2.4 of the 1999 edition of the *Code*, so that their implementation now requires more attention to avoid errors. This article is confusing and untoward, and its suppression would be appropriate.

Both terms ***synonymy**** and ***homonymy**** designate widely different situations: ***isonymy*** (objective synonymy) and ***(primary) homonymy*** depend only on nomenclature and are permanent and irreversible, whereas ***doxisonymy*** (subjective synonymy) and ***secondary homonymy*** (which exists only in the species-series) depend on the ergotaxonomy adopted and are labile and reversible so that the mere mention, e.g., in a synonymic list or in a database, that two nomina are 'synonyms' (or, worse, their simple connection by the sign =) or 'homonyms', without additional information, is uninformative and may be misleading.

The rules of the *Code* regarding the correct spellings of nomina, some of which are quite complex or delicate to use (e.g., regarding the endings of species-series ***epithets**** dedicated to persons), are not discussed here (see e.g. Dubois, 2018), but should be given attention by all users of nomina.

## Priority vs. 'Usage'

In order to be operational, any system regulating the validity of nomina must rely on a single fundamental Principle, to which ancillary Principles may be added for particular situations (e.g., Melville, 1958, p. 1249; Hołynski, 1994, p. 12; Dubois, 2010c, 2011, p. 35). This was the case in all editions of the *Code* until 1999: the basic and automatic Principle for validity was priority of publication, which could be tempered by a few other ones in special situations (see 'Zygoidy and Reversal of Precedence' above). However, the 1999 *Code* introduced a new criterion of 'prevailing usage' (defined differently in different parts of the work), to 'protect' some 'well-known' nomina and spellings.

The main change in this respect is the new Article 23.9 on reversal of priority whenever [1] a senior homonym or synonym has not been used as valid after 1899, while [2] its junior homonym or synonym has been used as valid for the same taxon in at least 25 works published by at least 10 authors for at least 10 years in the immediately preceding 50 years. The first one is called *nomen oblitum* and the junior one *nomen protectum*. This Article is intended to avoid useless nomenclatural instability not justified by taxonomic changes, but it raises several problems (see e.g., Dubois and Ohler, 2018). First, this condition is based on the fact that the senior nomen has been ***considered*** as ***valid***, not only as ***available***, so that a nomen that has been mentioned hundreds of times in the literature as available but invalid (e.g., for being a junior synonym) can be rejected as a *nomen oblitum*, which is fully illogical and should no doubt be modified. If it has not been done, some authors will feel entitled to introduce new nomina for taxa for which nomina are already available, instead of 'resurrecting' the latter, adding to nomenclatural inflation. Second, condition [1] must be taken strictly so that if even a single usage of a nomen as valid after 1899 has been overlooked, the use of this Article is not *Code*-compliant. Third, the simple fact that two competing nomina comply with the conditions of Article 23.9 does not by itself result in the invalidation of the senior one: this occurs only if an author points to the fact that these conditions are met and provides a list of 25 references showing that condition [2] is complied with. Finally, the formula *nomen oblitum* had been used in a different sense in previous editions of the *Code*, abandoned and finally 'resurrected', but for a different concept. This is a potential source of confusion which has resulted in errors that could have been avoided by the use of a new term such as ***distagmonym****.

Beside Article 23.9, the current *Code* allows in some other cases the Principle of Priority among nomina or spellings to be ignored. However, the Articles at stake (29.5, 33.2–33.3, 35.5, 40.2 and 75.6) rely on a deficient, confusing and non-operational concept of 'prevailing usage,' and should be modified or suppressed (for details see Dubois and Aescht, 2019, p. 37). Relying upon 'usage' when the latter is not *Code*-compliant amounts to encouraging superficial and hasty nomenclatural work and echoes Gould's (1990) famous sentence: 'What other field defines its major activity by the work of the least skilled?'. The old question of 'usage vs. priority' has long been, and still is, an important source of problems and conflicts in zoological nomenclature. It should be bluntly reconsidered.

Four practical aspects of this question have a direct negative impact on taxonomy in the century of extinctions: [1] it weakens the binding value and strength of the *Code* by encouraging careless nomenclatural work that, in the long run, may threaten the universality of nomenclature by causing chaos and indirectly lending support to the development of alternative nomenclatural systems; [2] it encourages personal (and sometimes very bitter) debates among taxonomists, giving undue importance to the medieval 'argument of authority' in nomenclatural decisions, and conveying a poor image of taxonomic research among the biological sciences; [3] it sends a wrong message to non-taxonomists as regards completion of taxonomic work, understating considerably the importance of the taxonomic gap; [4] it weakens the status of onymophoronts in zoological nomenclature, and thus acts as a threat against natural history museums and other permanent collections, which are indispensable for a proper survey of biodiversity (Dubois, 2010c; see 'Type Specimens' above). Most so-called *nomina oblita* derive from the fact that taxonomists have failed to examine the type specimens of subjective synonyms, which are copied uncritically from publication to publication (Dubois 2010b, c). Dubois (2010c, 2011) suggested that 'protection' of nomina 'threatened' by rules of the *Code* should be limited strictly to nomina well-known ***outside the small world of taxonomy***, such as *Caenorhabditis elegans*, *Drosophila melanogaster* or *Archaeopteryx*. This would require new rules for the *Code* to clearly define ***categories of usage*** on the basis of objective criteria (see Dubois, 2015).

## Taxonomic and Nomenclatural Databases

Recent decades have seen the flourishing of online taxonomic and nomenclatural databases. Not all of them were created or are managed by competent taxonomists aware of nomenclatural rules. This results in the presence in some of them of basic confusions, imprecisions or errors that may mislead non-taxonomists who tend to trust these databases instead of consulting the scientific taxonomic literature. These are not 'imprecisions' or random factual errors but the result of misguided conceptions of these databases, concerning the kind of information documented, the terminology used, and whether original taxonomic publications have been examined and mentioned, or information only copied from other databases. This is even more irritating when editors or referees of journals require authors to follow these databases as though they were 'official' or 'authoritative.' The citation of the type species of genus-series nomina without providing the source of their designation, and the mere statement that two nomina are 'synonyms' or 'homonyms' without distinguishing the different categories of these concepts, are among many problems found in such databases. Many of them fail to use terms of the *Code* (e.g., 'available', 'valid' or 'preliminary/final version'), instead using terms borrowed from other fields and not recognised by this text (e.g., 'authority', 'basionym', 'legitimate', 'accepted' or 'version of record'). Few of them distinguish properly what depends on taxonomy alone, on nomenclature alone and on interactions between both fields. For example, the statement that a given nomen is 'valid' is not useful information if no reference is given to at least one published work where this validity is supported and justified. Authors and managers of databases should pay attention to the various pieces of information that may be provided by them (see Dubois, 2017a, b, 2022b; Dubois and Ohler, 2018; Dubois and Frétey, 2023) in order to avoid missing data, confusion and errors.

## Deliberate Ignorance of the *Code* and of the Commission's Decisions

For a long time, zoological nomenclature has attracted little attention, this domain being entrusted to the specialists of taxonomy, mostly museum curators and researchers. This situation has changed drastically in recent decades, to the point that many people now think themselves competent to propose changes to the *Code*, even if it is clear from their writings that they base their ideas on preconceived erroneous beliefs. The well-known, long-standing disinterest in, and even contempt for, taxonomic matters by the community of biologists (Dubois, 2003; Wheeler, 2004, 2024; Carvalho et al., 2005, 2007, 2008; Crisci 2006a, b; Valdecasas et al., 2008) has a new manifestation: after having ignored this domain for decades, destructive proposals regarding taxonomy are published in 'highly rated' journals (see 'Nomenclatural Availability' above) which refuse to publish responses from professional taxonomists (e.g., Jiménez-Mejías et al., 2024).

Many such proposals will disappear as fast as they appeared, with little attention or impact. More worrying are taxonomists tempted by this approach. Citing published papers showing a poor mastery of taxonomic principles and nomenclature, they propose to police taxonomy by imposing 'best practices', so-called ethical considerations, and dictating the acceptable taxonomic paradigm. The unclear concept of 'taxonomic vandalism' is part of this game which aims to ignore some works instead of discussing and refuting them. The history of science is filled with episodes during which ideological, political, financial or other personal motivations have distorted the functioning of science, sometimes with long-lasting damage. In order to avoid such nuisances, zoological nomenclature should stick to the statement of the *Code*'s Preamble about 'the freedom of taxonomic thought or actions' (see 'Some Basic Characteristic of the *Code*' above). Like all scientific disciplines, taxonomy should progress toward consensus through free debate and argument, not by administrative rules or lobbies.

This being said, it is clear that some works are unacceptable, breaching all scientific practices and causing major problems that cannot be easily solved by the standard nomenclatural rules. A solution exists. It is not to ignore these works, but to request that the Commission invalidate them under its Plenary Power on the grounds that they 'disturb stability and universality or cause confusion' (Article 81.1). In such cases, the Commission should not show signs of weakness or complacency as it could contribute to major crises as it is the case following one of its recent decisions (Anonymous, 2021). This matter would merit a thorough open international debate which has not yet taken place.

## HIGHER NOMENCLATURE

Many taxonomists believe that the *Code* does not regulate the nomenclature of taxa at ranks above superfamily (the nomenclatural ***class-series***; Dubois, 2000), which is wrong. As stressed in Article 1.2.2, it does provide basic rules for them, but only deals with their availability and correctness, not with their taxonomic allocation and validity. The *Code* does not require them to follow the basic nomenclatural concepts such as onomatophores, homonymy, synonymy and priority. It therefore abandons this domain to 'consensus' and 'usage', which does not work: in the literature, countless higher taxa bear alternative nomina, and the same nomina are often used for widely distinct taxa that do not even overlap in intension and extension. This failure of the *Code* opens the door to alternative and incompatible (e.g., for being intensional instead of ostensional) nomenclatural systems, such as the *Phylocode*, to claim authority for the regulation of these nomina, a serious threat to the unity and universality of nomenclature.

During the history of nomenclature, several systems of nomenclature for higher taxa have been proposed (Dubois, 2006, 2011), but all of them have been shown to pose major problems, mostly because of the long period of absence of such rules and of the need of compatibility with the *Code*'s system, in particular regarding the use of ostension for the taxonomic allocation of nomina. Recently however, a comprehensive nomenclatural system for class-series nomina, which is ostensional and compatible with the rest of the *Code*, has been proposed (Dubois, 2005b, 2006, 2015) and applied with success to the higher nomenclature of recent amphibians (Dubois et al., 2021). It would be worth scrutiny and testing by taxonomists working on other groups.

## CONCLUSIONS

After a long period of stasis, the domain of zoological nomenclature is currently in turmoil, being submitted to contradictory centrifugal forces. In recent decades, various proposals have been made to modify the *Code* or to replace it with new nomenclatural systems (such as the *Phylocode* or the *Biocode*). However, as testified by the contents of journals such as *Zootaxa*, *Zookeys* and *Zoosystema*, most zootaxonomists still adhere to the *Code* and are not ready to replace it with another set of rules based on widely different rationales. As discussed above, the *Code* needs clarification and improvement, involving in particular a considerable reduction in the number of exceptions and an increased automaticity. In use, the *Code* should seek to avoid to the extent possible the necessity of time-consuming investigation to resolve nomenclatural issues; these have long kept taxonomists busy but do not address real scientific questions.

Various proposals to modify the current *Code* have been published recently. Two important aspects deserve particular attention: 'simplification' of the rules and proliferation of exceptions to the general rules.

The widespread idea that zoological nomenclature should be 'simplified' in order to be used 'easily' by laypersons without making the effort to understand its functioning, rules and language, is misguided and will lead, if uncritically pursued, to growing nomenclatural chaos. As shown in detail above, most of the complexity of the *Code* is not due to the whims of its founders and improvers, but to the need to address a wide range of basic questions in order to ensure an automatic, unambiguous and universal functioning of the rules. In fact, two different ideas are behind the wish to 'simplify' the *Code*.

As we have seen, the *Code* could indeed benefit from some simplifications alleviating the work of taxonomists. Two examples of possible improvements were discussed above. One would be to 'de-emphasize' the role and importance of authorship in zoological nomenclature. Authorship does not play any role in the Nomenclatural Process leading to nomina of taxa. The late discovery that the 'real' author of a nomen, as defined in Article 50, is different from that previously believed, in a work otherwise well-known, has no consequence for the availability, allocation or validity of this nomen. Removing the need to establish the 'real authors' of nomina in complex situations and replacing the mention of the name(s) of author(s) in the nominal-complex by that of its effective year of promulgation, would be a real improvement of the *Code* which would solve two problems discussed above under 'The Role of Characters in the Availability of Nomina'. The second example is that of the 'defetishisation' of some old syntypes through the methodology of virtual lectotype designation of lost syntypes described above under 'Type Specimens', which avoids laborious enquiries trying to establish the status of defective syntypes of doubtful taxonomic allocation.

In contrast, some recently proposed or implemented 'simplifications' of the *Code* in fact increase confusion and ambiguity. This includes, for example, proposals to abandon the strict criteria of availability of nomina to abide by 'common usage,' either to reject some nomina for so-called 'ethical' reasons or to allow validation of nomina or nomenclatural acts currently in use even though they had not been properly promulgated (e.g., published online without *Zoobank* promulgation; see e.g., Krell, 2015 and Dubois et al., 2015). The reluctance of the Commission to replace the confusing 'common language terminology' of the *Code* with well-defined unambiguous terms is also a consequence of this attitude favouring 'simplification'.

A second recent worrying tendency, well-illustrated in the 1999 edition of the *Code*, is that of multiplication of exceptions to the rules of the *Code* in particular cases. This is well exemplified by the new wording of Articles that removes the distinctions between well-defined categories of subsequent spellings (see 'Emendations vs. Incorrect Spellings' above) or the non-respect of the Principle of Priority in some cases in the family-series (see 'Zygoidy and Reversal of Precdence' above). Such exceptions to the general rules make it difficult for zoologists who are not well acquainted with the taxonomy of a zoological group to understand why, for example, the nomen of a family is junior to that of one of its subfamilies, or why an incorrect subsequent spelling or an unjustified emendation is treated as the valid nomen of a taxon. Furthermore, the increase of such exceptions will render more difficult the computerisation of the rules of the *Code*, which would be very useful for taxonomists of the future (see 'Terminology' above).

The Commission is working on a new edition of the Code, but little information is published about the progress of its work. In contrast, the Linz Zoocode

Committee publishes progressively proposals for improvements to the *Code* (e.g., Dubois et al., 2019, 2022b) which are open to public discussion long before their potential implementation. Hopefully, solutions will be proposed for nomenclatural problems confronting taxonomists but, if some of these proposals are not accepted by the community, the current trend towards implosion of the nomenclatural system might increase, which could have dramatic consequences on communication about taxa and thus on biology as a whole. Therefore, before its adoption and implementation, the new *Code* should be submitted to a fully open discussion leaving enough time for real debates in the community, followed by a democratic decision involving all interested taxonomists, not only a small group of persons.

## NOTES

1 The terms in ***bold italics*** and followed by an asterisk * in this work are defined in Saliba et al. (2021) <https://skosmos.loterre.fr/th63/fr/index/C?clang=en>.
2 The definition of the term 'type' in the Glossary of the *Code* ('A term used alone, or forming part of a compound term, to denote a particular kind of specimen or taxon') is fully ambiguous, as it applies both to nomenclatural and taxonomic concepts. The ambiguity is removed for the nomenclatural use of this concept only when combined with another root (e.g., in 'holotype') or term (e.g., in 'name-bearing type'). Many biologists believe that the function of type specimens is to provide 'typical characters'. This testifies to a typological approach to nomenclature and has been used repeatedly to argue that zoological nomenclature was an 'obsolete' discipline.
3 Exceptional reasons may be justified, but should require action by the Commission under Plenary Power and are not discussed here.
4 Which was established 150 years later!
5 The mere publication of such statements points to deficiencies not only in the writing of the paper but also in its reviewing, and to a more general problem which is that some journals should be avoided by taxonomists (Dubois, 2003; Dubois et al., 2024).
6 Even when 'crossability' is used as a criterion for the ***taxonomic*** recognition of distinct taxa (e.g., Félix et al., 2014), ***nomenclaturally*** the availability of their nomina is not provided by this criterion but may be so by the molecular sequences considered diagnostic of these taxa (e.g., Kiontke et al., 2011).
7 Particular problems, not discussed here, occur when the original onomatophore is lost, incomplete or damaged and as a consequence cannot be taxonomically allocated.
8 The only exception in this respect is the new Article 24.2.4 introduced in the 1999 edition (see 'Zygoidy and Reversal of Precedence').
9 In cases where the onomatophore is currently missing, the taxonomic allocation of the nomen can be done by 'consensus,' without subsequent fixation of an onomatophore. However, more and more the subsequent designation of an onomatophore becomes necessary to restrict the use of a nomen to one among several taxa that had been kept for two centuries or more under a single nomen but distinguished recently, for example by molecular studies.
10 Such subsequent designations were possible until 1999, but under the current edition of the *Code* a new genus nomen proposed for several nominal species without fixation among them of a type species is unavailable, even if these different species are considered (at a given moment of the history of taxonomy at least) to belong in the same taxonomic genus.
11 In exceptional situations, in which the preservation of specimens is impossible, it is necessary to have recourse to alternative solutions (Dubois and Nemésio, 2007; Ceríaco et al., 2016; Mills and Dubois, 2023).
12 This expression is misleading, as no 'second reviser action' ever exists.

## REFERENCES

Anonymous [International Commission on Zoological Nomenclature] (1999) *International Code of Zoological Nomenclature.* Fourth edition, International Trust for Zoological Nomenclature, London.

Anonymous [International Commission on Zoological Nomenclature] (2012) *Bulletin of Zoological Nomenclature*, 69: 161–169.

Anonymous [International Commission on Zoological Nomenclature] (2017) *Bulletin of Zoological Nomenclature*, 73: 96–97.

Anonymous [International Commission on Zoological Nomenclature] (2021) *Bulletin of Zoological Nomenclature*, 78: 42–45.

Bauer, A.M., Parham, J.F., Brown, R.M., et al. (2011) Availability of new Bayesian-delimited gecko names and the importance of character-based species descriptions. *Proceedings of the Royal Society* (B), 278: 490–492. https://doi.org/10.1098/rspb.2010.1330

Blanchard, R., ed (1905) *Règles internationales de la nomenclature zoologique adoptées par les Congrès Internationaux de Zoologie*, Rudeval, Paris.

Bruun, A.F. (1950) The *Systema Naturae* of the twentieth century. *Science*, 112: 342–343. https://doi.org/10.1126/science.112.2908.342-a

Cantino, P.D. and Queiroz, K. de (2020) *International code of Phylogenetic Nomenclature (PhyloCode).* CRC Press, Boca Raton.

Carvalho, M.R. de, Bockmann, F.A., Amorim, D.S., et al. (2005) Revisiting the taxonomic impediment. *Science*, 307: 353. https://doi.org/10.1126/science.307.5708.353b

Carvalho, M.R. de, Bockmann, F.A., Amorim, D.S., et al. (2007) Taxonomic impediment or impediment to taxonomy? A commentary on systematics and the cybertaxonomic-automation paradigm. *Evolutionary Biology*, 34: 140–143. https://doi.org/10.1007/s11692-007-9011-6

Carvalho, M.R. de, Bockmann, F.A., Amorim, D.S., et al. (2008) Systematics must embrace comparative biology and evolution, not speed and automation. *Evolutionary Biology*, 35: 150–157. https://doi.org/10.1007/s11692-008-9018-7

Ceríaco, L.M.P., Aescht, E., Ahyong, S.T., Ballerio, A., Bouchard, P., Bourgoin, T., Dmitriev, D., Evenhuis, N., Grygier, M.J., Harvey, M.S., Kottelat, M., Kluge, N., Krell, F., Kojima, J.-I., Kullander, S.O., Lucinda, P., Lyal, C.H.C., Pyle, R.L., Rheindt, F.E., Scioscia, C.L., Welter-Schultes, F., Whitmore, D., Yanega, D. Zhang, Z.-Q., Zhou, H.-Z. and Pape, T. (2023) Renaming taxa on ethical grounds threatens nomenclatural stability and scientific communication: Communication from the International Commission on Zoological Nomenclature. *Zoological Journal of the Linnean Society*, 197: 283–286, https://doi.org/10.1093/zoolinnean/zlac107

Ceríaco, L.M.P., Gutiérrez, E.E., and Dubois, A. (2016) Photography-based taxonomy is inadequate, unnecessary and potentially harmful for biological sciences. *Zootaxa*, 4196: 435–445. https://doi.org/10.11646/zootaxa.4196.3.9

Cheng, S.J., Gaynor, K.M., Moore, A.C., Darragh, K., Estien, C.O., Hammond, J.W., Lawrence, C., Mills, K.L., Baiz, M.D., Ignace, D., Khadempour, L., McCary, M.A., Rice, M.M., Tumber-Dávila, S.J. and Smith, J.A. (2023) Championing inclusive terminology in ecology and evolution. *Trends in Ecology & Evolution*, 38: 381–384. https://doi.org/10.1016/j.tree.2022.12.011.

Crisci, J.V. (2006a) One-dimensional systematists: Perils in a time of steady progress. *Systematic Botany*, 31: 217–221. http://www.jstor.org/stable/25064142

Crisci, J.V. (2006b) Making taxonomy visible. *Systematic Botany*, 31: 439–440. https://doi.org/10.1600/036364406777585829

Dayrat, B. (2005) Towards integrative taxonomy. *Biological Journal of the Linnean Society*, 85: 407–415. https://doi.org/10.1111/j.1095-8312.2005.00503.x

Dubois, A. (1999) Miscellanea nomenclatorica batrachologica. 19. Notes on the nomenclature of Ranidae and related groups. *Alytes*, 17: 81–100.

Dubois, A. (2000) Synonymies and related lists in zoology: General proposals, with examples in herpetology. *Dumerilia*, 4: 33–98.
Dubois, A. (2003) The relationships between taxonomy and conservation biology in the century of extinctions. *Comptes Rendus Biologies*, 326: S9–S21. https://doi.org/10.1016/s1631-0691(03)00022-2
Dubois, A. (2005a) Proposed rules for the incorporation of nomina of higher-ranked zoological taxa in the *International Code of Zoological Nomenclature*. 1. Some general questions, concepts and terms of biological nomenclature. *Zoosystema*, 27: 365–426.
Dubois, A. (2005b) Proposals for the incorporation of nomina of higher-ranked taxa in the *Code*. *Bulletin of Zoological Nomenclature*, 62: 200–209.
Dubois, A. (2006) Proposed rules for the incorporation of nomina of higher-ranked zoological taxa in the *International Code of Zoological Nomenclature*. 2. The proposed rules and their rationale. *Zoosystema*, 28: 165–258.
Dubois, A. (2008a) A partial but radical solution to the problem of nomenclatural taxonomic inflation and synonymy load. *Biological Journal of the Linnean Society*, 93: 857–863. http://dx.doi.org/10.1111/j.1095-8312.2007.00900.x
Dubois, A. (2008b) Phylogenetic hypotheses, taxa and nomina in zoology. *Zootaxa*, 1950: 51–86. https://doi.org/10.11646/zootaxa.1950.1.7
Dubois, A. (2010a) Retroactive changes should be introduced in the *Code* only with great care: Problems related to the spellings of nomina. *Zootaxa*, 2426: 1–42. http://dx.doi.org/10.11646/zootaxa.2426.1.1
Dubois, A. (2010b) Nomenclatural rules in zoology as a potential threat against natural history museums. *Organisms, Diversity and Evolution*, 10: 81–90. http://dx.doi.org/10.1007/s13127-010-0015-1
Dubois, A. (2010c) Zoological nomenclature in the century of extinctions: Priority vs. 'usage'. *Organisms, Diversity & Evolution*, 10: 259–274. http://dx.doi.org/10.1007/s13127-010-0021-3
Dubois, A. (2011) The *International Code of Zoological Nomenclature* must be drastically improved before it is too late. *Bionomina*, 2: 1–104. https://doi.org/10.11646/bionomina.2.1.1
Dubois, A. (2015) The duplostensional nomenclatural system for higher zoological nomenclature. *Dumerilia*, 5: 1–108.
Dubois, A. (2017a) A few problems in the generic nomenclature of insects and amphibians, with recommendations for the publication of new generic nomina in zootaxonomy and comments on taxonomic and nomenclatural databases and websites. *Zootaxa*, 4237: 1–16. https://doi.org/10.11646/zootaxa.4237.1.1
Dubois, A. (2017b) The nomenclatural status of *Hysaplesia, Hylaplesia, Dendrobates* and related nomina (Amphibia, Anura), with general comments on zoological nomenclature and its governance, as well as on taxonomic databases and websites. *Bionomina*, 11: 1–48. http://dx.doi.org/10.11646/bionomina.11.1.1
Dubois, A. (2017c) The need for reference specimens in zoological taxonomy and nomenclature. *Bionomina*, 12: 4–38. http://dx.doi.org/10.11646/bionomina.12.1.2
Dubois, A. (2017d) Diagnoses in zoological taxonomy and nomenclature. *Bionomina*, 12: 64–88. http://dx.doi.org/10.11646/bionomina.12.1.8
Dubois, A. (2018) The correct spelling of the nomen *Nyctimystes cheesmani* Tyler, 1964 (Amphibia, Anura), with some comments on Latin grammar and the rules of the *Code*. *Alytes*, 35: 75–84.
Dubois, A. (2020a) Allocation of nomina to taxa in zoological nomenclature. *Bionomina*, 18: 1–43. http://dx.doi.org/10.11646/bionomina.18.1.1
Dubois, A. (2020b) The status regarding publication date and availability of taxonomic works published online without proper *Zoobank* registration. *Bionomina*, 18: 44–55. http://dx.doi.org/10.11646/bionomina.18.1.2

Dubois, A. (2022a) Nomenclatural acts in zoological nomenclature. *Bionomina*, 31: 1–23. http://dx.doi.org/10.11646/bionomina.31.1.1
Dubois, A. (2022b) Kalyptoidy: The nomenclatural status of new zoological nomina originally published as synonyms, with examples in herpetology and comments on taxonominal databases. *Bionomina*, 30: 1–82. http://dx.doi.org/10.11646/bionomina.30.1.1
Dubois, A. (2025) Zootaxonomy in the century of extinctions: Time for field work and collections. This volume.
Dubois, A. and Aescht, E. (eds) (2019) LZC Session 33. The principles of the *Zoocode*. 15. The principle of sozoidy. *Dumerilia*, 8: 133–142.
Dubois, A., Aneesh, P.T., Bauer, A.M., et al. (2022a) The Linz Zoocode project. Second report of activities (2020). Nomenclatural availability. 1. What is nomenclatural availability? *Bionomina*, 28: 1–17.
Dubois, A., Aneesh, P.T., Bauer, A.M., et al. (2022b) The Linz Zoocode project. Fifth report of activities (2022). Nomenclatural availability. 4. Electronic publication. *Bionomina*, 28: 71–119. http://dx.doi.org/10.11646/Bionomina.28.1.1
Dubois, A., Bauer, A.M., Ceríaco, L.M.P., et al. (2019) The Linz *Zoocode* project: A set of new proposals regarding the terminology, the principles and rules of zoological nomenclature. First report of activities (2014–2019). *Bionomina*, 17: 1–111. http://dx.doi.org/10.11646/bionomina.17.1.1
Dubois, A., Bour, R. and Ohler, A. (2015) Nomenclatural availability of preliminary electronic versions of taxonomic papers: In need of a clear definition. *Bulletin of Zoological Nomenclature*, 72: 252–265. http://dx.doi.org/10.21805/bzn.v72i3.a1
Dubois, A., Crochet, P.-A., Dickinson, E.C., et al. (2013) Nomenclatural and taxonomic problems related to the electronic publication of new nomina and nomenclatural acts in zoology, with brief comments on optical discs and on the situation in botany. *Zootaxa*, 3735: 1–94. http://dx.doi.org/10.11646/zootaxa.3735.1.1
Dubois, A., Denzer, W., Entiauspe-Neto, O., et al. (2024) Nomenclatural problems raised by the recent description of a new anaconda species (Squamata, Serpentes, Boidae), with a nomenclatural review of the genus *Eunectes*. *Bionomina*, 37: 8–58. http://dx.doi.org/10.11646/bionomina.37.1.2
Dubois, A. and Frétey, T. (2016) A new nomen for a subfamily of frogs (Amphibia, Anura). *Dumerilia*, 6: 17–23.
Dubois, A. and Frétey, T. (2023) Problems with the availability of zoological nomenclatural acts published online: The case of the *Journal of Herpetology*. *Bionomina*, 34: 9–34. http://dx.doi.org/10.11646/bionomina.34.1.2
Dubois, A., Frétey, T., and A. Ohler (2018) The *Relictus* case: It is high time that taxonomists follow the *Code*'s requirements for nomenclatural availability and validity of new zoological nomina. *Bionomina*, 13: 51–64. http://dx.doi.org/10.11646/bionomina.13.1.4
Dubois, A. and Nemésio, A. (2007) Does nomenclatural availability of nomina of new species or subspecies require the deposition of vouchers in collections? *Zootaxa*, 1409: 1–22. http://dx.doi.org/10.11646/zootaxa.1409.1.1
Dubois, A. and Ohler, A. (1997a) Early scientific names of Amphibia Anura I. Introduction. *Bulletin du Muséum National d'Histoire Naturelle*, 18(4): 297–320.
Dubois, A. and Ohler, A. (1997b) Early scientific names of Amphibia Anura II. An exemplary case: *Rana arborea* Linnaeus, 1758. *Bulletin du Muséum National d'Histoire Naturelle*, 18(4): 321–340.
Dubois, A. and Ohler, A. (2018) The *Hyla quoyi-Hyla prasina* case (Amphibia, Anura), with comments on bibliographic and taxonomic databases and on Article 23.9 of the *Code*. *Zoosystema*, 40: 501–506.
Dubois, A. and Ohler, A. (2021) *Bionomina*, a forum for biological terminology and nomenclature: The first 10 years. *Bionomina*, 21: 1–21. http://dx.doi.org/10.11646/bionomina.21.1.1

Dubois, A., Ohler, A. and Pyron, R.A. (2021) New concepts and methods for phylogenetic taxonomy and nomenclature in zoology, exemplified by a new ranked cladonomy of recent amphibians (Lissamphibia). *Megataxa*, 5: 1–738. http://dx.doi.org/10.11646/megataxa.5.1.1

Ereshefsky, M. (2001) *The Poverty of the Linnaean Hierarchy*. Cambridge University Press, Cambridge.

Félix, M.-A., Braendle, C. and Cutter, A.D. (2014) A streamlined system for species diagnosis in *Caenorhabditis* (Nematoda: Rhabditidae) with name designations for 15 distinct biological species. *PLoS One*, 9(4) [e94723]: 1–9. https://doi.org/10.1371/journal.pone.0118327

Gillman, L.N. and Wright, S.D. (2020) Restoring indigenous names in taxonomy. *Communications Biology*, 3(609): 1–3. https://doi.org/10.1038/s42003-020-01344-y

Gould, S.J. (1990) Bully for *Brontosaurus*. A minor victory snatched from the jaws of taxonomic triviality. *Bulletin of Zoological Nomenclature*, 47: 88–96. https://doi.org/10.5962/bhl.part.2673

Hammer, T.A. and Thiele, K.R. (2021) Proposals to amend Articles 51 and 56 and Division III, to allow the rejection of culturally offensive and inappropriate names. *Taxon*, 70: 1392–1394. http://dx.doi.org/10.1002/tax.12620

Hennig, W. (1966) *Phylogenetic Systematics*. University of Illinois Press, Urbana.

Hennig, W. (1974) Kritische bemerkungen zur frage 'Cladistic Analysis or Cladistic Classification.' *Zeitschrift für zoologische Systematik und Evolutionsforschung*, 12: 279–294.

Hołynski, R.B. (1994) Structure and function; or, what kind of nomenclatural regulations do we need? *Crystal*, Göd, Hungary, (ser. Zool.), 2: 1–50.

Jiménez-Mejías, P., Manzano, S., Gowda, V., et al. (2024) Protecting stable biological nomenclatural systems enables universal communication: A collective international appeal. *BioScience*, 74: 467–472. https://doi.org/10.1093/biosci/biae043

Kaiser, H., Crother, B.I., Kelly, C.M.R., et al. (2013) Best practices: In the 21st century, taxonomic decisions in herpetology are acceptable only when supported by a body of evidence and published via peer-review. *Herpetological Review*, 44: 8–23.

Kiontke, K.C., Félix, M.-A., Ailion, M., et al. (2011) A Phylogeny and molecular barcodes for *Caenorhabditis*, with numerous new species from rotting fruits. *BMC Evolutionary Biology*, 11(339): 1–18. https://doi.org/10.1186/1471-2148-11-339

Kluge, A.G. (2005) Taxonomy in theory and practice, with arguments for a new phylogenetic system of taxonomy, pp. 7–47. In *Ecology and Evolution in the Tropics: A Herpetological Perspective*. (eds M.H. Donnelly et al.). University of Chicago Press, Chicago.

Knapp, S., Vorontsova, M.S. and Turland, N.J. (2020) Indigenous species names in algae, fungi and plants: A comment on Gillman & Wright (2020). *Taxon*, 69: 1409–1410. http://dx.doi.org/10.1002/tax.12411

Knox, E.B. (1998) The use of hierarchies as organizational models in systematics. *Biological Journal of the Linnean Society*, 63: 1–49. https://doi.org/10.1111/j.1095-8312.1998.tb01637.x

Krell, F.-T. (2015) A mixed bag: When are early online publications available for nomenclatural purposes? *Bulletin of Zoological Nomenclature*, 72: 19–32.

Melville, R.V. (1958) Document 27/9. Objection to the introduction of a "Principle of Conservation" into the "Règles". *Bulletin of Zoological Nomenclature*, 15: 1247–1250.

Mills, C.E. and Dubois, A. (2023) The taxonominal status of the nomina *Lesueuria* Milne Edwards, 1831 and Lesueuriidae Chun, 1880, and introduction of a new genus and a new family for *Lesueuria pinnata* Ralph & Kaberry, 1950, as well as an additional new species of the new genus (Ctenophora, Lobata). *Bionomina*, 36: 1–35. http://dx.doi.org/10.11646/bionomina.36.1.1

Minelli, A. (2000) The ranks and the names of species and higher taxa, or a dangerous inertia of the language of natural history, pp. 339–351. In *Cultures and Institutions of Natural History: Essays in the History and Philosophy of Sciences* (eds M.T. Ghiselin and A.E. Leviton). California Academy of Sciences, San Francisco.

Mosyakin S.L. (2021) Proposals to amend Recommendation 7A on deposition of type material in institutions of countries of origin, and to add a new Recommendation 51A regarding avoiding potentially inappropriate or unacceptable names of taxa. *Taxon*, 70: 1379–1380. http://dx.doi.org/10.1002/tax.12606

Mosyakin S.L. (2022a) If "*Rhodes-*" must fall, who shall fall next? *Taxon*, 71: 249–255. https://doi.org/10.1002/tax.12659

Mosyakin S.L. (2022b) Defending Art. 51 of the Code: Comments on Smith & al. (2022). *Taxon*, 71: 1141–1150. http://dx.doi.org/10.1002/tax.12820

Ng, P.K.L. (1994) The citation of species names and the role of the author's name. *Raffles Bulletin of Zoology*, 42: 509–513.

Palma, R.L. and Heath, A.C.G. (2021) Science versus vernacular: Should some taxa of animals and plants be renamed according to 'indigenous' practices? *Bionomina*, 22: 1–7. https://doi.org/10.11646/bionomina.22.1.1

Pavlinov, I.Y. (2014) Taxonomic Nomenclature. Book 2. From Linnaeus to First Codes. *Zoologitsheskie Issledovaniya Zoological Research*, 15: 1–223. [In Russian].

Pethiyagoda, R. (2023) Policing the scientific lexicon: The new colonialism? *Megataxa*, 10: 20–25. https://doi.org/10.11646/megataxa.10.1.4

Pleijel, F. and Rouse, G.W. (2003) Ceci n'est pas une pipe: Names, clades, and phylogenetic nomenclature. *Journal of Zoological Systematics and Evolutionary Research*, 41: 162–174. https://doi.org/10.1046/j.1439-0469.2003.00236.x

Raikow, R.J. (1985) Problems in avian classification, pp. 187–212. In R.F. Johnston (ed.), *Current Ornithology*, Vol. 2, Plenum, New York.

Raposo, R.A., Kirwan, G.M., Lourenço, A.C.C., et al. (2021) On the notions of taxonomic 'impediment', 'gap', 'inflation', and 'anarchy', and their effects on the field of conservation. *Systematics & Biodiversity*, 19: 296–311. http://dx.doi.org/10.1080/14772000.2020.1829157

Rivas, J.A., Quintana, P. de la, Mancuso, M., et al. (2024) Disentangling the Anacondas: Revealing a new green species and rethinking yellows. *Diversity*, 16 (2) [127]: 1–29. https://doi.org/10.3390/d16020127

Saliba, E.M., Vignes-Lebbe, R., Dubois, A. and Ohler, A. (2021) ZooNom: An online thesaurus for alleviating ambiguity in the terminology of zoological nomenclature. *Zootaxa*, 5071: 42–50. https://doi.org/10.11646/zootaxa.5071.1.2

Simpson, G.G. (1940) Types in modern taxonomy. *American Journal of Science*, 238: 413–431.

Smith, A.B. (1988) Patterns of diversification and extinction in early Palaeozoic echinoderms. *Palaeontology*, 31: 799–828.

Smith, G.F. and Figueiredo, E. (2021a) Proposal to add a new Article 61.6 to permanently and retroactively eliminate epithets with the root *caf[e]r* or *caff[e]r* from the nomenclature of algae, fungi and plants. *Taxon*, 70: 1395–1396. https://doi.org/10.1002/tax.12622

Smith G.F. and Figueiredo, E. (2021b) "*Rhodes-*" must fall: Some of the consequences of colonialism for botany and plant nomenclature. *Taxon*, 71: 1–5. https://doi.org/10.1002/tax.12598

Smith G.F., Figueiredo, E., Hammer, T.A. and Thiele, K.R. (2022) Dealing with inappropriate honorifics in a structured and defensible way is possible. *Taxon*, 71: 933–935. https://doi.org/10.1002/tax.12742

Sundberg, P. and Pleijel, F. (1994) Phylogenetic classification and the definition of taxon names. *Zoologica Scripta*, 23: 19–25.

Thiele K.R., Smith, G.F., Figueiredo, E. and Hammer, T.A. (2022) Taxonomists have an opportunity to rid botanical nomenclature of inappropriate honorifics in a structured and defensible way. *Taxon*, 71: 1151–1154. http://dx.doi.org/10.1002/tax.12821

Valdecasas, A.G., Williams, D.M. and Wheeler, Q.D. (2008) 'Integrative taxonomy' then and now: A response to Dayrat (2005) *Biological Journal of the Linnean Society*, 93: 211–216. http://dx.doi.org/10.1111/j.1095-8312.2007.00919.x

Wheeler, Q.D. (2004) Taxonomic triage and the poverty of phylogeny. *Philosophical Transactions of the Royal Society of London*, 359: 571–583. https://doi.org/10.1098/rstb.2003.1452

Wheeler, Q. (2024) *Species, Science and Society. The Role of Systematic Biology*. Routledge, Abingdon.

# 10 The Survival of Taxonomy and the Digitization of Natural History Collections*

*Evgeny Mavrodiev, Manuel B. Crespo, and David M. Williams*

## INTRODUCTION

The massive digitization projects of Natural History collections of either living or extinct organisms, specifically those conserved in herbaria, are today considered an unambiguously positive phenomenon (e.g., Blagoderov et al., 2012, p. 143; Sweeney et al., 2018). The availability of many collections in the form of high-quality digital images will allow a significant increase in the efficiency of the work of a botanist. For example, in many cases, there will now be no need to request specific collections or to make long trips to work with herbarium specimens. Digitization of herbarium specimens, which will eventually combine images obtained from various herbaria into giant databases, will make it possible to include multiple collections in various analyses that are not directly related to taxonomic goals and thus will expand the functionality of collection repositories. Geo-referencing of the digitized samples will make it possible to utilize specimen collections in ecological, morphological and climatic bio-informatic studies. Digitized samples can now be the target of various artificial intelligence (AI)-based studies. Translating collections into a digital format will protect the actual specimens from accidental destruction, such as fire or other disasters, and simultaneously preserve our rapidly extinct biodiversity. The list of possible benefits of digitization goes on. Popular explanations of 'Why digitization matters' can be found today on the Internet, in the pages of scientific and popular journals, Museum exhibitions, etc.

Although numerous biological collections still need to be digitized, the process is already taking place on 'an industrial scale' following a 'high-throughput conveyor system' (Blagoderov et al., 2012, p. 133; Nelson et al., 2015; Sweeney et al., 2018, p. 166), even if it remains unclear how to proceed with the millions of digitized collections for global consumption (Moore, 2024). As summarized in Miralles et al., '91% of the 5.5 million plant specimens deposited in the world's largest herbarium (MNHN in Paris) have already been photographed at high resolution and made

* To Arthur Anker, taxonomist.


DOI: 10.1201/b22822-10

available online in less than a decade [...]' (Miralles et al., 2020, pp. 1234, 1241; for details on MNHN, Paris, see LeBras et al., 2017). Nevertheless, it is hard to discover the number of publications that have actually utilized these data, although some reviews are available (cited below).

The topic of massive digitization has been intensively promoted. For example, the paper 'Digitization of The New York Botanical Garden Herbarium' (Thiers et al., 2016) has received over 9,742 citations (Google Scholar accessed 24th March 2025). Still, less than 6% of these citations coincide with those found in the Web of Science (accessed 24th March 2025).

There are some doubts concerning the effectiveness of converting biological collections into electronic form. These doubts are largely based on the correctness of identification. Consider one example. Goodwin et al. evaluated 'the accuracy of names associated with 4,500 specimens of African gingers from 40 herbaria in 21 countries' and found '[...] that at least 58% of the specimens had the wrong name prior to a recent study' (Goodwin et al., 2015, p. R1066). As summarized in Konstantinova and Namyatova, a preliminary revision of the genus *Euscelidia* (Insecta, Diptera, Asilidae) showed that only 26% of the specimens in global collections were identified to species and a remarkable 83% were misidentified (Konstantinova and Namyatova, 2019, p. 341). Thus, misidentification is a fundamental issue.

However, misidentification is only part of a much larger problem that should have been discussed decades ago, before digitization began. Summarized in a few words, this is the lack of a positive relationship between taxonomy and large-scale digitization programs. Because of this, within the current digital reality, the science behind traditional taxonomy may not effectively survive. Below we outline arguments in favour of this proposition. Our first goal will be to avoid optimistic solutions. We believe that digitization is an irreversible phenomenon that will eventually place numerous biological disciplines, as we understand them today, in a new reality that must be considered in its proper context. Sketching this context is our second goal.

Of course, we do not want to suggest that taxonomists cannot use and benefit from the digital images in their work. Instead, we suggest that some of the protocols should be changed – that is, *limiting digitization solely to authentic collections would be a major improvement.*

## THE RELATIONSHIP BETWEEN TAXONOMISTS AND DIGITAL DATABASES

First, consider some words from the remarkable Russian taxonomist Nikolai Tzvelev (1925–2015; urn:lsid:ipni.org:authors:10947-1): 'Plant species, as they are in nature, are not sorted into folders as in the Herbarium'. This simple passage is hard to misunderstand: the language of traditional taxonomy does not cover the extent of observed variability. A fuller understanding is entirely different. The point being made is that any taxonomic work rarely achieves its stated goal, regardless of whether current taxonomy can cover the natural variation in formal terms (the latter point will be discussed later). In other words, for a given set of taxa there will always be some differences to those found in any set of herbarium folders. In general, the composition of herbarium folders and their labels has always been considered temporary, even if the illusion of constancy arises from the fact that nothing may actually change for decades. Taxonomic revisions are, in principle at least, in a state of continuously changing—revisions are the lifeblood of taxonomy.

Like many others, Miralles et al. have pointed out that ideally, taxonomic repositories should be able to handle universally unique "identifiers" that directly refer to specimens (e.g., Miralles et al., 2020, p. 1231, 1233-1234). They continue: '[…] For data reuse to be encouraged and facilitated in taxonomy and by its end users, emphasis should be on making data and metadata available in highly standardized formats, enhancing comparability across taxonomic studies. Metadata should thus include a specimen identifier in best-practice format for the respective group of organisms, in addition to a species-level name (accepted or candidate species) […]' (Miralles et al., 2020, p. 1245).

This passage begs a few questions: Who will establish 'comparability across taxonomic studies'? Who divided species into 'accepted' and 'candidates'? And even if today's large-scale efforts such as 'World Flora Online' (WFO, https://www.worldfloraonline.org/) or 'Plants of the World Online' (POWO, https://powo.science.kew.org/) (see Schellenberger Costa et al., 2023 for some comparisons) are already providing a role as World taxonomic authorities, it is not entirely satisfactory. It reflects nothing more than the requirement of someone or some institution acting as an authority, a person or persons that have already made 'infallible' final decisions. Thus, the assignment of a '(unique) specimen identifier' is an attempt to make something permanent from what is, or may be, temporary. Any currently 'accepted taxon' can, with time, be divided into many or synonymized, that is, entirely forgotten due to its inclusion in the circumscription of another taxon. Related sets of collections can be re-identified accordingly. Unique identifiers for the specimens can include georeferenced data or collection time, but not the 'final' taxonomy – because the latter simply does not exist; even within WFO or POWO. *Only the labels of effectively designated authentic collections can be technically assumed as something permanent* (e.g., for digitization purposes) – but even this assumption is not entirely realistic. There is always the possibility of taxonomic transfer of a species from one genus to another. It may seem that Miralles et al. understand and acknowledge the latter possibility: 'The assignment of species names to taxonomic data is secondary because these names are bound to change over time.' (Miralles et al., 2020, p. 1247). However, they continue: 'Yet, to facilitate their retrieval, data should be associated as much as possible with accepted and valid genus and species names' (Miralles et al., 2020, p. 1247). Such a link is actually strong, as in the Appendices of the cited paper. Our authors were even surprised that the International Code of Zoological Nomenclature does not require individual identifiers even for type specimens (Miralles et al., 2020, p. 1241).

Thus, when herbarium specimens are digitized, the temporary nature of the herbarium folders and the labels of the specimens are inevitably presumed permanent. Why so? First, taxonomists cannot be expected to go through and re-define (re-name) the thousands of digitized collections effectively. As Hussein et al. have indicated, 'The current challenges and the volume of specimens make it almost impossible for botanists to execute their tasks in a reasonable time' (Hussein et al., 2022, p. [2], they do go on to suggest that 'images of the specimen on the herbarium mounting sheet would greatly help botanists and taxonomists working in herbaria').

Second, there is a limited possibility of the acceptance of massive changes in the metadata of digitized specimens, even if these changes are technically possible.

In other words, after the digital revolution in Natural History, there is limited expectation that taxonomists will promote their conclusions through digital repositories.

Consider this example: as summarized by Alexeev and Mavrodiev (2017, p. 60), the first edition of *Flora europaea* (published between 1964 and 1980) lists nine species of the lovegrass (Angiosperms, Poaceae, *Eragrostis* Wolf). In a monograph of the genus, published 22 years later (Portal, 2002), for only Western European countries, 90 species of *Eragrostis* were listed (and some of these have two or three subspecies). According to recent accounts, more than 40 species of *Eragrostis* have been identified in the UK and more than 30 in Belgium (Alexeev and Mavrodiev, 2017, for references). Imagine, then, that in the 1980s, following the conclusions of the authoritative *Flora Europaea*, all the specimens in European herbaria of *Eragrostis* were carefully arranged into nine folders, each corresponding to one of the nine recognized species, their contents digitized, and the results placed all into one database that includes tens of thousands of specimen collections. How would it be possible to change the content of this database in 2002, when the number of species in Western Europe is recognized as 90? In other words, how would later taxonomic conclusions be included in the existing digital database? It is true that herbaria are not 'storage repositories' of studies finished some time ago, and the science of taxonomy is 'an ever-evolving discipline', as Edwards et al. (2024, p. 826) recently pointed out, but Linnaeus already made the same point. The question, however, is how taxonomists can rearrange thousands of digitized collections from nine virtual folders to 90.

## HOW DOES TAXONOMY WORK?

As far as we are aware, no comprehensive analyses of the structure of applied aspects of taxonomic practice have recently been made. One factor is true: most taxonomic treatments are based on a limited (albeit sometimes significant) number of specimens. Consider the current 'foundational' monograph on the genus *Ipomoea* L. (Angiosperms, Convolvulaceae), which is 'the most compelling example of modernizing and expediting monography via large collaborative teams' (Davis, 2023, p. 414). It is essentially based on (a) the molecular study of ca. 1600 herbarium specimens; and (b) primarily on the extensive collections of *Ipomoea* held at Kew Gardens (K) and the Natural History Museum, London (BM), even if data were requested from 72 herbaria worldwide (Muñoz-Rodríguez et al., 2019).

The taxonomic conclusions (e.g., the revised combinations and new binomials) may require decades to be accepted (or rejected) by the Natural History community. There is no uniformity of opinion: one can only speak of informal consensus regarding taxonomic names or systems, again (usually) formed over decades. In that sense, operating with specimen collections, taxonomists will still speak this non-formal language. But this non-formality results from the nature of Natural History collections, which systematists constantly strive for clarity of expression. In other words, taxonomy works primarily through multiple years of informal consensus of opinions that may or may not be reflected in the content and order of any particular biological collection. Digitization may destroy this non-formality and, in this way, inadvertently formalize an unrequired ambiguity. It happens simply because of the open architecture of current digital databases. Taxonomic solutions, many of which may be ambiguous, assume to be a final 'decision' by non-experts – even if the metadata of every collection can, in principle, be changed.

Rather than the taxonomic collections themselves, it is the taxonomic *conclusions* (presented in monographs, treatments, papers, floristic and faunistic inventories, etc.) that are the *primary* objects of future analyses (for use in the fields of ecology or biogeography, for example). The meaning behind most taxonomic works is to turn the rhetorical into the practical. Looking at the artificial example above, what would be the impetus to publish a monograph on *Eragrostis* in 2002 if (a) all the existing collections of 'love grass' are already digitized following the '*Flora europaea*', published in 1980, (b) are already in wide use (e.g., because they are georeferenced, the specimens are may now be involved in numerous niche modeling studies), and (c) there is no process to allow new taxonomic conclusions to be appended to the existing digital specimens?

Muñoz-Rodríguez et al. (2019) suggested a minimum estimate of the number of specimens in *Ipomoea* in the World's herbaria is ca. 200,000. It would be useful to understand the protocol implemented in propagating their taxonomic conclusions in these 200,000 specimens, if all the latter were digitized in the virtual Mega herbarium of the future.

## HOMONYMY AND NON-MONOPHYLY

> 'Are reports of the death of taxonomy an exaggeration?'
>
> *Wheeler, 2014*

True homonyms, as defined in the current *International Code of Nomenclature for Algae, Fungi, and Plants*, are taxa to which the same name has been validly and effectively published: 'A name spelled exactly like another name published for a taxon at the same rank based on a different type' (from the glossary in Turland et al., 2018). In addition to the term being defined this way, we have adopted a broader use of 'homonymy'. Misidentification being the first reason there is homonymy of taxonomic names. However, the lack of taxonomic revision and phylogenetic studies (i.e., the lack of demonstrated taxon monophyly) may also be a reason for any name ambiguity (either explicit or cryptic).

As Ebach and Williams pointed out

> 'A taxon is aphyletic when it is deemed to be non-monophyletic or unresolved [...] A problem arises in systematics when taxonomic names assigned to aphyletic taxa are treated as if they were natural groups. In the absence of a taxonomic and systematic revision, anomalous taxa should be labelled as aphyletic without recourse to phylogenetic inference (i.e., interpretation). Doing so avoids the validation of aphyletic names and the creation of dubious results in fields that rely on systematic and taxonomic data'
>
> *Ebach and Williams, 2010, p. 123*

Thus, non-monophyletic groups are taxonomically problematic and cannot be treated as if they were natural – and taxonomic revisions are frequently used to establish non-monophyly, even before any formal phylogenetic study has been undertaken.

Technically, taxonomic revisions can be performed by the elementary procedure of resorting herbarium folders, changing their order and composition (simply by shifting specimens from one folder to another) and re-writing herbarium labels.

The standard redefinition of specimens by the taxonomist implies the movement of folders and herbarium sheets as described above. However, any taxonomic revision is not necessarily the final word. It would be a catastrophe for botany if Engler's *Das Pflanzenreich: regni vegetablilis conspectus* (reviewed in Davis, 1957), implemented in numerous herbarium collections, was assumed to be final. The same is true of the *Flora of the USSR*, one of the most significant scientific projects in botany of the twentieth century (Sytin and Slastunov, 2021), if it framed all subsequent digital botanical studies of Eurasian flora. Thus, the composition of herbarium folders and the associated herbarium labels are always considered temporary. For example, in most herbaria, collections are repeatedly redefined by different taxonomists. Each of these redefinitions implies a change in the content of the folder, which, of course, does not always occur for technical reasons. Strictly speaking, even after the work of an experienced taxonomist (e.g., a monographer of any given genus or family), there is no reason that a herbarium manager will change the folder order or the labels content immediately.

In some cases, this may be done after the taxonomist redefines and re-names a certain number of specimens. In other cases, it will not be done at all simply because later another study will be published that conflicts with the redefinitions that have since been made. In effect, one herbarium may accept some taxonomic conclusions and rearrange their specimen folders accordingly, while others may place rearrangements on hold. Because of this, the meaning behind the same name frequently differs in different herbaria. For example, does the name '*Iris hexagona* Walter' (Angiosperms, Iridaceae) (Mavrodiev et al., 2021) mean the same in all collections in US herbaria?

Yet digitization projects will almost certainly imply that the union of digitized collections in these giant repositories (for example, the various virtual Consortia of Herbaria) integrates the data from multiple collection storages. Johnson et al. wrote, 'Integration of the world's natural history collections can provide a resource for decision-makers' (Johnson et al., 2023, p. 1192) – but logically, in what form are the collections really 'integrated'? It is only the digital images, as that is the primary goal of these projects (Johnson et al., 2023). *Therefore, there is no certainty that the same taxon names will have the same meaning across digital Consortia.*

Consider the lectotype of the almost forgotten Japanese cattail *Typha japonica* Miq. (Angiosperms, Typhaceae), mistakenly treated as a synonym of '*T. orientalis* C.Presl' (Mavrodiev et al., 2023). It has been digitized and stored in the Herbarium of Naturalis Biodiversity Center (L, Leiden, the Netherlands) under the name *T. angustifolia* L. (sect *Typha*) assigned by an unknown botanist, even if morphologically *T. japonica* appears more similar to *T. latifolia* L. (sect. *Ebracteolatae*) (see Kapitonova et al., 2024). Is the meaning of the name '*T. angustifolia*' the same in the digital collections of the Herbarium of Naturalis Biodiversity Center (L) and the Moscow State University Herbarium (MW), which likely contain no samples of *T. japonica* under any name (EVM, pers. obs.)? And if not, how can someone unite the digital images and metadata of both L and MW into a single database? The same applies to the binomials '*T. orientalis*' and '*T. angustifolia*', the meaning of which is different in various herbaria worldwide (Mavrodiev and Kapitonova, 2016; Mavrodiev et al., 2023). In other words, as in the case of *T. japonica*, various herbaria frequently store specimens with a different morphology, ecology, and distribution under the same name (under the binomial '*T. angustifolia*' in this example).

A further example is the binomial '*Stipa krylovii* Roshev.' (Angiosperms, Poaceae), which is now accepted among Russian botanists (Tzvelev and Probatova, 2019). It is related to both Siberian and Chinese plants. However, Paul Smirnov, an eminent *Stipa* taxonomist, pointed out that the type of this name is *S. sareptana* A.K. Becker, and therefore the Transbaikalian specimen of *S. krylovii*, should be named '*S. decipiens* P.A. Smirn.' (Smirnov, 1967). Thus, according to Smirnov, *S. krylovii* represents a mixture of two species: *S. sareptana* and *S. decipiens*. Tzvelev and Probatova (2019, p. 450) considered the latter name to be a synonym of *S. krylovii*. However, Chinese botanists circumscribed *S. krylovii* with *S. sareptana* at the rank of subspecies (or variety) but ignored the name '*S. decipiens*' (Wu and Phillips, 2006). *Stipa sareptana* A.K. Becker var. *krylovii* (Roshev.) P.C. Kuo and Y.H. Sun. is listed as the synonym of *S. krylovii* by Tzvelev and Probatova (2019, p. 450). From this complex of names, is the meaning of '*S. krylovii*', '*S. decipiens*', and '*S. sareptana*' the same across all Russian and Chinese herbaria? And if not, how can collections of their digital images be integrated into a single database? Similarly, the name '*Apocynum sibiricum*', widely used in the past by both Eurasian and North American botanists, was applied to distinct species from two genera and this taxonomic homonymy was resolved only recently (Mavrodiev et al., 2024). There are many more examples of this kind.

Of course, the results of any phylogenetic analysis should be considered preliminary. The number of imperfectly known phylogenies is negligible compared to the total number of genera and families of living organisms. It is an illusion to assume that every genus of living organisms can be the target of one or another phylogenetic analysis. The possibility of obtaining a resolved phylogeny for every group of organisms may indeed be part of the same illusion. Suffice to say, our phylogenetic knowledge is very limited. In addition, many taxonomic treatments are undertaken without any connection to phylogenetic analysis. Thus, today, most species, genera, and families etc. of different living organisms remain aphyletic (see above) – and, perhaps, will stay aphyletic for a very long time and their related collections cannot be meaningfully digitized.

Bebber et al. estimated that at least '[…] 70,000 species [are] still to be described, more than half already have been collected and are stored in herbaria' (Bebber et al., 2010, p. 22169). They also emphasize '[…] the need for widened access to global collections through the exchange and large-scale digitization of existing specimens' (Bebber et al., 2010, pp. 22170–22171). The first statement implies that numerous current botanic binomials that correspond to specimens in existing collections are treated as homonyms and, therefore, the corresponding species are likely to be non-monophyletic. But if so, how does the digitization of related collections help with future studies?

In summary, most taxa from all Kingdoms of Life are yet to be the object of a phylogenetic study or revised taxonomically. The described homonymy of the taxonomic names within digital collection databases intensifies this issue dramatically. It also suggests that the numerous analyses of non-monophyletic taxonomic entities place more than 60 years of phylogenetic systematics into question.

What does this mean to the working taxonomist? It is necessarily crucial if one or another herbarium keeps the order of their folders unchanged and ignores any recent monographs. But this will not be the case if a searchable digital repository includes numerous taxonomic homonyms.

## THE DIGITIZATION OF NATURAL HISTORY COLLECTIONS AND ARTIFICIAL INTELLIGENCE (AI)

> 'If we could capture in ordinary words, that is, in logical concepts, what is expressed by beauty, then we would precisely say it differently than aesthetically, thereby turning aesthetic experience into something unnecessary, into a simple duplicate of ordinary, logically expressible knowledge. Beauty expresses what is inexpressible logically and is expressible only aesthetically'
>
> *Frank, 1956*

Digitized Natural History collections inevitably involve the subject of specimen identification using artificial intelligence (AI) systems. Some earlier commentary on the subject of automatic identification can be found in Gaston and O'Neill (2004) and MacLeod et al. (2010); and in more recent contributions from Davis (2023), Hussein et al. (2022), Goëau et al. (2022), Karbstein et al. (2024), and Valdecasas (2024). As summarized by Davis, 'This is an intellectual frontier for biodiversity and computer scientists and for the large online amateur naturalist community. The promise of such synergistic efforts was demonstrated recently by the PlantCLEF2022 (https://www.imageclef.org/PlantCLEF2022, Goëau et al., 2022) challenge targeted at automating the identification of nearly 100,000 plant species' (Davis, 2023, p. 415, see Goëau et al., 2022). However, the application of names to specimens that require identification assumes a taxonomist was involved at some point with the identification of the material in the first place – moreover, not just an expert taxonomist, but a competent taxonomist in the particular group in question (Löbl et al., 2023). This statement would apply to many studies in ecology, climatology, morphology, and other fields that might require the use of digitized collections.

Therefore, what is essential in the procedures for AI machine learning is not the total number of processed images downloaded from one or another search engine, or from one or another Consortia of digital repositories, but the accuracy of the primary data. That is, the set of samples which the AI machines learnt from – for which only a taxonomist would be able to select appropriate specimens.

Before one is willing to perform the automated 'identification of nearly 100,000 plant species' (Davis, 2023; Goëau et al., 2022), a more pressing matter is: 'How many highly qualified experts-taxonomists will be involved in this procedure?' (no taxonomic expert was acknowledged in Goëau et al., 2022); additionally, 'How far can these experts go using the available material, not yet digitized?' These questions are rarely addressed perhaps because (a) the number of qualified taxonomists remains relatively small, (b) the material is frequently of poor quality, or (c) the herbarium is not actually accessible. It is worth stressing again that not all plant species can have their diagnostic characters successfully preserved through the standard drying and pressing procedures, and this is frequently unavailable to any AI machines (see above), etc. Often, instead of search engines (e.g., Google, Bing) websites (e.g., iNaturalist, https://www.inaturalist.org/) have been consulted (Davis, 2023; Goëau et al., 2022).

As calculated by Löbl et al. (2023, p. 4), 'From Hungary, a well-researched European country with a long taxonomic tradition, 35,650 animal species (excl. 'Protozoa') are recorded. For 15,250 of these (42.7%), there is no taxonomic expert in the country; for another 33.7% (12,010 species), there are only one or two, often retired experts. Currently, Hungarian taxonomists can reliably identify only 23.6% (8,410 species) of the Hungarian fauna (B. Páll-Gergely, pers. comm.) (Löbl et al., 2023, see also Páll-Gergely et al., 2024). But what is the number of qualified taxonomists worldwide? Miralles et al. summarized the situation: a 'substantial proportion of the previously estimated 30,000–40,000 taxonomists worldwide [...] in many cases are nonprofessional researchers' (Miralles et al., 2020, p. 1245). Referring to 'The Global Taxonomy Initiative 2010', MacLeod et al. (2010, p. 155) pointed out that there were likely no more than 4,000–6,000 taxonomists worldwide and, in 2018, David Shorthouse conducted 'A Compendium of Taxonomists on ORCID' (Shorthouse, 2018) referring to a personal communication from David J. Patterson (https://en.wikipedia.org/wiki/David_J._Patterson), that an estimate of active taxonomists was 10,000 – Shorthouse's 'starting' list of ORCIDs includes only ca. 1,500 taxonomists of unknown age. Even if there are three times more 'cryptic' taxonomic experts without ORCIDs, the total number would not exceed 4,500, which generally coincides with previous estimations of 'The global Taxonomy initiative' (see above). Thus, expressions such as 'Taxonomists – an endangered race' (Wägele et al., 2011) and 'An Extinction of Taxonomists' (Löbl et al., 2023) seem far from simple metaphors.

Is there a solution? The current option is to use the datasets '[...] with or without species labels provided and checked by human experts' (Goëau et al., 2022). Such datasets are now the target of automatic machine identification of plant species within the framework of PlantCLEF2022 project (Goëau et al., 2022). In other words, the primary material that is used to inform AI machines of the intensively advertised PlantCLEF2022 initiative (Goëau et al., 2022; Davis, 2023) has not necessarily been identified by, or even observed by, any taxonomist. Of course, iNaturalist, as well as the 'amateur naturalist community' (Davis, 2023) is now an alternative, or addition, to traditional taxonomic work.

Wide usage of AI-based systems in taxonomy and identification of natural history collections is based on a misunderstanding that the work of any taxonomist is always grounded, in part, in intuition and aesthetics (Mavrodiev and Yurtseva, 2017: Williams and Ebach, 2017; Ebach, 2022); a *human being* cannot, in principle, be replaced by an AI machine – this much is probably generally accepted (Valdecasas, 2024). Taxonomic work often requires the use of numerous 'non-formal' data: unpublished or electronically unavailable correspondence and manuscripts; resolving contradictory labels associated with the same specimens; photographs and drawings; even the smell of a collection (e.g., as in case of species from the genus *Seriphidium* (Besser ex Less.) Fourr. (Angiosperms, Asteraceae). As an example, based on personal observations, numerous herbarium specimens from the working collections of a well-known Russian taxonomist were only faintly annotated by pencil with a binomial and, in many cases, were rearranged in different folders, each corresponding to single species as that taxonomist had treated them. None of the numerous specimens were 'officially' determined by this taxonomist, so all the specimens examined were based solely on the old labels. A simple comparison between these pencil-written names and the published monograph showed that the latter is essentially based on the pencil-written names. These herbarium specimens, annotated in pencil, were frequently re-labelled with different taxonomic names from other 'official' sources. Sometimes, a single herbarium specimen (either authentic or not) may combine individuals from one to many taxa, and not all proper binomials are listed on the label(s) or (if listed) suitable in the herbarium collection. Recently, admixtures were detected in the type of *Scilla undulata* Desf. (*Squilla undulata* (Desf.) Mart. Azorín, M.B.Crespo and M.Á.Alonso; Angiosperms, Hyacinthaceae=Asparagaceae Scilloideae) (Crespo et al., 2019). It should be noted that proponents of digitising biological collections do sometimes emphasize the need for expert taxonomists (Little et al., 2020, Miralles et al., 2020, among others) – but in what capacity? For example, the digital data set of Little et al. (2020), curated by a taxonomist, '[…] consisted of 46,469 herbarium specimen images representing 683 species from the family Melastomataceae (Angiosperms) […] the data set was broken down into 75% training, 5% validation, and 20% test specimens that resulted in 34,225, 2679, and 9565 images, respectively' (Little et al., 2020, p. 3). 'In less than three months, the FGVC6 Herbarium 2019 Challenge drew 22 teams who entered 254 models for Melastomataceae species identification. The four best algorithms identified species with >88% accuracy' (Little et al., 2020, p. 1). In summary, 22 teams who dealt with 46,469 herbarium specimen images over a period of three months and generated 254 species identification models could not predict the future revision of a single species (section, genus) of the family involved with AI. No new taxonomic acts were suggested or any new species determined. In short, the AI Team of Little et al. (2020) mirrored the current order of the herbarium folders with >88% accuracy, reproducing the current taxonomy of the group, simply because, as shown above, the current taxonomy exists as if it was permanent. Human-based choices are not part of the machine-learning process.

Even if AI machines can perform taxonomic discovery (Badirli et al., 2023), who will verify it if an expert taxonomist is unavailable? Other AI machines? Not only the number of taxonomists but the number of collections of taxa is frequently limited. Little et al. noted that 'only species with a minimum of 20 specimens were included in the machine-learning dataset' (Little et al., 2020, p. 3). Thus, AI machines require prior estimates of content for their comprehensive training. But that does leave the uncountable species that are known only in one or few collections. Moreover, this is a trivial situation when authentic specimens of a given species are present in only a few collections (even in one). Authentic specimens are also not necessarily homogenous. Is there a way to teach AI machines how to operate with these limited authentic collections, e.g., to involve machines in essential aspects of taxonomic practice? So far, the basic solution seems to include combining authentic specimens with observations from the iNaturalist database in one dataset for future 'high-tech analyses' and machine 'learning' (see above). At present, the idea of identifying species of living organisms using AI machines might be considered 'naïve'.

## THE DIGITIZATION OF NATURAL HISTORY COLLECTIONS, AI AND RELIABLE TAXONOMIC PRACTICE

The digitization of Natural History collections and their replacement with digital images will not only make it difficult to undertake taxonomic revisions but will also affect numerous reliable practices for identifying living organisms from morphological and anatomical characters. For example, the detail in most digital images of animal anatomy (e.g., vertebrates) is not visible even when analyzing high-quality 3D images. Anatomical diagnostic characters are an information source for many herbarium specimens. Such information frequently cannot be obtained from computer analysis of electronic images. For example, in the taxonomy of *Podocarpus* Labill. (Gymnosperms, Podocarpaceae) and *Festuca* L. (Angiosperms, Poaceae), the transverse sections of leaf blades play a crucial role (Hackel, 1882; Orr, 1944; Smirnov, 1945, pp. 102–103; Buchholz and Gray, 1948; de Laubenfels, 1985; Mill, 2015; Martínez-Sagarra et al., 2017) with herbarium specimens being the primary source of material (e.g., Orr, 1944; Martínez-Sagarra et al., 2017). The number of nomenclatural acts in each of these two genera significantly exceeds the number of 'officially' recognized species, indicating the complexity of each taxon. The second example, *Festuca*, is probably non-monophyletic (Cheng et al., 2016). Because molecular systematic studies do not always make it possible to distinguish between closely related species from both genera, they must be undertaken in parallel with studies of the anatomy of leaf fragments (Martínez-Sagarra et al., 2017), which require special preparation. In addition to the study of transverse sections of leaf blades, in many plant groups, other parts of the anatomy are taxonomically significant (e.g., the transverse sections of fruits, seeds, stems or petioles (e.g., Das et al., 2023, and many others). How can petiole anatomy be comprehensively determined using only a digital copy of a herbarium specimen? How might one read Gaertner's *De fructibus et seminibus plantarum* (1788–1791, see Stafleu, 1969), and how might one interpret the studies of Nehemiah Grew (1641–1712) and Marcello Malpighi (1628–1694; reviewed in Robertson, 1906; Arber, 1942; Bolam, 1973, among others) after digitization? To avoid at least elementary mistakes in plant identification, ideally numerous digital images of herbarium specimens need supplementing with detailed images of the internal anatomy of leaves, stems, fruits, etc. – potentially hard to implement on an 'industrial scale' if the relevant experts are already 'extinct'.

Yet, even when analyzing 'standard' morphological characters, plain images of herbarium specimens are frequently insufficient. As summarized by Löbl et al.: 'X-ray microscopy, phase-contrast synchrotron X-ray microtomography, micro-CT scans, or genomic and other molecular technologies, often combined with morphological studies, provide excellent examples of technologically advanced taxonomic approaches' (Löbl et al., 2023, p. 7) – none of which can be applied to digital images.

Even the most straightforward diagnostic morphological features, such as the length of leaf ligulae of various species of grasses (Angiosperms, Poaceae) are not visible using digitizing herbarium specimens and require manipulation of the physical object. There is an informal expression used among botanists: 'boil a collection tissue'. Without such boiling, it is often impossible to estimate both flower and vegetative morphology accurately – for instance, working with herbarium collections of legumes or horned pondweeds (Angiosperms, Zannichelliaceae).

Accurate identification of cattails (*Typha* L.) is impossible without observing the fine morphology of female inflorescences and pollen (Mavrodiev and Kapitonova, 2016), which is not visible even with the best electronic images. The number of examples is numerous. Attempts to replace taxonomists with an AI machine and the original herbarium collection with its digital copy will lead to a significant loss of critical taxonomic information. If actual collections are not available or inaccessible (e.g., for example, the shifting of specimens from Paris to Dijon, about 300 km from Paris, see Rosier, 2023), after digitization, numerous anatomical and morphological aspects of herbarium specimens will be inaccessible thus potentially leading to future inaccuracies in the identification of collections and (in theory) subsequent machine learning.

It is a naïve expectation to believe that some formal model for an AI machine can somehow find unknown morphological external characters and effectively substitute the traditional internal ones listed above (e.g., the transverse sections of the leaf blades or almost the whole character arsenal of carpology). Such expectations seem to be based on irrational beliefs ('Situating methods in the magic of Big Data and AI', Elish and Boyd, 2018) that now can artificially replace past taxonomic achievements. Without basic acknowledgements of past achievement, such as respect for taxonomists and the traditional practices of taxonomy and specimen identification, current technological progress risks turning into what some philosophers have called the 'New Middle Ages'.

## MASSIVE DIGITIZATION AND THE FUTURE OF TAXONOMISTS

MacLeod et al. entitled their commentary, published in *Nature* as an 'Opinion' piece, 'Time to automate identification'. They wrote:

> 'Automating species identification using technologies developed by researchers in pattern recognition, artificial intelligence and machine learning *would transform alpha taxonomy from a cottage industry dependent on the expertise of a few individuals to a testable and verifiable science accessible to anyone needing to recognize objects.* Indeed, a concerted interdisciplinary research and development effort, within the next decade, could lead to automated systems capable of high-throughput identifications for hundreds or thousands of categories of living as well as non-living specimens'
>
> *MacLeod et al., 2010, p. 154, italics ours*

The presence of numerous non-taxonomists in the list of co-authors of many papers published on digitization-based topics is remarkable. Calls to 'Breathe new life into herbaria' (Davis, 2023, p. 415) and other repositories of Natural History collections, follow the implicit assumption that such collections, as well as the science of taxonomy that goes with it, are but imperfect appendices to something else – ecological or climatic analyses, for example. From that, the artificial 'revolution from above', which is now going through the digitization of biological collections, was initiated by scientific politicians for whom taxonomy is not seen as that valuable. That is, by people far from understanding what taxonomy actually is and what it has achieved. For these persons are, perhaps, unware that the word 'taxonomy' was coined by the elder de Candolle (Augustin Pyramus de Candolle, 1778–1841) who used it as a heading for a part of his '*Theorie Elementaire [...]*' (Candolle 1813) with the subheading 'Theorie de Classification', a phrase that presented a definition of taxonomy (Mason, 1950, p. 193; Drouin, 2001). From that humble beginning, traditional taxonomy is not, and was not, 'a cottage industry dependent on the expertise of a few individuals', as MacLeod et al. put it (MacLeod et al., 2010, p. 154) – but the foundation of much biological knowledge (Winsor, 2009). In other words, the destruction of taxonomy would be to detach from biological reality.

The idea of replacing Natural History repositories with digital copies implies the unrealistic assumption that traditional taxonomy has almost completed its task and now is the time to go forward. For example, only under this assumption can the digitization of herbarium folders with their future identification by AI machines without involving any taxonomic experts be logically understood, simply because some samples might be misidentified. The same is true of the irrational belief that AI, 'in itself', can replace professional taxonomists.

What is clear is that the attempt to sideline the intuitive, aesthetic and 'non-formal' input from taxonomy inevitably leads to the loss of taxonomists from Natural History collections. Indeed, if there is a digital copy of any herbarium, what is the need for an actual authentic herbarium? Why keep it, at least in its current form? Or at least in the buildings that now store it – especially if these buildings occupy prime real estate. Why not move Natural History collections elsewhere? Why not reduce the number of curators and collection managers? If taxonomic work is almost complete, any undergraduate student could dry and mount specimens following online instructions and place them on their proper shelves. All these questions relate to the massive rearrangements within the current infrastructure of the Museums of Natural History, especially keeping in mind, that '[…] more than a thousand Natural History museums today exist, with the largest ones located in Europe and North America' (Johnson et al., 2023, p. 1192), perhaps in business-attractive areas of many cities.

The digitization of collections has led to a dramatic decline in the prestige of Natural History Museums and their staff. Why do we need museum collections if there are their perfect digital copies, the transition to the various analyses of which promises such unprecedented scientific prospects? Reducing the number of taxonomists is one conclusion arising from this observation. Much of this is currently being challenged (e.g., 'Save the collections of the Paris National Museum of Natural History, an Open Letter and Petition: https://chng.it/sKR4pLZSdJ, 'Keep the Kew Herbarium at Kew: a Petition', https://chng.it/W5NPdxF2YN).

## A FRAMEWORK FOR THE FUTURE

'Mirrors reflect each other.
All reflections are mutually wrong.
I don't believe that evil is invincible,
But only in the imminence of my defeat'

*Ivanov, 1950*

While discussing 'priceless herbarium collections' (Davis, 2024), recent authors say nothing about their actual importance or about traditional taxonomy. It appears that specimen collections are thought of as material for other 'high-tech' research programmes ('Herbaria have long been a critical resource for ecological and evolutionary research but have recently become relevant to many more fields, including climate science, anthropology, genetics, computer science, chemistry, and medicine [...] Advances in genome sequencing and machine learning are guiding biodiversity monitoring efforts and revealing knowledge gaps where specimen sampling is needed [...] Collections increasingly have become the first line of genetic and genomic sampling for investigators who otherwise eschew conventional field work' (Davis, 2024, p. 1034).

The mass migration of biological collections from Natural History Museums illustrates some consequences of the digitization of biological collections, suggesting that the value of the original specimens is lower than that of the images. For example, AI experts can easily work with digital images, but not (necessarily) with physical collections. The opposite is found in Classical Art, for example, where the value of the original is always more than that of any copies: could we consider replacing the Louvre Museum with a digital image collection?

Biological collections are representations, snapshots of reality. Each individual specimen of a named species may represent the type or may be viewed as a kind of 'imprint' of the latter. The digital image of the collection is a second-order image: an image of the image. Scientific results using images of a collection is a third-order image. The important thing here is that a third-order representation is built from the second-order representation, not the first-order representation, the original. If each image of the subsequent order adequately represents the image of the previous order, no problems arise. But as discussed above, the adequacy of the images is only sometimes possible. For example, what does it mean when individuals of several different species are stored under the same binomial in a herbarium folder? It means that the type of a species is inadequately represented – some individuals taken from the folder will represent the correct species, others (maybe) not so. From that, the set of specimen images may be far from a correct representation of the species itself. The collection of images, in this case, will distort reality. Thus, when obtaining scientific results, specifically images of the third-order, there is a disconnection between that and the images of the first-order (the actual biological collection) and thus with reality. Third-order images will gain an independent and meaningful life of their own, obscuring reality, including images of the previous orders, down to the first. If the value of first-order images is not apparent, if only copies have actual value, then in the end, not only will taxonomy lose its meaning, but classical biology as well. What acquires meaning is something that is no longer associated with biological reality – the third-order images – and the algorithms behind their creation, the various models or programs that generate them.

Discussion of biological issues will be subtly replaced by a discussion of the semantic content of third-order images, which will be conducted in a biological context. That such a discussion is inadequate (simply because the source material does not correspond to its copies) is no longer a matter of debate. However, the third-order images can also be the subject of scientific work. Suppose one wants to summarize what is known today about the history of Earth's climate as reconstructed from biological collections. That scientist must translate many results, including the third-order images and produce another picture (the fourth-order image), which makes sense only to itself. Numerous manipulations with the metadata of collection images, recently proclaimed within the framework of the initiative Digitization 2.0 (Hedrick et al., 2020), are another example of how to create the images fourth-order from the images of second and third orders. All these observations – from the primacy of the value of a digital copy over the value of an original specimen to the concept of the false self-dependent images of the reality of the third and fourth orders – link to the philosophical work of Jean Baudrillard (1929–2007), one of the most prominent figures of European Postmodernism of the 20th century (Wolny, 2017, and others). Even if many details of his philosophy are yet to be reconstructed or clarified, it is perhaps now clear that what we refer to above as images of the third and fourth orders are, within established context, nothing but Baudrillard's 'Simulacra' of different degrees. Third- and fourth-order images tend to form what Baudrillard named a 'Hyper-reality' (Baudrillard, 1983; Wolny, 2017). Using Baudrillard's language, the disconnection between the real thing/object and its image encourages us to discuss replacing the 'representation' with a semantically opposite 'simulation'. For example, under the conditions discussed above, the set of digital images can now simulate the species, establishing the pictures of reality in future studies based on this set as simulacra.

The relevance of postmodernism for formalizing the results of modern floristic studies and biogeography has already been discussed (Mavrodiev et al., 2018). An analysis of what happened in biology in connection with the digitization of biological collections allows us to speak of the postmodernist connotations of what happened with much greater certainty, namely referring to the creative heritage of Baudrillard (as in Baudrillard, 1983). The potential 'demise' of taxonomy finalized by a digital stream means the birth of the biological 'Hyper-real', the semantics and aesthetics of which will now frame the context of most future studies based on digitized Natural History.

## REFERENCES

Alexeev, Y.E. and Mavrodiev, E.V. (2017) *Eragrostis frankii* C. A. Mey. ex Steud. - A new alien species of Lower Volga Region. *Bulletin of Moscow Society of Naturalists. Biological Series*, 122(5): 60–64 [In Russian].

Arber, A. (1942) Nehemiah Grew (1641–1712) and Marcello Malpighi (1628–1694): An essay in comparison. *Isis*, 34(1): 7–16, https://doi.org/10.1086/347742

Badirli, S., Picard, C.J., Mohler, G., Richert, F., Akata, Z. and Dundar, M. (2023) Classifying the unknown: Insect identification with deep hierarchical Bayesian learning. *Methods in Ecology and Evolution*, 14: 1515–1530, https://doi.org/10.1111/2041-210X.14104

Baudrillard, J. (1983) *Simulations* (P. Foss, P. Patton, and P. Beitchman, Trans.). Semiotext(e), Columbia University, USA, 159 pp.

Bebber, D.P., Carine, M.A., Wood, J.R.I., Wortley, A.H., Harris, D.J., Prance, G.T., Davidse, G., Paige, J., Pennington, T.D., Robson, N.K.B. and Scotland, R.W. (2010) Herbaria are a major frontier for species discovery. *Proceedings of the National Academy of Sciences of the United States of America*, 107(51): 22169–22171, https://doi.org/10.1073/pnas.1011841108

Blagoderov, V., Kitching, I.J., Livermore, L., Simonsen, T.J. and Smith, V.S. (2012) No specimen left behind: Industrial scale digitization of natural history collections. *Zookeys*, 209: 133–146, https://doi.org/10.3897/zookeys.209.3178

Bolam, J. (1973) The botanical works of Nehemiah Grew, FRS (1641–1712). *Notes and Records of the Royal Society of London*, 27(2): 219–231, https://doi.org/10.1098/rsnr.1973.0017

Buchholz, J.T. and Gray, N.E. (1948) A taxonomic revision of *Podocarpus*: I. the sections of the genus and their subdivisons with special reference to leaf anatomy. *Journal of the Arnold Arboretum*, 29: 49–63, http://www.jstor.org/stable/43781280

Candolle, de A.P. (1813) *Théorie élémentaire de la botanique, ou exposition des principes de la classification naturelle et de l'art de décrire et d'etudier les végétaux.* [2nd edn 1819]. Déterville, Paris.

Cheng, Y., Zhou, K., Humphreys, M.W., Harper, J.A., Ma, X., Zhang, X., Yan, H. and Huang, L. (2016) Phylogenetic relationships in the *Festuca-Lolium* complex (Loliinae; Poaceae): New insights from chloroplast sequences. *Frontiers in Ecology and Evolution*, 4, https://doi.org/10.3389/fevo.2016.00089

Crespo, M.B., Martínez-Azorín, M. and Alonso, M.Á. (2019) Lectotype designation for *Scilla undulata* Desf. (Asparagaceae) from among the mixed original material. *Phytotaxa*, 411: 89–92, https://doi.org/10.11646/phytotaxa.411.1.9

Das, P., Chettri, V., Ghosh, S. and Ghosh, C. (2023) Micromorphological studies of the leaf and stem of *Camellia sinensis* (L.) Kuntze with reference to their taxonomic significance. *Microscopy Research and Technique*, 86(4), 465–472, https://doi.org/10.1002/jemt.24287

Davis, C.C. (2023) The herbarium of the future. *Trends in Ecology and Evolution*, 38: 412–423, https://doi.org/10.1016/j.tree.2022.11.015

Davis, C.C. (2024) Collections are truly priceless. *Science*, 383(6687): 1035–1035, https://doi.org/10.1126/science.ado9732

Davis, M.T. (1957) A guide and an analysis of Engler's "*Das Pflanzenreich*". *Taxon*, 6: 161–182, https://doi.org/10.2307/1217029

de Laubenfels, D.J. (1985) A taxonomic revision of the genus *Podocarpus*. *Blumea*, 30(2): 251–278.

Drouin, J.-M. (2001) Principles and uses of taxonomy in the works of Augustin-Pyramus de Candolle. *Studies in History and Philosophy of Science Part C: Studies in History and Philosophy of Biological and Biomedical Sciences*, 32(2): 255–275, https://doi.org/10.1016/S1369-8486(01)00002-4

Ebach, M.C. (2022) *Goethe in the Age of Artificial Intelligence: Enlightened Solutions for a Modern Hubris.* Palgrave Macmillan, Singapore.

Ebach, M.C. and Williams, D.M. (2010) Aphyly: A systematic designation for a taxonomic problem. *Evolutionary Biology*, 37(2–3): 123–127, https://doi.org/10.1007/s11692-010-9084-5

Edwards, E.J., Mishler, B.D. and Davis, C.D. (2024) University herbaria are uniquely important. *Trends in Plant Science*, 29(8): 825–826, https://doi.org/10.1016/j.tplants.2024.06.002

Elish, M.C. and Boyd, D. (2018) Situating methods in the magic of Big Data and AI. *Communication Monographs*, 85(1): 57–80, https://doi.org/10.1080/03637751.2017.1375130

Frank, S.L. (1956) *Reality and Man: Metaphysics of Human Being*, YMCA-Press, Paris, 416 pp. [in Russian: С. Л. Франк. *Реальность и человек. Метафизика человеческого бытия*, YMCA-Press, Париж. 416 с.; also Frank, S. L. 1956. *Reality and man: An essay in the metaphysics of human nature* (translated by Natalie Duddington), Faber and Faber, London, 262 pp.]

Gaertner, J. (1788–1791) *De fructibus et seminibus plantarum.* Stutgardiae, Sumtibus Auctoris, Typis Academiae Carolinae.

Gaston, K.J. and O'Neill, M.A. (2004) Automated species identification: Why not? *Philosophical Transactions of the Royal Society of London. Series B: Biological Sciences*, 359(1444): 655–667, https://doi.org/10.1098/rstb.2003.1442

Goëau, H., Bonnet, P. and Joly, A. (2022) Overview of PlantCLEF 2022: Image-based plant identification at global scale, pp. 1916–1928. In *Proceedings of the Working Notes of CLEF 2022 - Conference and Labs of the Evaluation Forum*, Bologna, Italy, September 5th to 8th, 2022 (eds G. Faggioli, N. Ferro, A. Hanbury and M. Potthast) paper 153, https://ceur-ws.org/Vol-3180/paper-153.pdf

Goodwin, Z.A., Harris, D.J., Filer, D., Wood, J.R.I. and Scotland, R.W. (2015) Widespread mistaken identity in tropical plant collections. *Current Biology*, 25(22): R1066–R1067, https://doi.org/10.1016/j.cub.2015.10.002

Hackel, E. (1882) *Monographia Festucarum Europaearum*. Verlag von Theodor Fischer, Kassel.

Hedrick, B.P., Heberling, J.M., Meineke, E.K., Turner, K.G., Grassa, C.J., Park, D.S., Kennedy, J., Clarke, J.A., Cook, J.A., Blackburn, D.C. and Edwards, S.V. (2020) Digitization and the future of natural history collections. *BioScience*, 70(3): 243–251, https://doi.org/10.1093/biosci/biz163

Hussein, B.R., Malik, O.A., Ong, W.H. and Slik, J.W.F. (2022) Applications of computer vision and machine learning techniques for digitized herbarium specimens: A systematic literature review. *Ecological Informatics*, 69, Article 101641, https://doi.org/10.1016/j.ecoinf.2022.101641 [Hussein, B. R., Malik, O. A., Ong, W. H., and Slik, J. W. F. (2022). Corrigendum to "Applications of computer vision and machine learning techniques for digitized herbarium specimens: A systematic literature review". *Ecological Informatics* 69, 101641, ISSN 1574–9541, https://doi.org/10.1016/j.ecoinf.2022.101641, *Ecological Informatics*, 76, 2023, 102134, ISSN 1574-9541, https://doi.org/10.1016/j.ecoinf.2023.102134).

Ivanov, G.V. (1950) Mirrors Reflect Each Other […]. In the Collection: *Portrait without Resemblance: Poems*. Rhyme, Paris, 46 pp. [in Russian: Иванов, Г. В. (1950) Друг друга отражают зеркала […]. В Сборнике: *Портрет без сходства: Стихи*. Париж: Рифма. 46 с.]

Johnson, K.R., Owens, I.F.P. and The Global Collection Group (2023) A global approach for natural history museum collections. *Science*, 379(6638): 1192–1194, https://doi.org/10.1126/science.adf6434

Kapitonova, O.A., Belyakov, E.A. and Mavrodiev, E.V. (2024) Taxonomic status of the Far Eastern cattail *Typha japonica* Miq. (*T.* sect. *Ebracteolatae*) in the context of its first record in Russia. *Inland Water Biology*, 17(2): 229–237, https://doi.org/10.1134/S199508292402007X

Karbstein, K., Kösters, L., Hodač, L., Hofmann, M., Hörandl, E., Tomasello, S., Wagner, N.D., Emerson, B.C., Albach, D.C., Scheu, S., Bradler, S., de Vries, J., Irisarri, I., Li, H., Soltis, P., Mäder, P. and Wäldchen, J. (2024) Species delimitation 4.0: Integrative taxonomy meets artificial intelligence. *Trends in Ecology and Evolution,* 39: 771–784, https://doi.org/10.1016/j.tree.2023.11.002

Konstantinov, F. and Namyatova, A. (2019) Taxonomic revisions and specimen databases in the internet age: Dealing with a species rich insect taxon. *Entomological Review,* 99(3): 340–361, https://doi.org/10.1134/S0013873819030072

Le Bras, G., Pignal, M., Jeanson, M.L., Muller, S., Aupic, C., Carré, B., Flament, G., Gaudeul, M., Gonçalves, C., Invernón, V.R., Jabbour, F., Lerat, E., Lowry, P.P., Offroy, B., Pimparé, E.P., Poncy, O., Rouhan, G. and Haevermans, T. (2017) The French Museum national d'histoire naturelle vascular plant herbarium collection dataset. *Scientific Data*, 4: Article 170016, https://doi.org/10.1038/sdata.2017.16

Little, D.P., Tulig, M., Tan, K.C., Liu, Y., Belongie, S., Kaeser-Chen, C., Michelangeli, F.A., Panesar, K., Guha, R.V. and Ambrose, B.A. (2020) An algorithm competition for automatic species identification from herbarium specimens. *Applications in Plant Sciences* Jul 1, 8(6): e11365. https://doi.org/10.1002/aps3.11365

Löbl, I., Klausnitzer, B., Hartmann, M. and Krell, F.-T. (2023) The silent extinction of species and taxonomists—An appeal to science policymakers and legislators. *Diversity*, 15(10): 1053, https://doi.org/10.3390/d15101053

MacLeod, N., Benfield, M. and Culverhouse, P. (2010) Time to automate identification. *Nature*, 467(7312): 154–155, https://doi.org/10.1038/467154a

Martínez-Sagarra, G., Abad, P. and Devesa, J.A. (2017) Study of the leaf anatomy in cross-section in the Iberian species of *Festuca* L. (Poaceae) and its systematic significance. *Phytokeys* 83: 43–74, https://doi.org/10.3897/phytokeys.83.13746

Mason, H.L. (1950) Taxonomy, systematic botany and biosystematics. *Madroño*, 10(7): 193–208, https://www.biodiversitylibrary.org/page/47857404#page/237

Mavrodiev, E.V., Gómez, J.P., Mavrodiev, N.E., Melton, A.E., Martínez-Azorín, M., Crespo, M.B., Robinson, S.K. and Steadman, D.W. (2021) On biodiversity and conservation of the "*Iris hexagona* complex" (*Phaeiris*, Iridaceae). *Ecosphere*, *12*(1), Article e03331. https://doi.org/10.1002/ecs2.3331

Mavrodiev, E.V. and Kapitonova, O.A. (2016) Taxonomic composition of Typhaceae of the flora of European Russia. *Novitates Systematicae Plantarum Vascularium*, 46: 5–24, https://doi.org/10.31111/novitates/2015.46.5 [in Russian]

Mavrodiev, E.V., Sun, M., Schroder, L., Steadman, D.W. and Ebach, M.C. (2018) Moving from modern toward post-modern science: Comment on "An integrated assessment of the vascular plants of the Americas". *Phytotaxa*, 351(1): 96–98, http://dx.doi.org/10.11646/phytotaxa.351.1.10

Mavrodiev, E.V., Kapitonova, O.A. and Belyakov, E.A. (2023) New species of *Typha* (Typhaceae) from the Russian Far East. *Novitates Systematicae Plantarum Vascularium*, 54 (e07): 1–6.

Mavrodiev, E.V., Sytin, A.K., Laktionov, A.P., Vasjukov, V.M., Penton, J.S., Jr. and Panfilova, K.I. (2024) What is "*Apocynum sibiricum*"? A critical comment on the long-term taxonomic homonymy. *Taxonomy*, 4: 314–325, https://doi.org/10.3390/taxonomy4020015

Mavrodiev, E.V. and Yurtseva, O. (2017) "A character does not make a genus, but the genus makes the character": Three-taxon statement analysis and intuitive taxonomy. *European Journal of Taxonomy*, 377, https://doi.org/10.5852/ejt.2017.377

Mill, R. (2015) A monographic revision of the genus *Podocarpus* (Podocarpaceae): III. The species of the Central America and northern Mexico bioregions. *Edinburgh Journal of Botany*, 72: 243–341, https://doi.org/10.1017/S0960428615000050

Miralles, A., Bruy, T., Wolcott, K., Scherz, M.D., Begerow, D., Beszteri, B., Bonkowski, M., Felden, J., Gemeinholzer, B., Glaw, F., Glöckner, F.O., Hawlitschek, O., Kostadinov, I., Nattkemper, T.W., Printzen, C., Renz, J., Rybalka, N., Stadler, M., Weibulat, T., Wilke, T., Renner, S.S. and Vences, M. (2020) Repositories for taxonomic data: Where we are and what is missing? *Systematic Biology*, 69(6): 1231–1253. https://doi.org/10.1093/sysbio/syaa026

Moore, A. (2024) Digitizing Kew`s herbarium: Do we have a protocol for that? *Biodiversity Information Science and Standards,* 8: e137832, https://doi.org/10.3897/biss.8.137832

Muñoz-Rodríguez, P., Carruthers, T., Wood, J.R.I., Williams, B.R.M., Weitemier, K., Kronmiller, B., Goodwin, Z., Sumadijaya, A., Anglin, N.L., Filer, D., Harris, D., Rausher, M.D., Kelly, S., Liston, A. and Scotland, R.W. (2019) A taxonomic monograph of *Ipomoea* integrated across phylogenetic scales. *Nature Plants*, 5(11): 1136–1144, https://doi.org/10.1038/s41477-019-0535-4

Nelson, G., Sweeney, P., Wallace, L.E., Rabeler, R.K., Allard, D., Brown, H., Carter, J.R., Denslow, M.W., Ellwood, E.R., Germain-Aubrey, C.C., Gilbert, E., Gillespie, E., Goertzen, L.R., Legler, B., Marchant, D.B., Marsico, T.D., Morris, A.B., Murrell, Z., Nazaire, M., Neefus, C., Oberreiter, S.P.D, Ruhfel, B.R., Sasek, T., Shaw, J., Soltis, P.S.,

Watson, K., Weeks, A. and Mast, A.R. (2015) Digitization workflows for flat sheets and packets of plants, algae, and fungi. *Application in Plant Sciences*, 3(9): apps.1500065. Sep. 10, https://doi.org/10.3732/apps.1500065

Orr, M.Y. (1944) The Leaf Anatomy of *Podocarpus*. *Transactions of the Botanical Society of Edinburgh*, 34(1): 1–54, https://doi.org/10.1080/13594864409441551

Páll-Gergely, B., Krell, F.-T., Ábrahám, L., Bajomi, B., Balog, L.E., Boda, P., Csuzdi, C., Dányi, L., Fehér, Z., Hornok, S., Horváth, A., Kóbor, P., Koczor, S., Kontschán, J., Kovács, T., Majoros, G., Murányi, D., Németh, T., Pernecker, B., Puskás, G., Rózsa, L., Soltész, Z., Szita, É., Szűts, T., Tóth, B., Tőke, A., Vas, Z., Zsuga, K., Zsupos, V., Csabai, Z. and Móra, A. (2024) Identification crisis: A fauna-wide estimate of biodiversity expertise shows massive decline in a Central European country. *Biodiversity and Conservation*. https://doi.org/10.1007/s10531-024-02934-6

Portal, R. (2002) *Eragrostis de France et de l'Europe occidentale*. Imprimerie Jeanne-d'Arc, Le Puy-en-Velay, France.

Robertson, A. (1906) Nehemiah Grew and the study of plant anatomy. *Science Progress in the Twentieth Century (1906–1916)*, 1(1): 150–158.

Rosier, F. (2023) Moving the collections of the Paris Natural History Museum, a real headache. https://www.lemonde.fr/en/science/article/2023/05/27/moving-the-collections-of-the-paris-natural-history-museum-a-real-headache_6028159_10.html

Schellenberger Costa, D., Boehnisch, G., Freiberg, M., Govaerts, R., Grenié, M., Hassler, M., Kattge, J., Muellner-Riehl, A.N., Rojas Andrés, B.M., Winter, M., Watson, M., Zizka, A. and Wirth, C. (2023) The big four of plant taxonomy – a comparison of global checklists of vascular plant names. *New Phytologist*, 240: 1687–1702, https://doi.org/10.1111/nph.18961

Shorthouse, D.P. (2018) A compendium of taxonomists on ORCID. https://info.orcid.org/a-compendium-of-taxonomists-on-orcid/

Smirnov, P.A. (1945) On the taxonomy and geography of *Festuca ovina* L. s. ampliss. group in Central Russia. *Bulletin of Moscow Society of Naturalists. Biological series*, 50: 89–103 [in Russian].

Smirnov, P.A. (1967) On our understanding of the Transbaikalia "hair-like" species of *Stipa* L. *Bulletin of Moscow Society of Naturalists. Biological series*, 72(4): 81–88 [in Russian].

Stafleu, F.A. (1969) Joseph Gaertner and his carpologia. *Acta botanica neerlandica*, 18(1): 216–223, https://doi.org/10.1111/j.1438-8677.1969.tb00586.x

Sweeney, P.W., Starly, B., Morris, P.J., Xu, Y.M., Jones, A., Radhakrishnan, S., Grassa, C.J. and Davis, C.C. (2018) Large-scale digitization of herbarium specimens: Development and usage of an automated, high-throughput conveyor system. *Taxon*, 67(1): 165–178. https://doi.org/10.12705/671.10

Sytin A.K. and Slastunov D.D. (2021) V.L. Komarov and the authors of "Flora of the USSR". *Historical and Biological Studies*, 13(1): 24–71.

Thiers, B.M., Tulig, M.C. and Watson, K.A. (2016) Digitization of the New York Botanical Garden Herbarium. *Brittonia*, 68(3): 324–333, https://doi.org/10.1007/s12228-016-9423-7

Turland, N.J., Wiersema, J.H., Barrie, F.R., Greuter, W., Hawksworth, D.L., Herendeen, P.S., Knapp, S., Kusber, W.-H., Li, D.-Z., Marhold, K., May, T.W., McNeill, J., Monro, A.M., Prado, J., Price, M.J. and Smith, G.F. (eds) (2018) *International Code of Nomenclature for algae, fungi, and plants (Shenzhen Code) adopted by the Nineteenth International Botanical Congress Shenzhen, China, July 2017.* Regnum Vegetabile 159. Glashütten, Koeltz Botanical Books, https://doi.org/10.12705/Code.2018

Tzvelev, N. and Probatova, N. (2019) *Grasses of Russia* (ed T.I.V). KMK Scientific Press, Moscow, 646 pp. [In Russian].

Valdecasas, A.G. (2024) Can taxonomists think? Reversing the AI equation. *Taxonomy*, 4(4): 713–722, https://doi.org/10.3390/taxonomy4040037

Wägele, H., Klussmann-Kolb, A., Kuhlmann, M., Haszprunar, G., Lindberg, D., Koch, A. and Wägele, J.W. (2011) The taxonomist - an endangered race. A practical proposal for its survival. *Frontiers in Zoology*, 8, Article 25, https://doi.org/10.1186/1742-9994-8-25

Wheeler, Q. (2014) Are reports of the death of taxonomy an exaggeration? *New Phytologist,* 201(2): 370–371, https://doi.org/10.1111/nph.12612

Williams, D.M. and Ebach, M.C. (2017) What is intuitive taxonomic practice? *Systematic Biology*, 66(4): 637–643, https://doi.org/10.1093/sysbio/syw094

Winsor, M.P. (2009) Taxonomy was the foundation of Darwin's evolution. *Taxon*, 58: 43–49. http://www.jstor.org/stable/27756822

Wolny, R.W. (2017) Hyperreality and Simulacrum: Jean Baudrillard and European post-modernism. *European Journal of Interdisciplinary Studies*, 3(3): 75–79, https://doi.org/10.26417/ejis.v8i1.p76-80

Wu, Z.Y. and Phillips, S.M. (2006) *Stipa L.*, pp. 279–281. In *Flora of China* 22 (eds Z.Y. Wu, P.H. Raven, and D.Y. Hong). Beijing, Science Press; St. Louis, Missouri, Botanical Garden Press.

# 11 Taxonomy Positive

*Michelle J. Price*

'[...] ignorance more frequently begets confidence than does knowledge: it is those who know little, not those who know much, who so positively assert that this or that problem will never be solved by science'

*Darwin, 1871, p. 9*

## INTRODUCTION

Within the taxonomic community, the worth and significance of taxonomy is widely, or even unanimously, understood, even if frustration about the perceived lack of recognition of taxonomy as being important or relevant by those outside of the field persists in parallel (de Carvalho et al., 2007; Wheeler and Valdecasas, 2007; Ebach et al., 2011a, Wheeler et al., 2012; Dijkstra, 2016). Taxonomy as a scientific undertaking, the perceived slowness of progress in the field and barriers or challenges to be addressed to mitigate the factors that hinder taxonomy have been the focus of reflection, comment and criticism, both inside and outside of the taxonomic community (de Carvalho et al., 2007; Grimaldi and Engel, 2007; Ebach et al., 2011b; Coleman, 2015; Britz et al., 2020; Dupérrén, 2020; Raposo et al., 2020; Engel et al., 2021; Löbl et al., 2023; Sandall et al., 2023). The 'taxonomic impediment' was brought to the forefront under the auspices of the Convention on Biological Diversity to englobe the slowness of progress, gaps in taxonomic knowledge, the shortage of trained taxonomists and their uneven distribution as well as the global shortage of taxonomic infrastructure at the time (Secretariat of the Convention on Biological Diversity, 2006). Although well-intended, the term 'impediment', with its negative connotations, was perhaps not best chosen to achieve the desired result, namely, a sea-change over how taxonomy was perceived, funded and generally supported by society, policy makers and other scientists. In addition to the negative connotations associated with the use of the term 'impediment', taxonomy is often framed as being somewhat of a failure as taxonomists have still not managed to name all species on Earth and are not even capable of correctly estimating how many species there are (Mora et al., 2011; Costello et al., 2013; Larsen et al., 2017). This perceived failure implies that taxonomists, reduced to biodiversity naming service providers (de Carvalho et al., 2007), are far from reaching their goal, a goal that depends on an estimation that they are incapable of pinning down in the first place. Even the Global Taxonomy Initiative of the Convention for Biological Diversity (CBD, 2006), put in place to support and promote taxonomy on a global scale, was unable to frame taxonomy in a positive light:

'Unfortunately, taxonomic knowledge is far from complete. In the past 250 years of research, taxonomists have named about 1.78 million species of animals, plants and


DOI: 10.1201/b22822-11

micro-organisms, yet the total number of species is unknown and probably between 5 and 30 million'

As a community, we should question whether the naming of all known species within 250 years was the goal? If it was a suitable and suitably defined goal? A scientific goal? And if this rather perfunctory process still is the goal. It would seem to be both strategically unwise and self-defeating to propose and then adhere to a goal that expands over time, and for which estimates of where the goalposts lie vary by significant factors of magnitude.

A premise about taxonomy that is inherent to it, and obvious to practitioners, is that the understanding of past and present biological entities is constantly growing, changing and progressing. This inevitably, and correctly, leads to the ongoing reassessment of taxon concepts, circumscriptions and classifications as a result of new discoveries as well as the re-evaluation of our past understanding based on the expanding and densifying body of biodiversity knowledge. For many practitioners, the scientifically dynamic process of documenting, describing or circumscribing, and systematically organizing organisms according to their evolutionary relationships, combined with an ever more refined understanding of their evolutionary histories, biology, and phylogenetic affinities, over time, is exactly what attracted them to taxonomy in the first place. Ironically, the key dynamic processes inherent in taxonomy are the very things about taxonomy that appear to be most misunderstood by non-practitioners. Namely, that change should not happen, or that change is a bad sign as it is an indication that things were not done properly or correctly in the first place. Although taxonomy practitioners are inherently aware that the foundational body of knowledge on biological entities is in constant flux, with changes to our conceptual framework a normal part of the taxonomic process, this is not always obvious for all. Change is inherent in scientific processes, but a mis-messaging on its inevitability and significance in taxonomy seems to have taken place. Changes in names of taxa, the concepts, or circumscriptions of taxa or the position of them within the Tree of Life may be seen as an indication that the original science on which it was based was somewhat wanting or sub-standard, or that taxonomists are not up to par as they do not really know what they are doing. The view of change as a direct result of taxonomic endeavors is a question of perception, with perceptions based on the depth of understanding of the taxonomic process and the outcomes of it (Raposo et al., 2020). The misunderstanding of the processes that lead to dynamic taxonomy, and the misunderstanding of the reasons that changes take place need to be addressed in a more proactive and coherent way by taxonomy practitioners. Dynamic change in taxonomy is a direct consequence of taxonomic endeavors and represents a healthy indicator of the scientific process and scientific progress. As change in the fabric of taxonomy may not be an evident outcome to non-taxonomists or non-scientists, efforts are needed to communicate about taxonomic processes, outcomes and their implications in a positive and engaging way, and by encouraging taxonomic literacy across stakeholders (Evenhuis, 2007; Grimaldi and Engel, 2007; Dijkstra, 2016; Nanglu et al., 2023).

Based on ten years of leading the Consortium of European Taxonomic Facilities (CETAF), the unique exchange, coordination and collaboration platform for natural history institutions in and across Europe, several themes related to how we, as a taxonomic community, situate ourselves, communicate to others, set our goals and targets, and position within the complex and competitive scientific landscape that we are an integral part of, are explored. The aim is to stimulate thought and reflections on how adapting our approaches, attitudes, and dialogue toward a more taxonomy-positive point of view may strengthen our position. In full acknowledgement that the notions, ideas or propositions presented herein are neither novel nor innovative (see Wheeler, 2008a, b and chapters in Wheeler, 2008c), what is proposed is the collective exploration of undertaking a conscious change in our attitude, messaging, and approaches to presenting and communicating about our field. Looking at taxonomy from a positive perspective, behaving in a positive manner about it, while advocating for positive proactivity rather than passive acceptance, or worse, may well enable the transformation of a certain amount of our perceived problems into challenges to be tackled or opportunities to be seized. Taking direct responsibility for – positively – creating and curating our own surroundings will make it easier for us to quickly identify and profit from the huge number of openings that we have before us. Our opportunities are developing a rigorous scientific agenda, positioning ourselves correctly and honestly within global crises, exploiting technical and technological advances, compiling and continuously curating the taxonomic backbone to unify biodiversity knowledge and digital biodiversity data, continuing to develop collaborations across borders, focusing on strategic outreach, training, and capacity building efforts for taxonomy, among others. Efforts to exploit opportunities would ideally culminate in the creation of an ambitious vision for taxonomy that leverages the expertise, knowledge and impact of the exploration of life on our singular planet through taxonomic endeavors across the globe. Taking on a taxonomy-positive approach to life cannot entirely transform the circumstances in which we find ourselves, but it will, at the very least, make discussions less depressing, passive or defeatist, and the journey much more engaging (see Dijkstra, 2016 for a proposition to reintroduce the term 'Bionomy'). Changing the tone, content, and delivery of our discourse may also play a role in inspiring more confidence in taxonomy as a science. When practitioners rally behind taxonomy in a positive way, unite under a common vision and move from a defensive positioning to one of engagement the community gains a certain critical mass, as well as, incidentally, realizing a more complete state of taxonomic mindfulness (see Gerber et al., 2022, for an example in link with the biodiversity crisis).

To reclaim and then rewrite the message about taxonomy and its goals or the position of taxonomy and its roles, positive change is required from a sociological, cultural and community perspective. This positive change, or taxonomic positivity, must originate from its practitioners and be practiced by them as part of their belief and value system. A taxonomy-positive approach, based on strategic choices made by and implemented from within the taxonomic community, is a sociological-scientific issue that we should explore further, ideally in partnership with experts from the appropriate fields of sociology, psychology, and human behavior. Positive change, driven by the recognition of the reasons it is needed, should be paralleled with the culturing of an adequate and audience-adapted communication on the messaging, positioning, nature, identity, purpose, use, outcomes, and values of taxonomy.

## A TAXONOMY IDENTITY CRISIS?

Taxonomy alone does not relieve societal problems or resolve all our Anthropocene challenges. Taxonomy alone does not stop climate change. Taxonomy alone does not stop biodiversity loss. To create the necessary circumstances for positive change, truthfully and clearly communicating what taxonomy is and is not, why it is done, what its outcomes are and are not, and what the scientific findings that result from it are used for, directly and indirectly, is essential. To communicate unequivocally about taxonomy, the essence of it should not be diminished by framing it in terms of what people might want to hear, what is perceived to give it validity, or what appears to give its outcomes scientific credibility. By doing this, taxonomy itself is weakened, with the risk that it is viewed and treated simply as a service industry for other disciplines (see in de Carvalho et al., 2007), or one without much scientific weight (Wheeler and Valdecasas, 2007).

Taxonomy practitioners should be able to ask themselves and answer three simple questions, What is taxonomy for? Why is taxonomy needed? What would happen if there were no taxonomy and no taxonomists? We should also strive to find out what the answers would be when the same questions are addressed to the diverse stakeholders of taxonomy. The answers to these questions form the foundations from which to re-message, re-purpose, and re-think how we position taxonomy, communicate on taxonomy, and show the role and value of its outcomes in the biodiversity landscape and beyond. Framing taxonomy as a solution to global problems creates further goals that are not our own and that cannot realistically be fulfilled by ourselves alone. Further perceptions about taxonomy, and its relative failure or success, may be created that are divergent from the reality of the nature, practice and outcomes of taxonomy, and from the taxonomic process itself. Just as taxonomists should uphold high scientific standards, ensure the reproducibility of taxonomic information, and embrace new technologies to promote excellence in the field (see Wheeler and Valdecasas, 2005), effectively we should be promoting what taxonomy is, not 'selling' what it is not and then crossing our fingers in the hope that taxonomy's lot will improve.

The global taxonomic community has the enviable position of being at the forefront of species discovery and the leading 'distributed expert' on past and present biological organisms. It is best placed to identify and address knowledge gaps and to advise on questions that fall within its remit as the foremost distributed expert on the components of biodiversity. By adopting and promoting the identity of taxonomy, the role of taxonomy and the outcomes of taxonomy in a taxonomy positive way a phase that replaces the taxonomic impediment with the taxonomic imperative can be reached. The taxonomic imperative would express the necessity and importance of taxonomy and taxonomic expertise, while evoking the urgency of assuring its scientific, financial, infrastructural and technical health. The taxonomic imperative, perhaps to be renamed the bionomy imperative, would pave the way to realizing more ambitious thinking and big-science ideas that consolidate the importance of planetary biodiversity discovery and documentation with the necessary scientific experts and expertise.

## UNREALISTIC VERSUS REALISTIC GOALS

The perceived failure of taxonomy and taxonomists to describe all species on Earth over the past 250 years, or to be able to correctly estimate the number of species, is only a failure if a complete global species inventory within this time frame was the goal in the first place. Did the taxonomic community collectively decide upon and set this goal for itself? An inventory *per se* is not an awe-inspiring activity, neither is it a particularly scientifically stimulating process or a headline making endeavor. Did any taxonomist engage in the field of taxonomy because attaining the global inventory of all species was a motivating factor? Taxonomy is more than the identification of biological 'things' so the apparent linking of it as a scientific endeavor to the making of a global list of biological organisms needs to be set in a broader evolutionary context. If a global inventory supported by descriptions of all known species really is our destination, then we need to become much more adept at telling the story about the complexities of the journey, give indications on how long the journey will take, and explain why. The journey evokes the discovery and documenting of the unknown, across time and geographical space, combined with the ongoing development of a better scientific understanding of known biological entities and biological diversity in an evolutionary context. It is the taxonomic journey that is scientifically stimulating, that leads to discovery and progress, that changes our perceptions of biological entities on a continual basis. Whereas the destination is just that, an endpoint.

Setting the goals for our own field, based on our own expertise and from our own informed perspective, allows us to decide upon our destination and more importantly permits us to own the journey. If we truly aim to describe all biological entities on Earth within a given time frame then we rapidly need to work on accurately outlining these ambitions while scaling them up to the appropriate level as well as communicating on them in the appropriate, and positive, terms. In addition, some care is needed to separate the urgency over the need to know, name and document species before they are lost from the perceived failure of taxonomists to name things fast enough. The former is down to a systematic failure to care for the well-being of our planet whereas the latter reflects an external pressure that is caused by the former, combined with varying perceptions of goals, needs, priorities and scientific competence. If a global species inventory done as rapidly as possible is not our primary taxonomic goal then we should see this as an opportunity to redefine our purpose, directions, aims and ambitions. Owning the dialogue on the role of taxonomy, the goals of taxonomy, the successes of taxonomy and the outcomes of taxonomy is a necessary first step in outlining our ambitions, perhaps even pushing us out of our traditional rather modest 'taxonomy comfort zone' into 'big ambition' thinking (see Wheeler et al., 2004; Wheeler, 2008a).

## MESSING UP THE MESSAGE

The linking of taxonomy and its outcomes to global crises, whether intentional or non-intentional, should be re-evaluated from a strategic perspective within the community. The climate change and biodiversity crises, as evoked in the well-known Kunming-Montreal Global Biodiversity Framework agreed by the Conference of the Parties to the Convention on Biological Diversity, the United Nations Sustainable

Development Goals, climate change mitigation efforts such as the European Green Deal, as well as reporting activities of the Intergovernmental Science-Policy Platform on Biodiversity and Ecosystem Services, are global issues with an urgent need for global responses. There is growing concern that a general fatigue about the crises and their implications has set in with the emphasis on a continual crisis mode giving rise to feelings of uselessness, disinterest, disengagement and disbelief (Pihkala, 2020; de Lange et al., 2022; Bird et al., 2024). Attempts to ameliorate, restore, monitor, report on and manage biodiversity, and use the world's resources in a climate-neutral, sustainable fashion, are all underpinned by the need for taxonomic information on biological entities, where they live and with which other entities they interact, cohabit, or depend on in some way. However, taxonomic information is but one contribution, among many, that is needed to have any impact on biodiversity loss or climate change on a global scale. Promoting taxonomy as a solution to global crises due to the importance of taxonomic information in resolving them is a dangerous precept. The input from taxonomy alone will have no impact on the achievement of global or regional goals and targets, as their ultimate success depends on a multitude of other inter-dependent factors, as well as massive change in individual, community, societal, financial and political spheres, including the implementation of the necessary policies and legislation.

Messaging on the role of taxonomy in underpinning global efforts to stem global crises should be done in realistic terms. The following questions should be explored: What does taxonomy concretely contribute to resolving global crises? How can the understanding of it, and its role(s), be increased? How should taxonomy be framed and positioned? What do stakeholders expect or want from taxonomy? Taxonomic outcomes and taxonomic information, namely species circumscriptions, the organizational structure for ordering biological entities according to their genetic affinities, the provision of unique scientific names that apply to entities, and information on what these entities are in physical and biological terms, where they live and in which types of habitats they are found, are widely acknowledged to underpin the majority of efforts to preserve, restore, manage, and sustainably use biodiversity. Taxonomy contributes taxonomic information that it is essential at a foundational level, amongst others, but it is not the vector by which measures are implemented or actions are taken on the ground. Taxonomic information, as an essential tool for implementing actions on the ground, or policy measures, should not be presented as separated from the scientific processes of the practice of taxonomy that led to the obtention of it. Correctly positioning taxonomy, taxonomic expertise, taxonomic outcomes and taxonomic information within the framework of efforts to tackle global challenges transforms into an opportunity to engage in the process, at the right levels, based on the universal understanding of the factual inputs from taxonomy to the issues in hand, and through the provision of tangible contributions that can be taken-up, implemented and measured, on their own terms.

## OPPORTUNITY DOES NOT KNOCK TWICE

Opportunities for the taxonomic community are multiple, ranging from the use of new tools, advances in computing, genetics and genomics, innovative analytical techniques, including artificial intelligence, and the availability of ever-increasing amounts of data on biological specimens, with a focus herein on the latter. The taxonomic community is one of the bastions of knowledge on biodiversity and is one of the main contributors to the enrichment of natural history collections through their collecting activities. This same community is also the main scientific user of collections on an ongoing basis as part of their research activities. The importance of natural history collections, their enrichment and scientific value and use are well-documented (Bebber et al., 2010; Card et al., 2021; Davies, 2023). Taxonomic expertise applied to the scientific study of natural history specimens in turn contributes to the ongoing curation and scientific accuracy of the physical specimens as well as to the growing pool of digital biodiversity data.

The mobilization of digital data on species occurrences from observations or physical specimens from collections in natural history institutions has accelerated over the last 10–15 years (Page et al., 2015; Nelson and Ellis, 2019), even if the amount of data that has been mobilized represents only a small proportion of what remains as 'dark data' on biodiversity (Johnson et al., 2023), with the scientific value of both physical and digitized specimens recognized (Suarez and Tsutsui, 2004; Lister, 2011; Krell et al., 2016; Soltis, 2017; Funk, 2018; Meineke et al., 2019; Hedrick et al., 2020; Card et al., 2021; Moreau and Ware, 2021; Johnson et al., 2023). In support of digitization efforts, biodiversity data standards and identifiers have been implemented, under the FAIR data principles that ensure that digital data is findable, accessible, interoperable and reusable (Wilkinson et al., 2016). Open access to aggregated digital data on species and specimens, combined with analytical and technical advances, has opened unprecedented opportunities to analyse biodiversity knowledge in innovative ways, to address innovative questions, using ever growing sets of data and ever-increasing capacities to handle and analyse large and/or complex data sets. Species occurrence data obtained from specimen labels is increasingly enriched with meta-data derived from the physical specimens themselves, such as genetic or chemical composition, phenology, morphological-anatomical traits, diseases or parasite load, digital images or scans, among others, with the obvious benefit of providing open access to multi-dimensional data sets on biological entities, at least for the well-known and well-documented taxa that are readily available within natural history collections. The aggregation of unparalleled amounts of digital data on species and specimens has raised the profile of the growing body of global biodiversity knowledge and its impact, highlighting the importance of unlocking and interpreting data from within the world's natural history collections. Open access to unprecedented amounts of biodiversity data provides opportunities for the taxonomic community, as users or contributors of data as well as the key curators of taxonomic/nomenclatural information and the taxonomic backbone that links it all together.

Biodiversity data mobilization and aggregation efforts rely on taxonomic information, with the taxonomic backbone of life on Earth forming the framework within which data is organized and data are linked to it via scientific names. The names and nomenclatural structure used to group them allow for the aggregation, interconnection and retrieval of information. Data derived from specimens ensure long-term reproducibility, although the accuracy of the digital data is directly linked to the accuracy of the taxonomic identification of the specimens themselves. Taxonomists and natural history collection curators know that specimens are not always correctly identified by the persons that collected and/or determined them (Costa et al., 2015; Goodwin et al., 2015). In addition to this, the name applied to a specimen at one point in time may change as taxonomic concepts are revisited and revised, which progresses our knowledge, but implies the need for taxonomic verification of specimens to correctly assign them as a consequence of taxonomic dynamics. The taxonomic verification of specimens is something that is outside of the scope and capacity of normal curatorial tasks where nomenclatural changes resulting from revisions may be managed as part of routine nomenclatural updating, but the re-determination of large numbers of accessioned specimens is not. The juncture between the physical specimen and its digital equivalent is one of the challenges that is being currently addressed from an informatics perspective by international data aggregators, such as the Global Biodiversity Information Facility or Distributed System of Scientific Collections ESFRI, with the need to coordinate information, for example, annotations, determinations or verifications, between the two representations of the same entity, the physical specimen and its digital twin. Taxonomy and taxonomists have a key role to play in shaping the tools needed to ensure the taxonomic component of two-way linking, making the annotations of the physical or digital specimens based on their taxonomic expertise, and curating the nomenclatural and classification frameworks used for structuring and linking data. Inherent in this process is safeguarding the correct citation of scientific names, taxonomic information and taxonomic works. But more important still is the provision of the conceptual framework needed to understand the natural world, the definition of the entities that it contains, and the establishment of a single name for each that is then used to communicate to share and organize the vast body of biodiversity data and knowledge.

The scientific processes involved in the discovery, description, documentation and phylogenetic classification of living entities have become increasingly collaborative and technological with much emphasis placed on developing integrative approaches to taxonomy (Dayrat, 2005; Padial et al., 2010). Technological advances have also provided opportunities to develop distributed networks of taxonomic experts as well as trans-disciplinary actions or projects that involve taxonomic expertise. Technical and technological advances, access to huge numbers of digital specimens, networking capacity, and trans-disciplinary collaborations are all opportunities for the taxonomic community. To fully profit from the opportunities in hand as well as to be well-positioned to tackle future emerging issues, proactivity, foresight, and inventiveness, combined with a positive attitude, are key. These should be supported by increased efforts to engage, listen, discuss, react, collaborate and interact from a positive rather than negative perspective, both within the community and with the stakeholders in taxonomic information. Ideally, with a slightly utopian twist, the taxonomic community should initiate, co-develop, implement, and monitor a *Global Strategy for Taxonomy* to coordinate activities and fully profit from the current and future opportunities in technology, analytical capacities, tool development as well as to develop research possibilities and outreach, education or training activities as well as realize across-borders capacity building.

## A FOCUS ON THE SOLUTIONS AND NOT THE PROBLEMS

Reflections are needed from within and across the taxonomic community on whether previous efforts to expound the taxonomic impediment and thus ensure that the situation for taxonomy improved or that taxonomy and taxonomists were less misunderstood have succeeded or failed, and why. To explain what taxonomy is and why it is important, to communicate the objectives and goals of taxonomy, to reinforce the position of taxonomists as competent, hypothesis-driven scientists and to strengthen the recognition of taxonomy as underpinning biodiversity knowledge are not new or in any way innovative propositions. What appears to be needed is a fresh approach to how we do any one or all of the above. How we communicate our science and our expertise, how we position ourselves within the scientific and biodiversity landscapes, how we value the taxonomic information we produce, and how we explain what we are doing and why it is important, is just as important as what we are saying. What we are saying is likely to become more impactful if framed in a positive manner. With this in mind, we also need to focus on analyzing how taxonomy is framed by ourselves and others, what we are saying about taxonomy and why, what it is useful for and used for, why we really do it, who really needs it, what stakeholders really want as outcomes, and what the community goals for taxonomy really are.

Discourse with taxonomists, from personal experience, often leads to exchange on what is lacking, not right, not working or otherwise unsatisfactory. The lack of taxonomic experts with a North-South bias, an aging cohort, taxonomy teaching non-existent in schools and disappearing at universities, and limited career options are frequently evoked. The general lack of funding, with the funding base disappearing, pressures on institutions to raise their own revenues increasing, and competition growing with other disciplines, cited alongside the notion that the image of taxonomy is rather decrepit and staid. Concerns are often raised about the lack of due recognition of taxonomy as a science and lack of understanding of the importance of its outcomes. Contextual problems like the lack or depauperate state of infrastructure, the lack of adequate tools, the 'slowness' of describing species, issues with nomenclature and the realization of a taxonomic backbone, challenges linked with taxonomic publishing, especially of taxonomic revisions or monographs, the negative effects of regulatory or legislative issues and the implementation of policies are also often evoked. Problem-orientated thinking may lead to the creation of an in-looking vision, a rather depressing messaging and a discourse that is not very engaging. The identification and elaboration of problems should be complemented by the search for and presentation of innovative, dynamic and novel solutions for the problems in hand to ensure that efforts bear tangible fruit. A problem-orientated focus in communications creates a dialogue that is difficult for an audience to hear and creates a set of circumstances that are difficult to for anyone to do anything about. In the light of the knowledge that 'nobody likes a whinger,' the sea-change that the taxonomic community is still waiting for may come closer to being realized if we are able to focus on finding, creating and proposing solutions (see the 'eight steps' in Evenhuis, 2007).

Taxonomic positivity requires sociological, behavioral and attitude changes to how we present and address our perceived problems, and on how we combine our soothsaying with ideas, options and solutions that advance our ambitions. The taxonomic community has little to lose and much to gain by adopting a taxonomy-positive attitude. Taking a no-whinging, solution-orientated approach to taxonomy and the challenges linked to taxonomic endeavors, backing this up with increased efforts to engage, listen, discuss, react, collaborate and interact from a positive rather than negative standpoint, both within the community and with the stakeholders in taxonomic information at large, may well open up our horizons.

## THINKING AND ACTING TAXONOMY POSITIVE

Changing the dialogue on taxonomy from a 'crisis mode' format (Agnarsson and Kuntner, 2007; Dubois, 2010; Löbl et al., 2023) to something other should be considered within the community as how we present our field and its complexities may well have an influence how it is perceived. Making a conscious choice to adopt a taxonomy-positive outlook is an important step in reclaiming taxonomy and reframing it. The taxonomic community, at different scales and at different levels, is able to contribute to positive change by collaborating, cooperating and communicating across the community, across borders and across languages, on common issues, projects and initiatives. The development of commonly agreed, realistic and tangible goals and targets will help frame future collective efforts. The community, as a distributed network, may focus on addressing taxonomic literacy and taxonomy education as well as fostering taxonomy training through teaching and mentoring, at whatever level is possible and realistic. If the community speaks with one voice, advocacy, interactions with stakeholders and dialogues at the science-policy interface can be done in a coordinated and strategic way, especially if undertaken as part of a global taxonomy strategy.

Key roles for taxonomy practitioners in effecting taxonomic positivity are assuring the promotion of the fascination for living species and nature objects, promoting taxonomy literacy, engaging in education or training activities and becoming involved at the science-public, science-stakeholders or science-policy interfaces. People's general curiosity for the natural world is what led to the creation of natural history institutions and the vast global distributed natural history collection in the first place (Johnson et al., 2023). The public continues to be fascinated by nature and the natural world, retaining curiosity about biological entities, where they are and how they live. It is at this interface where there is an exciting opportunity to address taxonomic literacy, to explain the dynamic nature of taxonomy, and the necessity for the ongoing reassessment of what is known about biological entities, what they are, where they live and how they are related. Effectively, combining curiosity with science and scientific observations goes some way to reinforcing the notion that taxonomy is all around us and that we use it all the time. It is through outreach activities that the public can gain a fuller appreciation of the scientific processes involved in taxonomy, especially if taxonomists are able to transmit the scientific challenges or the intrigues they have encountered, the dynamism of the scientific processes they engage in and the joy they experience about the natural world.

The exploration of our living planet, the description and documentation of past and present organisms and the ongoing development of biological, genetic or other sorts of knowledge on past or present entities is one of the most profound, exciting and scientifically engaging endeavors imaginable. It is one that we, the taxonomic community, should capitalize on. Stories, experiences, passion for the living world, and curiosity about nature are our currency for outreach and communications. Reconnecting with our own roots within our field, rekindling the passion(s) that led us there and expressing our scientific curiosity about nature are some of the holistic ways of communicating with the public, school children or students. Showing the joy of our discoveries and demonstrating the different methods we use to understand species and establish how they are related provide a window into the scientific processes involved in taxonomy. Coupled with genuine passion for what we do and stories about the outcomes or revelations we encounter along the way, the science of taxonomy becomes more tangible, more real, more understandable. The scientific drivers that stimulate taxonomists to explore, describe and re-evaluate biodiversity to build up biodiversity knowledge, and adapt interpretations based on new scientific findings, are often what attracted scientists to the field of taxonomy in the first place. They are also the key ingredients in outreach activities that allow for human connection and that can stimulate an interest in the science of taxonomy and processes behind our discoveries and findings. The sharing of the emotions experienced when working with natural history specimens, undertaking fieldwork, describing or circumscribing species, exploring evolutionary hypotheses about them or finding organisms that are new to science is powerful. Combine this with a taxonomy-positive approach and we can make small, individual steps toward changing our situation and surroundings for the better.

## ACKNOWLEDGMENTS

I would like to thank Quentin Wheeler and David Williams who invited me to participate in the New Taxonomy Symposium in Reading, England, on the 20th June 2024, and to the Systematics Association for hosting this second symposium on the topic. Ana Casino is thanked for our many conversations on the European natural history institution scene, the strategic positioning of the Consortium of European Taxonomic Facilities (CETAF), the elaboration of its research agenda, especially around the topic of supporting taxonomy, and the development of a taxonomy-orientated project at the heart of the CETAF community, TETTRIS. I would also like to acknowledge the many, many discussions I have had with Erik Smets and Thierry Bourgoin on science in natural history institutions, taxonomy and the challenges facing taxonomy and taxonomists, big-thinking or out of the box thinking, and on evoking change. I would like to acknowledge the CETAF community and its members who participated in a series of workshops on the role of natural history collections, the role of taxonomy and the most pressing scientific questions that European natural history institutions should address with the view of elaborating a Research Agenda for European natural history collections. The Conservatory and Botanical Garden of Geneva is acknowledged for providing an excellent environment in which to be a collections curator and undertake taxonomic research, with my colleagues thanked for stimulating discussions along the way.

## REFERENCES


Agnarsson, I.and Kuntner, M. (2007) Taxonomy in a changing world: Seeking solutions for a science in crisis. *Systematic Biology*, 56(3): 531–539, https://doi.org/10.1080/10635150701424546.

Bebber, D.P., Carine, M.A., Wood, J.R., Wortley, A.H., Harris, D.J., Prance, G.T., Davidse, G., Paige, J., Pennington, T.D., Robson, N.K. and Scotland, R.W. (2010) Herbaria are a major frontier for species discovery. *Proceedings of the National Academy of Sciences U. S. A.*, 107(51): 22169–22171, https://doi.org/10.1073/pnas.1011841108.

Bird, L.H., Thomas, E.F., Wenzel, M. and Lizzio-Wilson, M. (2024) Thinking about the future: Examining the exacerbating and attenuating factors of despair-induced climate burnout. *Journal of Environmental Psychology,* 98: 102382, https://doi.org/10.1016/j.jenvp.2024.102382.

Britz, R., Hundsdörfer, A. and Fritz, U. (2020) Funding, training, permits—the three big challenges of taxonomy. *Megataxa*, 2020; 1: 49–52, https://doi.org/10.11646/megataxa.1.1.10.

Card, D.C., Shapiro, B., Giribet, G., Moritz, C. and Edwards, S.V. (2021) Museum Genomics. *Annual Review of Genetics*, 55: 633–659, https://doi.org/10.1146/annurev-genet-071719-020506.

CBD (2006). Global Taxonomy Initiative - What is Taxonomy? Secretariat of the Convention of Biological Diversity https://www.cbd.int/gti/taxonomy.shtml. Accessed: 18 June 2024.

Coleman, C.O. (2015) Taxonomy in times of the taxonomic impediment – Examples from the community of experts on Amphipod Crustaceans, *Journal of Crustacean Biology*, 35(6): 729–740, https://doi.org/10.1163/1937240X-00002381.

Costa, H., Foody, G.M., Jiménez, S., and Silva, L. (2015) Impacts of species misidentification on species distribution modeling with presence-only data. *ISPRS International Journal of Geo-Information*, 4(4): 2496–2518, https://doi.org/10.3390/ijgi4042496.

Costello, M.J., May, R.M., Stork, N.E. (2013) Can we name earth's species before they go extinct? *Science*, 339, 413–416, https://doi.org/10.1126/science.1230318.

Darwin, C.R. (1871) *The Descent of Man, and Selection in Relation to Sex.* Volume 1 (1st edition). John Murray, London.

Davies, C.C. (2023) The herbarium of the future. *Trends in Ecology and Evolution*, 38(5): 412–423, https://doi.org/10.1016/j.tree.2022.11.015.

Dayrat, B. (2005) Towards integrative taxonomy. *Biological Journal of the Linnean Society*, 85: 407–415, https://doi.org/10.1111/j.1095-8312.2005.00503.x.

de Carvalho, M.R., Bockmann, F.A., Amorim, D.S., Brandão, C.R.F., de Vivo, M., de Figueiredo, J.L., Britski, H.A., de Pinna, M.C.C., Menezes, N.A., Marques, F.P.L., Papavero, N., Cancello, E.M., Crisci, J.V., McEachran, J.D., Schelly, R.C., Lundberg, J.G., Gill, A.C., Britz, R., Wheeler, Q.D., Stiassny, M.L.J., Parenti, L.R., Page, L.M., Wheeler, W.C., Faivovich, J., Vari, R.P., Grande, L., Humphries, C.J., DeSalle, R., Ebach M.C. and Nelson, G. (2007) Taxonomic impediment or impediment to taxonomy? A commentary on systematics and the cybertaxonomic-automation paradigm. *Evolutionary Biology*, 34: 140–143, https://doi.org/10.1007/s11692-007-9011-6.

de Lange, E., Sharkey, W, Castelló y Tickell, S., Migné, J., Underhill, R., and Milner-Gulland, E.J. (2022) Communicating the biodiversity crisis: From 'warnings' to positive engagement. *Tropical Conservation Science*, 15, https://doi.org/10.1177/19400829221134893.

Dijkstra, K.D. (2016) Natural history: Restore our sense of species. *Nature*, 533: 172–174, https://doi.org/10.1038/533172a.

Dubois, A. (2010) Taxonomy in the century of extinctions: Taxonomic gap, taxonomic impediment, taxonomic urgency. *Taprobanica,* 2(1): 1–5, https://doi.org/10.4038/tapro.v2i1.2702.

Dupérrén, N. (2020) Old and new challenges in taxonomy: What are taxonomists up against? *Megataxa*, 001(1): 059–062, https://doi.org/10.11646/megataxa.1.1.12.

Ebach, M.C., De Carvalho, M.R. and Nihei, S.S. (2011a) Saving our science from ourselves: The plight of biological classification. *Revista Brasileira de Entomologia*, 55(2): 149–153, https://doi.org/10.1590/S0085-56262011005000005.

Ebach, M.C., Valdecasas, A.G., and Wheeler, Q.D. (2011b) Impediments to taxonomy and users of taxonomy: Accessibility and impact evaluation. *Cladistics*, 27(2011): 550–557, https://doi.org/10.1111/j.1096-0031.2011.00348.x.

Engel, M.S., Ceríaco, L.M.P., Daniel, G.M., Dellapé, P.M., Löbl, I., Marinov, M., Reis, E.R., Young, M.T., Dubois, A., Agarwal, I., Lehmann P.A., Alvarado, M., Alvarez, N., Andreone, F., Araujo-Vieira, K., Ascher, J.S., Baêta, D., Baldo, D., Bandeira, S.A., Barden, P., Barrasso, D.A., Bendifallah, L., Bockmann, F.A., Böhme, W., Borkent, A., Brandão, C.R.F., Busack, S.D., Bybee, S.M., Channing, A., Chatzimanolis, S., Christenhusz, M.J.M., Crisci, J.V., D'elía, G., Da Costa, L.M., Davis, S.R., De Lucena, C.A.S., Deuve, T., Fernandes Elizalde, S., Faivovich, J., Farooq, H., Ferguson, A.W., Gippoliti, S., Gonçalves, F.M.P., Gonzalez, V.H., Greenbaum, E., Hinojosa-Díaz, I.A., Ineich, I., Jiang, J., Kahono, S., Kury, A.B., Lucinda, P.H.F., Lynch, J.D., Malécot, V., Marques, M.P., Marris, J.W.M., Mckellar, R.C., Mendes, L.F., Nihei, S.S. Nishikawa, K., Ohler, A., Orrico, V.G.D., Ota, H., Paiva, J., Parrinha, D., Pauwels, O.S.G., Pereyra, M.O., Pestana, L.B., Pinheiro, P.D.P., Prendini, L., Prokop, J., Rasmussen, C., Rödel, M.-O., Trefaut Rodrigues M., Rodríguez, S.M., Salatnaya, H., Sampaio, Í., Sánchez-García, A., A Shebl, M., Santos, B.S., Solórzano-Kraemer, M.M., Sousa, A.C.A., Stoev, P., Teta, P., Trape, J.-F., Van-Dúnem Dos Santos, C., Vasudevan, K., Vink, C.J., Vogel, G., Wagner, P., Wappler, T., Ware, J.L., Wedmann, S., and Kusamba Zacharie, C. (2021) The taxonomic impediment: A shortage of taxonomists, not the lack of technical approaches. *Zoological Journal of the Linnean Society*, 193(2): 381–387, https://doi.org/10.1093/zoolinnean/zlab072.

Evenhuis, N.L. (2007) Helping solve the "other" taxonomic impediment: Completing the eight steps to total enlightenment and taxonomic Nirvana. *Zootaxa*, 1407: 3–12

Funk, V.A. (2018) Collections-based science in the 21st century. *Journal of Systematics and Evolution*, 56: 175–193, https://doi.org/10.1111/jse.12315.

Gerber, L.R., Reeves-Blurton, Z., Gueci, N., Iacona, G.D., Beaudette, J.A. and Pipe, T. (2022) Practicing mindfulness in addressing the biodiversity crisis. *Conservation science and practice*, 5(7), e12945, https://doi.org/10.1111/csp2.12945.

Goodwin, Z.A., Harris, D.J., Filer, D., Wood, J.R.I., and Scotland, R.W. (2015) Widespread mistaken identity in tropical plant collections. *Current Biology*, 25(22): R1066–R1067, https://doi.org/10.1016/j.cub.2015.10.002.

Grimaldi, D.A., and Engel, M.S. (2007) Why descriptive science still matters. *BioScience*, 57(8): 646–647, https://doi.org/10.1641/B570802.

Hedrick, B.P., Heberling, J.M., Meineke, E.K., Turner, K.G., Grassa, C.J., Park, D.S., Kennedy, J., Clarke, J.A., Cook, J.A., Blackburn, D.C., Edwards, S.V. and Davis, C.C. (2020) Digitization and the future of natural history collections. *BioScience*, 70(3): 243–251, https://doi.org/10.1093/biosci/biz163.

Johnson, K.R. Owens, I.F.P. and Global Collection Group (2023) A global approach for natural history museum collections in the 21st century. *Science*, 379(6638): 1192–1194, https://doi.org/10.1126/science.adf6434.

Krell, F.-T., Klimeš, P., Rocha, L.A., Fikáček, M., and Miller, S.A. (2016) Preserve specimens for reproducibility. *Nature*, 539: 168, https://doi.org/10.1038/539168b.

Larsen, B.B, Miller, E.C., Rhodes, M.K., and Weins, J.J. (2017) Inordinate fondness multiplied and redistributed: The number of species on Earth and the new pie of life. *The Quartely Review of Biology*, 92: 229–265, https://doi.org/10.1086/693564.

Lister, A.M. (2011) Natural history collections as sources of long-term data sets. *Trends in Ecology and Evolution*, 26: 153–154, https://doi.org/10.1016/j.tree.2010.12.009.

Löbl, I., Klausnitzer B., Hartmann M., Krell F.-T. (2023) The silent extinction of species and taxonomists—an appeal to science policymakers and legislators. *Diversity*, 15(10): 1053, https://doi.org/10.3390/d15101053.

Meineke, E.K., Davies, T.J., Daru, B.H. and Davis, C.C. (2019) Biological collections for understanding biodiversity in the Anthropocene. *Philosophical Transactions of the Royal Society B, Biological Sciences*, 374: 20170386, https://doi.org/10.1098/rstb.2017.0386.

Mora, C., Tittensor, D.P., Adl, S., Simpson, A.G.B. and Worm, B. (2011) How many species are there on earth and in the ocean? *PLoS Biology*, 9: e1001127, https://doi.org/10.1371/journal.pbio.1001127.

Moreau, C.S., and Ware, J.L. (2021) Fund natural-history museums, not de-extinction. *Nature*, 598(7879): 32, https://doi.org/10.1038/d41586-021-02710-4.

Nanglu, K., de Carle, D., Cullen, T.M., Anderson, E.B., Arif, S., Castañeda, R.A., Chang, L.M., Iwama, R.E., Fellin, E., Manglicmot, R.C., Massey, M.D. and Astudillo-Clavijo, V. (2023) The nature of science: The fundamental role of natural history in ecology, evolution, conservation, and education. *Ecology and Evolution*, 23;13(10): e10621, https://doi.org/10.1002/ece3.10621.

Nelson, G. and Ellis, S. (2019) The history and impact of digitization and digital data mobilization on biodiversity research. *Philosophical Transactions of the Royal Society B, Biological Sciences*, 374: 20170391, https://doi.org/10.1098/rstb.2017.0391.

Padial, J.M., Miralles, A., De la Riva, I., and Vences, M., (2010) The integrative future of taxonomy. *Frontiers in Zoology*, 7: 1–14, https://doi.org/10.1186/1742-9994-7-16.

Page, L.M., MacFadden, B.J., Fortes, J.A., Soltis, P.S. and Riccardi G. (2015) Digitization of biodiversity collections reveals biggest data on biodiversity. *BioScience*, 65: 841–842, https://doi.org/10.1093/biosci/biv104.

Pihkala, P. (2020) Anxiety and the ecological crisis: An analysis of eco-anxiety and climate anxiety. *Sustainability*, 12(19), 7836.

Raposo, M.A., Kirwan, G.M., Lourenço, A.C.C., Sobral, G., Bockmann, F.A. & Stopiglia, R. (2020) On the notions of taxonomic 'impediment', 'gap', 'inflation' and 'anarchy', and their effects on the field of conservation. *Systematics and Biodiversity*, 19(3): 296–311, https://doi.org/10.1080/14772000.2020.1829157.

Sandall, E.L., Maureaud, A.A., Guralnick, R., McGeoch, M.A., Sica, Y.V. Rogan, M.S., Booher, D.B., Edwards, R., Franz, N., Ingenloff, K., Lucas, M., Marsh, C.J., McGowan, J., Pinkert, S., Ranipeta, A., Uetz, P., Wieczorek, J., and Jetz, W. (2023) A globally integrated structure of taxonomy to support biodiversity science and conservation. *Trends in Ecology & Evolution*, 38(12): 1143–1153, https://doi.org/10.1016/j.tree.2023.08.004.

Secretariat of the Convention on Biological Diversity (2006) Guide to the global taxonomy initiative. *CBD Technical Series*, 30: 1–195.

Soltis, P.M. (2017) Digitization of herbaria enables novel research. *American Journal of Botany*, 104(9): 1281–1284, https://doi.org/10.3732/ajb.1700281.

Suarez A.V. and Tsutsui, N.D. (2004) The value of museum collections for research and society. *BioScience*, 54: 66–74, https://doi.org/10.1641/0006-3568(2004)054[0066:TVOMCF]2.0.CO;2.

Wheeler, Q.D., Knapp, S., Stevenson, D.W., Stevenson, J., Blum, S.D., Boom, B.M, Borisy, G.G., Buizer, J.L., De Carvalho, M.R., Cibrian, A., Donoghue, M.J., Doyle, V., Gerson, E.M., Graham, C.H., Graves, P., Graves, S.J., Guralnick, R.P., Hamilton, A.L., Hanken, J., Law, W., Lipscomb, D.L., Lovejoy, T.E., Miller, H., Miller, J.S., Naeem, S., Novacek, M.J., Page, L.M., Platnick, N.I., Porter-Morgan, H., Raven, P.H., Solis, M.A., Valdecasas, A.G., Van Der Leeuw, S., Vasco, A., Vermeulen, N., Vogel, J., Walls, L.,

Wilson, E.O. and Woolley, J.B. (2012) Mapping the biosphere: Exploring species to understand the origin, organization and sustainability of biodiversity. *Systematics and Biodiversity*, 10: 1–20, https://doi.org/10.1080/14772000.2012.665095.

Wheeler, Q.D., Raven, P. and Wilson, E.O. (2004) Taxonomy: Impediment or expedient? *Science*, 303: 285, https://doi.org/10.1126/science.303.5656.285.

Wheeler Q.D. (2008a) Taxonomic shock and awe, pp. 211–226. In: *The New Taxonomy* (ed Q.D. Wheeler). Boca Raton: CRC Press; 2008.

Wheeler, Q.D. (2008b) Introductory: Toward the new taxonomy, pp. 1–17. In *The New Taxonomy*. Volume 76. (ed Q.D. Wheeler). Boca Raton: CRC Press.

Wheeler, Q.D. (ed.) (2008c) *The New Taxonomy*. Systematics Association Special Volumes Series. Volume 76. Boca Raton: CRC Press.

Wheeler, Q.D. and Valdecasas, A.G. (2007) Taxonomy: Myths and misconceptions. *Anales del Jardín Botánico de Madrid*, 64(2): 237–241.

Wheeler, Q.D. and Valdecasas, A.G. (2005) Ten challenges to transform taxonomy. *Graellsia*, 61(2): 151–160.

Wilkinson, M., Dumontier, M., Aalbersberg, I., Appleton, G., Axton, M., Baak, A., Blomberg, N., Boiten, J.-W, Bonino da Silva Santos, L., Bourne, P.E., Bouwman, J., Brookes, A., Clark, T., Crosas, M., Dillo, I., Dumon, O., Edmunds, S., Evelo, C.T., Finkers, R., Gonzalez-Beltran, A., Gray, A.J.G., Groth, P., Goble, C., Grethe, J.S., Heringa, J., 't Hoen, P.A.C., Hooft, R., Kuhn, T., Kok, R., Kok, J., Lusher, S.J., Martone, M.E., Mons, A., Packer, A.L., Persson, B., Rocca-Serra, P., Roos, M., van Schaik, R., Sansone, S.-A., Schultes, E., Sengstag, T., Slater, T., Strawn, G., Swertz, M.A., Thompson, M., van der Lei, J., van Mullingen, E., Velterop, J., Waagmeester, A., Wittenburg, P., Wolstencroft, K., Zhao, J. and Mons, B. (2016) The FAIR guiding principles for scientific data management and stewardship. *Scientific Data*, 3(1): 160018, https://doi.org/10.1038/sdata.2016.18.

# 12 A Single Authoritative List of the World's Species

## *Background and Road Map*

*Frank E. Zachos, Stijn Conix, Les Christidis, Aaron M. Lien, and Stephen T. Garnett*

## INTRODUCTION – WHY TAXONOMY MATTERS AND WHY TAXONOMISTS STILL OFTEN DISAGREE

Taxonomy, or systematics,[1] is perhaps the most fundamental biological discipline – more or less all other realms of the life sciences, from molecular biology and genetics to physiology, morphology and ecology, depend in one way or another on taxonomic information. Among the first of the biological sciences to develop modern standards – to this day, we by and large follow the procedures standardised by Linnaeus in the 18th century – taxonomy was for a long time the core discipline of what is today called biology. However, taxonomy's appeal has waned. Other disciplines have taken the lead in biology, and most of the public funding with it. While support for these disciplines is well justified, taxonomy's wallflower image is not. Many biological disciplines still rely heavily on the classifications (and related understanding of the evolutionary relationships among taxa) provided by taxonomists. While less popular than it was, taxonomy remains a cutting-edge scientific discipline that uses the latest theoretically and technologically rigorous methods to bring order to the Tree of Life.

Not only does taxonomy serve as the backbone to many biological disciplines, it also plays a crucial role in conservation and management, which are increasingly and rightfully seen as disciplines combining life and social sciences. Species names (or sometimes names at other ranks such as the subspecies) are at the core of multilateral environmental agreements such as the Convention on Biological Diversity (CBD), the Convention on International Trade in Endangered Species (sic!) of Wild Fauna and Flora (CITES) and the Convention on Migratory Species (CMS). Because of this, taxonomy has important legal dimensions, and taxonomic decisions have consequential ramifications 'downstream' within biology and beyond. The quip 'taxonomy as destiny' (May, 1990) is no exaggeration, and taxonomic definitions and decisions are subject to legal debate (e.g., Wheeler, 2014 on orcas *Orcinus orca* and George and Mayden, 2005 more generally). There have even been petitions made by law firms to have government authorities define the terms species and subspecies so that binding legal decisions can be made (Schiff and Wood, 2017).

DOI: 10.1201/b22822-12

Given that taxonomy, and particularly the species level in our classifications, has such outstanding relevance – from enumerating biodiversity to understanding evolutionary biology and palaeontology (Thomson et al., 2021), creating Red Lists of species at risk of extinction and regulating wildlife trade (CITES) – it might surprise non-experts that there are still challenges within the science of taxonomy that complicate the use of taxonomic information for end-users. Some difficulties are routine housekeeping: species lists for many taxa are outdated, incomplete, challenging to access, or confined to limited geographic regions. Taxonomic lists are also plagued by practical problems related to nomenclature. Many species are known by multiple names (synonyms), and sometimes a single name is employed to refer to multiple species. Consequently, it is not always straightforward to determine the group of organisms to which a specific name refers, or whether it corresponds to another group of interest. Resolving such ambiguities is both time-consuming and challenging, as uncovering the history of name usage can be difficult, and the practice of anchoring name references to type specimens allows for interpretive flexibility (Kennedy et al., 2006; Franz et al., 2008; Lepage et al., 2014).

In addition to these practical problems, there is also a deeper issue on which taxonomists often disagree – the fundamental question of what a species is. In one survey there was a 79% chance that two taxonomists surveyed about species concepts would have different definitions (Stankowski and Ravinet, 2021). There is often uncertainty about whether two populations comprise a single species or two. This is the heart of the species problem – the lack of a universally accepted and universally applicable species concept that ensures that what we call a species in, say, bacteria, is commensurable with what we call a species in mammals or insects. There is a huge body of literature on the species problem and the resulting taxonomic 'disorder' (e.g., Zachos, 2016; Wilkins, 2018; Pavlinov, 2023; Maddison and Whitton, 2023), but it is not within the scope of this chapter to discuss this in detail. Very briefly, the fundamental problem is that the discrete classification scheme of taxonomy is imposed on the (often ephemeral) results of a continuous process (evolutionary diversification) and that these results are of a fractal nature – there are taxa within taxa within even more inclusive taxa. This means that taxonomy is fundamentally a two-step enterprise (see Zachos et al., 2020; Zachos, 2022). While taxa can be objectively delimited – i.e., they are discovered – the decision which groups to recognise as species is not entirely an objective matter: taxa are real, and ranks are human constructs.

There are two reasons for this. First, there is not always one single best way to group organisms into taxa. There are many biological properties of interest in delimiting evolutionary groups, and sometimes these cross-classify organisms into groups (Ereshefsky, 2000). While all these groups may be real, and delimiting them thus an objective matter, settling on a single correct classification to use requires an executive decision that is not in the same way fully determined by the data.

Secondly, and more importantly, judgement on which of the many hierarchical and encaptic levels should be recognised as the species level is not entirely objective. An analogy with mountains, which just like species taxa are natural phenomena with fuzzy boundaries, might help. Altitudinal data within a mountain range are a matter of rigorous scientific quantification, but the question of how many mountains there are in, say, the Alps is not. Accordingly, Thiele et al. (2021, p. 618) use map-making and mountain ranges to show why taxonomy is as good a scientific discipline as any despite the species problem:

> Consider two cartographers independently tasked with drawing a map of a complex mountainous landscape from a set of primary data (gridded height measurements). Both are likely to agree on the pattern of topography (they will draw equivalent topographic lines). But as soon as they start marking peaks on their maps [how many mountains are there?], they may diverge: one may mark a single peak where the other marks two peaks separated by a small saddle. The topographic map is a first-order abstraction from the primary data, but the delimitation of peaks is a second-order abstraction. Neither map is false, they are just differently marked.
>
> In the same way, the delimitation of taxa from primary data (the patterns of variation among organisms) is a high-order abstraction. And just as the cartographers can still draw highly useful maps even without a rigorous, objective, shared definition of 'peak', so too can taxonomists usefully 'map' biodiversity using concepts (such as species) that defy rigorous, objective definition. The failure to agree on a single, objectively determined species definition is not a failure of taxonomy; rather, it reflects the inadequacy of the question 'what exactly is a species?' [...] There will always be friction caused by the exigencies of abstraction, no matter what species concept one chooses.

This is the underlying, ultimate reason for fundamental taxonomic disagreement – disagreement that cannot be overcome by more data or better, more rigorous analyses. Organisms are grouped into taxa based on data, but ranking includes an executive decision, and one taxonomist's species often turns out to be another taxonomist's subspecies. While grouping, therefore, is a testable hypothesis, ranking is not, and the species problem as we know it is here to stay.

Of course, there are clear cases where two taxa have been separated for so long that no two taxonomists disagree. But boundaries are fuzzy in a grey area of relatively recent lineage sundering. For example, whether there is one species of red deer (*Cervus elaphus*) or several (as has been suggested, e.g., by Groves and Grubb, 2011), is not a disagreement over the biological properties of the different red deer populations but about how these properties are to be translated into ranks. It is a matter of lumping or splitting given the same data, in other words: a matter of taxonomic taste. As a result, well-studied groups in which many taxonomists work will produce many such disagreements. Sometimes this even results in several different global species lists for the same taxon that are used by different stakeholders (e.g., four such lists in birds or vascular plants, see Sandall et al., 2023 and Schellenberger Costa et al., 2023). In a study of avian raptors, only 68% of taxa were consistent across four global lists (McClure et al., 2020).

The ranking problem and the resulting incommensurability ('apples and oranges') of the species rank assigned by different authors and in different taxonomic groups cannot easily be solved, and perhaps not at all under the present codes of nomenclature. As a result, different stakeholders use different classifications, and non-expert users are forced to make arbitrary choices or delve into the taxonomic literature to make substantiated ones. In an ideal world, we might have a rich bioinformatics system that makes all competing taxonomic information available to users in an easy way, while allowing us to map between different classifications that users with different aims might choose to use (Sterner et al., 2020). However, we are currently nowhere near such a system, and it is unclear whether we will ever get there (Conix et al., 2022). In that light, it would clearly be progress if a certain standardisation and a single list of accepted species for each taxon could be achieved. In these lists, taxonomic specialists could make the (sometimes arbitrary) contested decisions to get to a single list that non-specialist users can rely on. These taxon-specific lists combined would then be the single authoritative list of the world's species.

Apart from a consensus within each taxonomic community that includes scientific debates, such a global list would need some level of governance. Garnett et al. (2020, see next section and also Conix et al., 2022) laid out potential principles that would maximise legitimisation and acceptance of such a list. Importantly, governance in this sense should not interfere with the taxonomic science as carried out and discussed by the various experts within their communities, but rather create quality standards for accepting sublists (birds, mammals, vascular plants etc.) into the global list. These standards do not refer to the content of the list but to its consistency and the procedure of its generation.

The benefits of a single global list would include, among other things, consistent usage by stakeholders (for example, so that the same species name in two different lists really refers to the same organisms), increased transparency about list quality, and a reduction in duplicated work. However, a single list is just, and can only ever be, a distillation of presently available taxonomic knowledge. It is by no means a replacement for the desperately needed boost to taxonomic research in the face of the ongoing extinction crisis (see Wheeler, 2024).

## ORIGIN OF THE SINGLE-LIST PROJECT

### Catalogue of Life

Creating a list of the world's species has long been an aspiration for those with an interest in taxonomy. As recounted by Hobern et al. (2021), the most advanced approach has been the Catalogue of Life (COL). In the 1990s, two groups had taken on the task of creating a global list. One, Species 2000, was established in 1996 as a task group under the Taxonomic Databases Working Group (now Biodiversity Information Standards) which had been set up in 1985 to develop standards for the exchange of biodiversity data and is responsible for the Darwin Core standards. Species 2000 was also sponsored by the Committee on Data for Science and Technology (a committee of the International Science Council), the International Union of Biological Sciences (IUBS), and the International Union of Microbiological Societies. In the same year that Species 2000 was initiated, the Integrated Taxonomic Information System was established at the behest of the White House Subcommittee on Biodiversity and Ecosystem Dynamics and became a long-lived partnership between government agencies in the USA, Canada, and Mexico to create a database with information on global species names and their hierarchical classification. In 2001, these two initiatives were combined to form COL. The original partnership was upgraded in 2017 as Catalogue of Life Plus through the Catalogue of Life Plus project (Bánki et al., 2019).

COL itself has a relatively simple structure, aggregating lists from authoritative sources into a single list within an agreed taxonomic backbone. The list is updated monthly as well as there being an annual Checklist which provides a dated, verifiable reference for the usage of names, which was a collaborative project between COL and the Global Biodiversity Information Facility (GBIF) among others. Among the important roles of COL is as a taxonomic backbone to GBIF, an international network and data infrastructure funded by governments globally which aims to provide open access to data about all life forms that was created in 1999 following a recommendation from the Biodiversity Informatics Subgroup of the Organization for Economic Cooperation and Development's Megascience Forum. As of 2024, GBIF agreed to serve as the administrative host for the Species 2000 Secretariat that runs COL. Among the initiatives that have already emerged from the GBIF/COL partnership is ChecklistBank, an online repository for species checklist datasets which allows the ready comparison of lists to help identify synonyms and variations in taxonomic treatment. COL is also used by the Biodiversity Heritage Library, the Barcode of Life Data System, the Encyclopedia of Life, and GBIF and, in 2023, was accepted as one of the Global Core Biodata Resources by the Global Biodata Coalition, an organisation that promotes biocuration and fosters support and funding for sustaining biological data resources.

## Global Species List Working Group

Independently of COL, Garnett and Christidis (2017) published a paper challenging the coherence of governance in taxonomy. There were two issues about which they expressed concern. The first was how lists of species are developed for taxonomic groups, the second was how these lists are compiled into a single global list of all life. The issue arose because, although groups, including COL, can assemble lists of species, the basis for decisions on which lists to accept is almost invariably opaque. This matters little for the vast majority of species – most of which are invertebrates – because there are so few taxonomists and so little dispute. However, for the small number of species, like vertebrates and plants, that have high social salience, taxonomists are numerous and dispute a frequent, and important, part of the taxonomic process (Thiele et al., 2021).

The paper by Garnett and Christidis (2017) attracted much criticism (e.g., Raposo et al., 2017; Thomson et al., 2018), partly because it was published in *Nature* and its title suggested the governance of taxonomy was anarchic, and partly because it was interpreted as a plea to control the ways in which taxonomy is conducted. The independence of taxonomy, however, was never challenged. Rather, the point being made was that inconsistent approaches to how species were defined and a lack of quality control in the rigour of treatments meant that conservation could be misdirected. The importance of keeping taxonomy informed by other considerations, but not dictated or influenced by them, was dealt with in a subsequent paper on stakeholder engagement (Conix et al., 2021).

As acknowledged by Conix (2019), Garnett and Christidis (2017) were raising important questions about quality control in taxonomy. Reviewers of journal articles may assess the quality of data and how it was analysed but rarely question the taxonomic decisions being made – it is assumed that such decisions are the prerogative of the authors. Furthermore, few journal editors or reviewers have the necessary taxonomic knowledge to assess such decisions because they are usually chosen for their technical expertise (e.g., molecular and/or phylogenetic methodologies) rather than their views on taxonomic hierarchies (explicitly taxonomic journals are an exception, of course).

In some cases, such as birds and mammals, it can be important to assess the number of contributors to the global list. As an example, the bovids (antelopes, cattle and relatives) were split into numerous species by two authors (Groves and Grubb, 2011), an approach widely criticised but adopted in the influential series *Handbook of the Mammals of the World* (HMW, Wilson and Mittermeier, 2011). Many of the species acknowledged by Groves and Grubb were so poorly defined that they would not even be considered subspecies by many other authors. Nevertheless, this major revision was taken up by others (Castelló, 2016), and poorly justified species were being recognised. It was not until a more representative group of mammalian taxonomists reviewed the Bovidae that their taxonomy was returned to pre-HMW days in their publication *All the Mammals of the World* (Lynx Nature Book, 2023, see also the ASM Mammal Diversity Database, https://www.mammaldiversity.org/). Moreover, many of the species recognised by Groves and Grubb were not then accepted as subspecies in this latest treatment. This bovid example illustrates poignantly why Garnett and Christidis (2017) advocated for some form of governance to provide the quality control and taxonomic rigour that is not always applied in journals.

Subsequent to the debate raised by Garnett and Christidis (2017), it became apparent to many of those involved that they shared concerns about list governance (https://theconversation.com/we-need-a-single-list-of-all-life-on-earth-and-most-taxonomists-now-agree-on-how-to-start-216006). As a result, the International Union for Biological Sciences was approached to support collaboration to discuss the governance of lists. In February 2020, on the birthday of Charles Darwin at the Charles Darwin University in the city of Darwin, Australia, a workshop was held that led to the creation of the Global Species List Working Group (GSLWG). This group published a list of ten principles for the global governance of species lists shortly afterwards (Garnett et al., 2020). This was followed first by six papers that explored different aspects of the process (Conix et al., 2021; Hobern et al., 2021; Lien et al., 2021; Pyle et al., 2021; Thiele et al., 2021; Thomson et al., 2021) and then with the results of a global survey of taxonomists to understand the extent to which they supported creation of such a list and their ideas on how it should be governed (Lien et al., 2023).

## TAXONOMIC LIST GOVERNANCE AND WHAT THE BIOLOGICAL COMMUNITY THINKS

Governance, generally speaking, is the system of rules, norms and strategies used by people to guide behaviour and decision-making (Robichau, 2011). The larger field of taxonomy has many governance institutions already in place. For example, the *International Code of Zoological Nomenclature* (ICZN), the *International Code of Nomenclature for algae, fungi, and plants* (ICN or ICNafp), the *International Code of Nomenclature for Cultivated Plants* (ICNCP), the *International Code of Nomenclature of Prokaryotes* (ICNP, in addition, the *Code of Nomenclature of Prokaryotes Described from Sequence Data* (SeqCode), https://registry.seqco.de/page/seqcode), and the *International Code of Virus Classification and Nomenclature* (ICVCN) establish clearly defined rules, or governance institutions, for the naming of new species. However, these codes only address the process of naming a species once a taxonomist has made a decision about its ranking. They are silent on the *process* of making these decisions and also on the process of aggregating species names into lists of agreed taxa. The goal of taxonomic list governance is to provide a clear set of rules and norms to guide the aggregation of species lists into a single list of life on Earth on the one hand, while avoiding interference with the science of taxonomy and taxonomic decisions on the other (Garnett et al., 2020).

How then can this balance be achieved, and is the taxonomic community likely to accept the outcome? Garnett et al. (2020) began by establishing a clear set of principles to guide the development of taxonomic list governance. These principles are designed to set boundaries on the scope of the governance effort such that list governance does not drift into the governance of taxonomic science. As such, these principles are explicit about the need for list governance not to interfere with the scientific freedom of individual taxonomists, that list decisions be grounded in science and not on other considerations, and that the process of list governance is distinct from other scientific activities undertaken by taxonomists, such as species description and species naming. Building from these principles, Pyle et al. (2021) suggested in detail the potential structure and content of a global list focused on the sole task of aggregating accepted species. This included proposing exactly what a global list would contain (taxonomic name usages, TNUs, indicating a specific treatment of a taxon), the breadth and granularity of a global list, and the information needed about each of the entries to be included in the global list (Pyle et al., 2021).

These principles and proposed list contents begin to answer the question of *what* a global list may look like but do not answer the question of *how* a global list is ultimately developed and maintained over time. This latter question is one of governance – how can the system of rules to regulate the development of a global list be developed, what do these rules consist of, and who is responsible for implementing them. Lien et al. (2023) took up this question by conducting a survey of the taxonomic community, as well as other scientists and users of taxonomic information. In total, a convenience sample of 1,134 responses was collected, comprising 560 taxonomists, 409 scientists from other disciplines, 50 users of taxonomy, and 115 respondents who selected more than one category. On the whole, the survey results show a remarkable degree of agreement within the community of taxonomists, as well as scientists in other disciplines and users of taxonomy.

Overall, 77% of all respondents agreed that a single global list would be beneficial. While all categories of respondents showed strong majorities supportive of a global list, users of taxonomic information showed the strongest support (88%), while taxonomists showed the least support (73%). When asked about challenges respondents currently experience with the use of taxonomic lists, the most commonly reported were that lists are outdated (64%), have nomenclatural problems

(55%) and that there are competing lists (48%). These common problems help to explain some of the reasons why support for a single list may be so high; a global list would need to address each of these challenges in order to create a list that is used and useful for taxonomists and other users of taxonomic information. Respondents also largely agreed on the list contents. Using the information about list entries suggested by Pyle et al. (2021), there was general agreement that entries should include unique, persistent identification numbers; treatment authors; classification at genus-level and higher; and information about any uncertainty in taxonomic classification or rank. Respondents were most ambivalent about the inclusion of non-Linnaean names and inclusion of fossil taxa, though this was manifested as a large number of neutral opinions rather than opinions that inclusion of this information would be undesirable.

Governance of a global list would involve the establishment of rules, or criteria, for inclusion of entries into the global list and the designation of an existing, or creation of a new organisation, to manage the list and implement these criteria. Of the criteria for inclusion of entries into a global list evaluated by survey respondents, the strongest support was for the authors of items submitted for inclusion to be acknowledged (92%), followed by standardisation of open-access licensing (90%), that items for aggregation into a global list should be based on science alone (89%), and that there should be a quality control process in place to evaluate submissions before inclusion in a global list (88%).

To implement these criteria, respondents were generally supportive of creation of a new, independent oversight board or commission (72%) or creating a new subcommittee of an existing aggregator of species lists such as COL (68%). There was relatively little support for empowering an existing non-governmental organisation that does not currently have a role in taxonomy or list aggregation (40%). Respondents generally supported an oversight body with representation from taxonomists (95%) who represent a wide diversity of taxa (91%) and geography (72%). Representation by nomenclatural commissions was also supported (69%). There was relatively weak support for inclusion of users of taxonomy, especially amongst taxonomists (38%). This may reflect concerns about inclusion of non-scientific considerations in decision-making if non-taxonomists are involved in oversight. Preferences for selection of representatives to the oversight body were more ambiguous, with small majorities supporting the ability for taxonomic speciality groups to develop their own selection process or a nomination process. Small majorities also supported a requirement that individuals voting for representatives to the oversight body should themselves be involved in the creation of taxonomic lists in some way.

## THE WAY FORWARD

Progress towards a single authoritative list of the world's species is proceeding apace, much of it now within the Catalogue of Life. In 2023, many of those involved with the GSLWG also joined a working group within COL to form the Species List Governance Group (https://www.catalogueoflife.org/about/governance) within COL's Global Team as well as participating in the Taxonomic Working Group that assesses and chooses lists for inclusion in COL. The aim of the Species List Governance Group is to consider the long-term governance of global species lists and to establish ways in which the quality of list governance and list contents can be measured. There are two independent lines of thinking.

The first has been a review of the governance of global scientific organisations, assessing how they are governed through the lens of scientific knowledge as a global commons. A set of ten design principles for the Global Scientific Commons has been drafted and the extent to which taxonomic organisations comply with these principles is being tested. These principles will then form the basis of recommendations to COL on a potential oversight group for the governance of global species lists. Options on potential governance arrangements for such an oversight body will also be tested with the global taxonomic community.

At the same time, a set of metrics has been developed based on Garnett et al. (2020) for governance and Pyle et al. (2021) for list contents. These have been trialled and refined with the creators of lists for a wide range of life forms to remove ambiguities and try to ensure they are universally applicable to the diversity of taxonomic lists available. The metrics are being constructed so that each list will have separate scores for different elements of governance (12 metrics) and contents (24 metrics) each of which is weighted by the importance given to them in the survey of Lien et al. (2023) so that lists will have single scores for governance and contents. The aim of the scoring is fourfold. First, it allows list creators to assess where they are weakest and provides an incentive to improve. Secondly, it provides list users with a transparent and largely objective measurement of list quality so they can assess whether to use the list. Thirdly, it allows COL to decide whether to adopt a list, especially where there are competing treatments. Finally, it will allow measurement of trends and patterns in list quality, both across different branches of the Tree of Life and also over time, helping to focus investment to remedy deficiencies.

## NOTE

1 Taxonomy and systematics are often assigned different meanings such that taxonomy refers to the description and naming of taxa and is thus part of systematics which also includes phylogenetics, the analysis of (evolutionary) relationships among taxa. Other authors use taxonomy and systematics synonymously, and there may be good reasons to do so. In line with the title of this book, we follow the synonymous usage in this chapter.

## REFERENCES

Bánki, O., Hobern, D., Döring, M. and Remsen, D. (2019) Catalogue of Life Plus: A collaborative project to complete the checklist of the world's species. *Biodiversity Information Science and Standards*, 3: e37652. https://doi.org/10.3897/biss.3.37652

Castelló, J. R. (2016) *Bovids of the World. Antelopes, Gazelles, Cattle, Goats, Sheep, and Relatives*. Princeton University Press, Princeton.

Conix, S. (2019) Radical pluralism, classificatory norms and the legitimacy of species classifications. *Studies in History and Philosophy of Science*, 73: 27–34. https://doi.org/10.1016/j.shpsc.2018.11.002

Conix, S., Garnett, S. T., Thiele, K. R., Christidis, L., van Dijk, P. P., Bánki, O. S., Barik, S. K., Buckeridge, J. S., Costello, M. J., Hobern, D., Kirk, P. M., Lien, A., Nikolaeva, S., Pyle, R. L., Thomson, S. A., Zhang, Z.-Q. and Zachos, F. E. (2021) Towards a global list of accepted species III. Independence and stakeholder inclusion. *Organisms Diversity and Evolution*, 21: 631–643. https://doi.org/10.1007/s13127-021-00496-x

Conix, S., Garnett, S. T., Zachos, F. E. and Christidis, L. (2022) Taxonomic order, disorder, and governance, pp. 161–173. In *Species Problems and Beyond – Contemporary Issues in Philosophy and Practice* (eds J. S. Wilkins, F. E. Zachos and I. Ya. Pavlinov). CRC Press, Boca Rotan.

Ereshefsky, M. (2000) *The Poverty of the Linnaean Hierarchy: A Philosophical Study of Biological Taxonomy*. Cambridge University Press, Cambridge.

Franz, N. M., Peet, R. K. and Weakley, A. S. (2008) On the use of taxonomic concepts in support of biodiversity research and taxonomy, pp. 63–86. In *The New Taxonomy* (ed Q. Wheeler). CRC Press, Boca Raton.

Garnett, S. T. and Christidis, L. (2017) Taxonomy anarchy hampers conservation. *Nature*, 546: 25–27. https://doi.org/10.1038/546025a

Garnett, S. T., Christidis, L., Conix, S., Costello, M. J., Zachos, F. E., Bánki, O., Bao, Y., Barik, S. K., Buckeridge, J. S., Hobern, D., Lien, A., Montgomery, N., Nikolaeva, S., Pyle, R. L., Thomson, S. A., van Dijk, P. P., Whalen, A., Zhang, Z.-Q. and Thiele, K. R. (2020) Principles for creating a single authoritative list of the world's species. *PloS Biology*, 18(7): e3000736. https://doi.org/10.1371/journal.pbio.3000736

George, A. L. and Mayden, R. L. (2005) Species concepts and the endangered species act: How a valid biological definition of species enhances the legal protection of biodiversity. *Natural Resources Journal*, 45: 369–407. http://www.jstor.org/stable/24888981

Groves, C. and Grubb, P. (2011) *Ungulate Taxonomy*. The Johns Hopkins University Press, Baltimore.

Hobern, D., Barik, S. K., Christidis, L., Garnett, S. T., Kirk, P., Orrell, T. M., Pape, T., Pyle, R. L., Thiele, K. R., Zachos, F. E. and Bánki, O. (2021) Towards a global list of accepted species VI. The Catalogue of Life checklist. *Organisms Diversity and Evolution*, 21: 677–690. https://doi.org/10.1007/s13127-021-00516-w

Kennedy, J., Hyam, R., Kukla, R. and Paterson, T. (2006) Standard data model representation for taxonomic information. *OMICS: A Journal of Integrative Biology*, 10: 220–230. https://doi.org/10.1089/omi.2006.10.220

Lepage, D., Vaidya, G. and Guralnick, R. (2014) Avibase–a database system for managing and organizing taxonomic concepts. *ZooKeys*, 420: 117–135. https://doi.org/10.3897/zookeys.420.7089

Lien, A. M., Conix, S., Zachos, F. E., Christidis, L., van Dijk, P. P., Bánki, O. S., Barik, S. K., Buckeridge, J. S., Costello, M. J., Hobern, D., Montgomery, N., Nikolaeva, S., Pyle, R. L., Thiele, K., Thomson, S. A., Zhang, Z.- Q. and Garnett, S. T. (2021) Towards a global list of accepted species IV. Overcoming fragmentation in the governance of taxonomic lists. *Organisms Diversity and Evolution*, 21: 645–655. https://doi.org/10.1007/s13127-021-00499-8

Lien, A. M., Bánki, O., Barik, S. K., Buckeridge, J. S., Christidis, L., Cigliano, M. M., Conix, S., Costello, M. J., Hobern, D., Kirk, P. M., Kroh, A., Montgomery, N., Nikolaeva, S., Orrell, T. M., Pyle, R. L., Raz, L., Thiele, K., Thomson, S. A., van Dijk, P. P., Wambiji, N., Whalen, A., Zachos, F. E., Zhang, Z.-Q. and Garnett, S. T. (2023) Widespread support for a global species list with a formal governance system. *Proceedings of the National Academy of Sciences of the United States of America*, 120(45): e2306899120. https://doi.org/10.1073/pnas.2306899120

Lynx Nature Books. (2023) *All the Mammals of the World*. Lynx Edicions, Barcelona.

Maddison, W. P. and Whitton, J. (2023) The species as a reproductive community emerging from the past. *Bulletin of the Society of Systematic Biologists*, 2: 1–35. https://doi.org/10.18061/bssb.v2i1.9358

May, R. M. (1990) Taxonomy as destiny. *Nature*, 347: 129–130. https://doi.org/10.1038/347129a0

McClure, C. J., Lepage, D., Dunn, L., Anderson, D. L., Schulwitz, S. E., Camacho, L., Robinson, B. W., Christidis, L., Schulenberg, T. S., Iliff, M. J., Rasmussen, P. C., and Johnson, J. (2020) Towards reconciliation of the four world bird lists: Hotspots of disagreement in taxonomy of raptors. *Proceedings of the Royal Society B*, 287: 20200683. https://doi.org/10.1098/rspb.2020.0683

Pavlinov, I. Ya. (2023) *The Species Problem. A Conceptual History*. CRC Press, Boca Raton.

Pyle, R. L., Barik, S. K., Christidis, L. *et al.* (2021) Towards a global list of accepted species V. The devil is in the detail. *Org Divers Evol*, 21: 657–675. https://doi.org/10.1007/s13127-021-00504-0

Raposo, M. A., Stopiglia, R., Brito, G. R. R., Bockmann, F. A., Kirwan, G. M., Gayon, J. and Dubois, A. (2017) What really hampers taxonomy and conservation? A riposte to Garnett and Christidis (2017). *Zootaxa*, 4317(1): 179–184. http://dx.doi.org/10.11646/zootaxa.4317.1.10

Robichau, R. W. (2011) The mosaic of governance: Creating a picture with definitions, theories, and debates. *Policy Studies Journal*, 39(s1): 113–131. https://doi.org/10.1111/j.1541-0072.2010.00389_8.x

Sandall, E. L., Maureaud, A. A., Guralnick, R., McGeoch, M. A., Sica, Y. V., Rogan, M. S., et al. (2023) A globally integrated structure of taxonomy to support biodiversity science and conservation. *Trends in Ecology and Evolution*, 38: 1143–1153. https://doi.org/10.1016/j.tree.2023.08.004

Schellenberger Costa, D., Boehnisch, G., Freiberg, M., Govaerts, R., Grenié, M., Hassler, M., Kattge, J., Muellner-Riehl, A. N., Rojas Andrés, B. M., Winter, M., Watson, M., Zizka, A. and Wirth, C. (2023) The big four of plant taxonomy – a comparison of global checklists of vascular plant names. *New Phytologist*, 240: 1687–1702. https://doi.org/10.1111/nph.18961

Schiff, D. M. and Wood, J. (2017) Petition of Pacific Legal Foundation et al., for rule-making under the Administrative Procedure Act, https://zenodo.org/records/1067325

Stankowski, S. and Ravinet, M. (2021) Quantifying the use of species concepts. *Current Biology*, 31: R428–R429. https://doi.org/10.1016/j.cub.2021.03.060

Sterner, B., Witteveen, J. and Franz, N. (2020) Coordinating dissent as an alternative to consensus classification: Insights from systematics for bio-ontologies. *History and Philosophy of the Life Sciences*, 42: 1–25. https://doi.org/10.1007/s40656-020-0300-z

Thiele, K. R., Conix, S., Pyle, R. L., Barik, S. K., Christidis, L., Costello, M. J., van Dijk, P. P., Kirk, P., Lien, A., Thomson, S. A., Zachos, F. E., Zhang, Z.-Q. and Garnett, S. T. (2021) Towards a global list of accepted species I. Why taxonomists sometimes disagree, and why this matters. *Organisms Diversity and Evolution*, 21: 615–622. https://doi.org/10.1007/s13127-021-00495-y

Thomson, S. A., Pyle, R. L., Ahyong, S. T., Alonso-Zarazaga, M., Ammirati, J., Araya, J. F., [...] and Segers, H. (2018) Taxonomy based on science is necessary for global conservation. *PLoS Biology*, 16(3), e2005075. https://doi.org/10.1371/journal.pbio.2005075

Thomson, S. A., Thiele, K., Conix, S., Christidis, L., Costello, M. J., Hobern, D., et al. (2021) Towards a global list of accepted species II. Consequences of inadequate taxonomic list governance. *Organisms Diversity and Evolution*, 21: 623–630. https://doi.org/10.1007/s13127-021-00518-8

Wheeler, Q. (2024) *Species, Science and Society: The Role of Systematic Biology*. Routledge, Abingdon.

Wheeler, T. (2014) The concept of species with constant reference to Killer Whales. *Washington Journal of Environmental Law and Policy*, 4: 250–284. https://digitalcommons.law.uw.edu/wjelp/vol4/iss1/10

Wilkins, J. S. (2018) *Species: The Evolution of the Idea*. Second Edition. CRC Press, Boca Raton.

Wilson, D. E. and Mittermeier, R. A. eds (2011) *Handbook of the Mammals of the World.* Vol. 2. Hoofed Mammals. Lynx Edicions, Barcelona.

Zachos, F. E. (2016) *Species Concepts in Biology. Historical Development, Theoretical Foundations and Practical Relevance.* Springer International Publishing.

Zachos, F. E. (2022) Critique of taxonomic reason(ing): Nature's joints in light of an 'honest' species concept and Kurt Hübner's historistic philosophy of science, pp. 315–340. In *Species Problems and Beyond – Contemporary Issues in Philosophy and Practice* (eds J. S. Wilkins, F. E. Zachos and I. Ya. Pavlinov). CRC Press, Boca Raton.

Zachos, F. E., Christidis, L. and Garnett, S. T. (2020) Mammalian species and the twofold nature of taxonomy: A comment on Taylor et al. 2019. *Mammalia*, 84: 1–5. https://doi.org/10.1515/mammalia-2019-0009

# 13 Species Descriptions Go Digital

*Peter Uetz and Donat Agosti*

More than 2 million species have been described since Carl von Linné founded biological taxonomy in 1753. The descriptions of these species use an estimated 500 million pages, of which about 50 million pages have been digitized in the Biodiversity Heritage Library (BHL, https://www.biodiversitylibrary.org/) and other sites such as JSTOR (https://www.jstor.org/). However, species are not just described – they are redescribed and redefined, split into multiple species, or multiple species are merged into one, all documented in scientific publications. Hence, our understanding of species constantly changes. In other words, descriptions go through versions that need to be stored and linked to each other via citations which help scientists to constantly track and update this knowledge.

Over the past 20 years, a substantial information infrastructure has been developed to index most species known to science. Foremost and most complete among them is the Catalogue of Life (CoL, https://www.catalogueoflife.org/), a project that started in 2001 and has produced a list of the aforementioned 2 million species. However, various studies have suggested that there may be millions, even billions, more species to be discovered (Li and Wiens, 2022), and that many species have multiple synonymies and changed names.

While the CoL is an impressive achievement, it remains an aggregate of about 160 different species databases, each of them specializing in a different taxonomic group, such as plants, spiders, molluscs and reptiles. That is, the CoL is a meta-database that simply merges the species from their source databases. The problem with this approach is that each source database collects different types of information. While all of them collect species names, there is little common beyond that: some provide only the currently accepted name, while others include detailed synonymies; some collect references to the taxonomic literature, and links to BHL or digital object identifiers (DOIs), and others provide the actual papers as PDFs, some collect photos of species, others do not. Many have detailed locations of species, but some have only country names.

Most do not collect actual species descriptions, even if they have indexed the literature that does. In other words, users often have to unravel the citation, and then search tediously for the often obscure papers for the description of a species. The sources may be behind paywalls or in different formats, languages, or illustrating different body parts, if any. Older papers may not even be available online.

It is surprising that in the new age of artificial intelligence (AI) the need for digitized species descriptions remains largely unmet. One reason may be that the most popular groups of species, such as mammals and birds, have excellent and largely

DOI: 10.1201/b22822-13

stable reference literature, such as field guides or even massive encyclopedias, like the Lynx editions for birds and mammals (del Hoyo et al., 2011; Wilson et al., 2019). But they are not easily accessible and certainly not free. Another is that many taxonomists work in small groups, where the exchange of publications is routinely done by email without a need to share more widely.

Even if a description is available online, **only a small fraction of species descriptions is actually available in a structured digital format** that is findable, accessible, interoperable, and reusable (FAIR). This is critical if you want to compare, let's say, all the species of a genus or family in a specific area, or compare the descriptions of the same species in different regions, now increasingly facilitated by AI applications like ChatGPT.

Databases have started to manually collect that data from published papers. For instance, the Reptile Database has descriptions of more than 8,000 (out of 12,000) species (Uetz et al., 2023). When we surveyed the 30 largest CoL source databases, namely those with more than 10,000 species each, most of them did not contain descriptions, or more generally, taxonomic **'treatments'** – that is, chunks of information that are used to describe species, such as a diagnosis, a detailed description of the type specimen, or description of variation or even images (Catapano, 2010).

Manual digitization is time-consuming and expensive. One solution would be the use of AI to extract data from published papers. We have used ChatGPT to convert published descriptions into tables of characters, so that species can be compared to each other (Uetz et al., 2023). ChatGPT is reasonably good at text mining, but it is far from perfect. However, it will certainly improve over the near future and, most importantly, it can be trained on existing treatments, automated, and, in theory, applied to the whole body of literature – if it were accessible (see above).

An automated workflow to **extract taxonomic treatments, tables and figures** from published papers and making them accessible in TreatmentBank (https://plazi.org/tags/treatment-bank/) and the Biodiversity Literature Repository (BLR, https://biolitrepo.org/) has been developed by the Swiss non-profit association Plazi (Bénichou et al., 2021; Fawcett et al., 2022). This data is, after quality control steps, made accessible as FAIR data, currently providing access to 1,000,000 treatments and 680,000 figures via TreatmentBank and BLR (Agosti et al., 2022). These are routinely used by the Global Biodiversity Information Facility (GBIF, https://www.gbif.org/), ChecklistBank (https://www.checklistbank.org/) and the fledgling BiodiversityPMC (https://biodiversitypmc.sibils.org/), allowing the use of AI to explore it as part of the PMC and PubMed (Pasche et al., 2023) corpora. Importantly, data is always linked to the original publication. Some publishers such as Pensoft and Muséum national d'Histoire naturelle Paris, have already adopted this approach and offer 'semantic publishing', that is, they provide semantic annotations and fully automated dissemination.

Once the publications are annotated with taxonomic names, treatments and cited specimens, the taxonomic data can be automatically integrated via ChecklistBank into an extended CoL. This provides databases a gateway to new species, records of existing taxa, and other taxonomic data in human and machine-readable formats. In other words, speeding up the long-awaited reference system to link all data about biodiversity in a knowledge graph and expose the biodiversity knowledge to the rapidly growing AI world.

A limitation of this approach is the large stack of publications only available as scanned pages without optical character recognition (OCR) standards and resources to digitize them. Even with digital publishing, the thousands of journals with biodiversity content use different formats and thus parsers have to be tweaked to each specific journal. Finally, paywalls are a huge obstacle with publishers and authors still unwilling or unable to publish open access and thus moving to semantically enhanced publishing.

Based on the aforementioned limitations of the current publication system, and the advent of AI and other technologies, we recommend the following for taxonomic and scientific publishing (see Figure 13.1):

- Scientists and funders should aim for **open digital access knowledge**.
- Authors should publish in a **digitally accessible format** that funders can support.
- **Publications**, especially taxonomic treatments and names, and figures **should be available immediately** and citable as digitally accessible biodiversity knowledge (FAIR data) in globally available databases such as GBIF, BiodiversityPMC, Biodiversity Literature Repository, Wikidata, ChecklistBank or equivalent. Cited publications should be treated similarly.

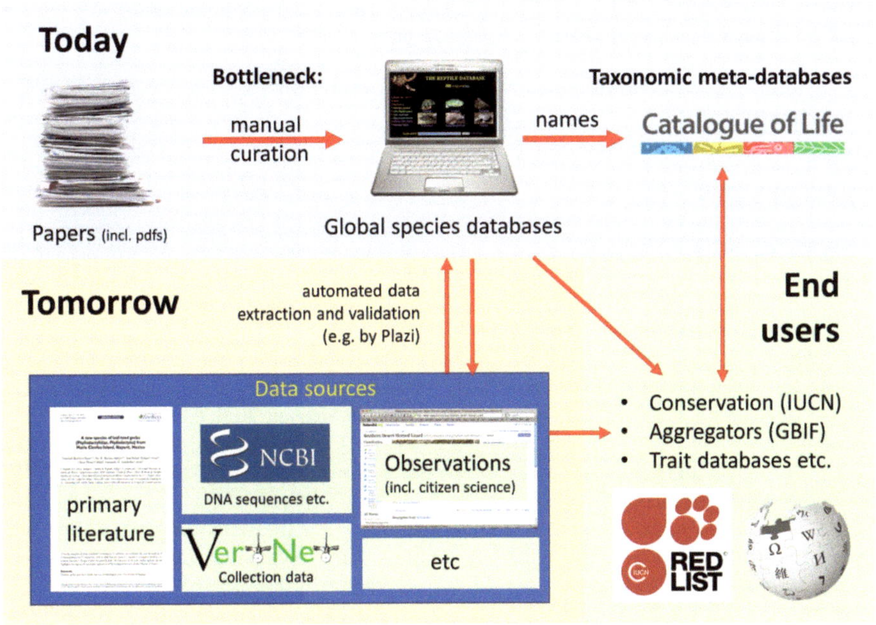


**FIGURE 13.1** Species descriptions, traditionally published on paper or in PDFs need to be made computer-readable to make them fully accessible and useful. That would allow the integration of traditionally published information with other data sources and provide integrated information to end users.

- Digital **knowledge** and its sources should be **accessible as training data** for AI applications to open up newly published unstructured data as well as legacy literature.
- A minimal **shared vocabulary** should be established to annotate biodiversity terms complementing global vocabularies such as schema.org or TDWG terminology and to map to others such as Wikidata to exchange and link data.
- The fledgling **infrastructure** to convert, annotate, and provide access to FAIR data, and long-term storage needs to be expanded and **funded** as a global shared infrastructure.
- A **knowledge graph based on taxonomic names** should be made accessible to the global community as a reference system to link and jointly explore data to serve as a digital infrastructure.

Recent developments to convert publications into FAIR digital objects and their reuse can potentially open up the huge corpus of taxonomic literature to modern data science and AI. We have outlined above a path ahead towards this goal. Modernization will help protect the discipline as much as the biosphere itself, and provide a global reference system to link and explore biodiversity across domains, all the more urgent given the rapid biodiversity loss due to overpopulation, overconsumption and climate change.

## ACKNOWLEDGMENTS

We thank Laurence Bénichou, Willi Egloff, and Puneet Kishor for their helpful comments.

## REFERENCES

Agosti D., Ruch, P., Benito Gonzalez Lopez, J. and Penev, L. (2022) Enabling published taxonomic data to be used to address the biodiversity crisis: Biodiversity Literature Repository and TreatmentBank. *Biodiversity Information Science and Standards*, 6: e91167. https://doi.org/10.3897/biss.6.91167

Bénichou, L., Guidoti, M., Gérard, I., Agosti, D., Robillard, T. and Cianferoni, F. (2021) European Journal of Taxonomy: A deeper look into a decade of data. *European Journal of Taxonomy*, 782(1): 173–196. https://doi.org/10.5852/ejt.2021.782.1597

Catapano, T. (2010) TaxPub: An extension of the NLM/NCBI journalpPublishing DTD for taxonomic descriptions. *Proceedings of the Journal Article Tag Suite Conference 2010.* https://doi.org/10.5281/zenodo.3484285

Fawcett, S., Agosti, D., Cole, S.R. and Wright, C.F. (2022) Digital accessible knowledge: Mobilizing legacy data and the future of taxonomic publishing. *Bulletin of the Society of Systematic Biologists*, 1(1): 1–12. https://doi.org/10.18061/bssb.v1i1.8296

del Hoyo, J., Elliott, A., Christie D.A., et al. eds (2011) *Handbook of the Birds of the World*, 16 volumes. Lynx Nature Books, Barcelona. Available from: https://www.lynxeds.com/product-category/by-category/encyclopedias/hbw/

Li, X. and Wiens, J.J. (2022) Estimating Global Biodiversity: The Role of Cryptic Insect Species. *Systematic Biology*, 72(2): 391–403. https://doi.org/10.1093/sysbio/syac069

Pasche, E., Agosti, D., Penev, L., Groom, Q., Flament, A., Gobeill, J. and Ruch, P. (2023) Towards "Biodiversity PMC". *Biodiversity Information Science and Standards*, 7: e111647. https://doi.org/10.3897/biss.7.111647

Uetz, P., Darko, Y.A. and Zeliff, D. (2023) Towards digital descriptions of all extant reptile species. *Megataxa*, 10 (1): 027–042. https://doi.org/10.11646/megataxa.10.1.6

Wilson D.E., Mittermeier, R.A. and Lacher, T.E. eds (2019) *Handbook of the Mammals of the World*. Lynx Edicions, Barcelona. https://www.lynxeds.com/product/handbook-of-the-mammals-of-the-world-volumes-1-9/

# 14 Saving Systematics
## *Taxonomy's Identity, Traditions, and Great Expectations*

*Quentin D. Wheeler*

## INTRODUCTION

Science, applied to solving an immediate problem, like identifying species, is limited by current knowledge and technology. Science done for its own sake, in contrast, creates new knowledge, unexpected possibilities, and, as a fringe benefit, discoveries that, sooner or later, are found to have practical applications. With both outcomes in mind, we support and encourage scientists to pursue pure, curiosity-driven research in ecology, astronomy, physics, and all basic fields.

In contrast, taxonomists are being asked to abandon the exploration of species and details of evolutionary history to simply focus on making species identifiable. When identifying species is the goal, rather than a by-product of knowing species, we invite shortcuts and compromises in the quality and quantity of knowledge, and we are tempted to replace rigorous scholarship and testable hypotheses with efficient, rote procedures. Further, taxonomists are encouraged to sacrifice multiple sources of evidence—traditionally including comparative morphology, ontogeny, and fossils—for the convenience of relying primarily upon a single one (i.e., molecular). In both the short and long terms, science is better served by letting taxonomists create as much knowledge as possible and applying it, as appropriate, to problem-solving.

Taxonomy is unsurpassed in its record of service to the life sciences. Its hypotheses, identifications, classifications, scientific names, and growing body of knowledge of species, characters, clades, and relationships, make credible biology possible. The value of taxonomy—measured in the amount and reliability of knowledge and information—is determined by the quality of taxonomy itself. Yet, information-rich, corroborated hypotheses based on detailed comparative studies and multiple lines of evidence are being abandoned to accelerate the identification of species, most of which we know very little about.

We evaluate science by advances in knowledge and impacts. In spite of centuries of progress in taxonomy, including about two million species discovered, described, and named; empowering biology, agriculture, medicine, commerce, and society with accurate identifications of the most common, useful, and impactful species; theoretical advances second to none; integration of generations of new technologies; synthesis of diverse evidence; and creation of a conceptual framework for biology, in

 DOI: 10.1201/b22822-14

the form of phylogenetic classifications (Hennig, 1966), cynics nonetheless take the measure of taxonomy in what it has not accomplished (Scotland et al., 2003):

> The collective achievement of taxonomists over the last 200 years is every bit as impressive as the contributions made by related fields of ecology and evolution. Despite this the spotlight remains firmly on what we still do not know about biodiversity, rather than the massive achievements already made. Taxonomy is not in crisis but merely insufficiently resourced to complete the inventory within the time frame that is being imposed by those concerned about the real crisis, which is the biodiversity crisis.

Colleagues who, for decades, have belittled systematics and natural history as 'stamp collecting' (Johnson, 2007), and denied positions and funding to taxonomists, now whinge about the dearth of taxonomic information. Faced with mass extinction and ignorance, created by years of neglecting taxonomy, they complain that millions of species cannot be identified. With a stunning lack of self-awareness, they blame the same taxonomy that they consistently malign and neglect for this paucity of information. Having learned no lessons from past mistakes, they continue to ignore the needs of taxonomists and seek shortcuts, like DNA barcodes and minimalist taxonomy (Sharkey et al., 2021); expediencies that promise identifications without having to share research funds with systematics.[1]

The emphasis on molecular data and identifying species represents a fundamental change in the identity of systematics. The goals of knowing species, clades, characters, and history are replaced by far less ambitious ones: demonstrating that species exist and telling them apart. This impoverished taxonomy is conceived by and for non-taxonomists and accompanied by changes in the goal of cladistic analyses, too. Rather than mapping character transformations on cladograms and informing classifications with knowledge of history, 'trees' are published depicting only relationships among species, stripped of the characters that would make them most interesting. It is important that taxonomy enables identifications and creates a cladistic framework for biology, of course, but these are only as informative, useful, and reliable as the quality of taxonomy behind them. The description, comparison, and analysis of characters are core interests of taxonomists, and anyone else wishing to understand biodiversity and evolutionary history. Molecular taxonomy focuses on species, not characters, because that is the superficial level of information needed immediately by ecologists, conservationists, and other users of taxonomy.

Make no mistake; molecular tools are an exciting and powerful addition to taxonomy. However, they should augment, not replace, a traditional approach that includes comparative, revisionary, and descriptive studies; synthesis of diverse evidence; and explicit hypotheses about homology, character polarity, species status, and cladistic relationships. The preference for a molecular-based taxonomy is driven by funding, popularity, and needs of other disciplines, not by what is in the best interest of taxonomy itself. Those eager to make identifications easy should be careful what they wish for. The most reliable identifications come from the best taxonomy. And the best taxonomy is informed by theory, inspired by curiosity, pursued for its own sake, cognizant of all relevant evidence, and conducted to its own high standards. There are no shortcuts without compromises.

The need for identifications, magnified by extinction rates and environmental problems, should be cause for a renewal, not replacement, of taxonomic traditions. Studies based entirely on molecular data represent compromises that should be unacceptable to everyone. Focusing on identifications, it is easy to lose sight of the broader roles of taxonomy in exploring species, characters, and evolutionary history, and organizing knowledge in a phylogenetic classification. Species and characters are two sides of the same coin (Nelson and Platnick, 1981). Therefore, it is impossible to fully understand one without comparably detailed attention to the other. Identifying species while ignoring characters is an exercise in futility, putting names on species about which almost nothing is known.

Similarly, cladograms are only as informative and interesting as the attributes of species and clades that can be understood by virtue of their branching pattern. Without knowledge of characters, there is little for a cladogram to explain. Unless we reassert the identity, goals, and mission of taxonomy, and reengage Hennig's revolution (Wheeler et al., 2013), we risk degrading the science of systematics to little more than an identification service and imposing permanent ignorance of the diversification and organization of life. This is a disservice, including to those seeking to identify species, and needlessly undermines what we ultimately know of the history of life.

Because of the number, severity, and magnitude of environmental problems, the urgent necessity to identify species is self-evident. That need is best met, however, by maintaining the excellence, integrity and information content of taxonomy—not by cutting corners. The suggestion that molecular data is the only, or best, way to accelerate identifications is false. The option exists to make identifications faster, more reliable and more informative while, at the same time, advancing fundamental taxonomy to a high level of excellence. This requires recognition that the highest quality identifications are by-products of taxonomy, not an end in themselves; and, ultimately, a result of a commitment to inventory, describe, and classify all species,[2] maintaining the traditional aims and integrity of taxonomy (Wheeler et al., 2012). Given that we can have excellent taxonomy *and* multiple ways to rapidly identify species, from DNA barcodes (e.g., Hebert et al., 2003; but see Meier et al., 2006) to image recognition and artificial intelligence (e.g., Ness, 2008; Russell et al., 2007), why would we choose to limit ourselves to molecular data and less knowledge?

Taxonomy has always been 'big science' in concept, if not in practice. In the past, the ambition to know all species was stifled by limitations on travel, communication, and the means to handle vast amounts of data. These former constraints, and the sheer number of species, led people to conclude, prematurely, that the discovery, description, and classification of all species was a practical impossibility. Advances have minimized or eliminated such constraints, yet we persist in accepting limitations on taxonomy that need not apply.

The rate of extinction dictates that we think big, accept no limitations, compromises or shortcuts, act on a planetary scale, and insist on gathering as much information, creating as much knowledge and preserving as much evidence, including preserved specimens, of as many species as possible. To do so, taxonomy must be pursued for its own sake and natural history museums must be ever cognizant of why their collections exist, living up to the responsibility to protect, grow, develop, curate and use them for their highest purposes: advancing taxonomic knowledge and documenting life on earth. The good news for users of taxonomy is that this will result in more, and more reliable, information, and identifications.

Our choices are simple: Return support to taxonomy, defend its identity, modernize its infrastructure, rebuild its workforce and create a legacy of knowledge of the diversity and history of life; or lower expectations, ignore taxonomy's unique mission and accept permanent ignorance of species, the organization of the biosphere and the story of evolution. The role of molecular data in taxonomy is important; but it is rightfully a supporting, not a leading, role. Top billing belongs to studies of characters, homology, and apomorphy. To this end, it is necessary to explore all relevant evidence—morphological, fossil, ontogenetic, and molecular—with equal vigor. Stated another way, taxonomy's aim to know species *and* characters, the pattern of relationships among them, and to create an information-rich, phylogenetic classification requires more than molecular evidence can provide.

The ideal of a molecular-based taxonomy, useful, and technologically impressive as it could be, is misguided. Although molecular data can be parsed as characters, too, understanding more complex characters, and their origins and diversification, requires knowledge beyond the molecular. Species are not fascinating because there are so many of them; simply enumerating species is only marginally more interesting than counting grains of sand. Their exploration is not compelling because they differ from one another by some percentage of DNA. After all, individuals and populations differ in percentages of DNA as well. Species are interesting foremost because of diverse and improbable characters, unexpected novelties, and complexity that include emergent properties not reducible to the molecular level and only fully appreciated when described and compared.

Ancient Minoans used knowledge of the night sky to navigate, but guiding sailors has never been understood as the only, primary or ultimate purpose of astronomy. Its potential is much greater, to discover what exists in the universe and understand how it came to be. The same is true of taxonomy. Its applications, while profoundly important, should not detract from its mission as a fundamental science. General biologists ask how life works, from intracellular mechanisms to biosphere functions, but taxonomy is different. It asks what kinds of living things exist, what characters they possess and how both species and characters are related; it is descriptive, comparative and historical in perspective; it seeks to summarize, synthesize, and make sense of all we know of species in a phylogenetic context and classification. Far from a shortcoming, its non-experimental methods, and all-or-nothing predictions about character distributions, make its hypotheses among the most elegant and strenuously testable in science (Popper, 1959; Gaffney, 1979; Nelson and Platnick, 1981; Wheeler, 2024).

## IDENTITY

Taxonomy does not exist to identify species; species are identifiable because taxonomy exists. In their introduction, Williams and Ebach (2020) wrote 'With this book we hope, at the very least, to suggest why we believe taxonomy is not just exceedingly important, not just central to all biology, but has a vibrant research agenda of its own' (p. 8). Taxonomists must clarify this agenda, reclaim their science, concede nothing that dilutes or distracts from its mission and make no apologies for its theories and methods, or the importance of natural history collections as a primary research resource. That some are content with minimal knowledge is no reason to reduce the standards or information content of taxonomy. They are welcome to know as little as they wish, of course, but they do so at their own peril. Applied taxonomy should be just that: the application of knowledge and information created by fundamental research. Making identifications an end, rather than a means to access knowledge, invites compromises that limit what we ultimately learn about biodiversity and its origins.

The following mission statement, included in the report from a workshop (Wheeler et al., 2012), expresses taxonomy's fundamental and applied aspirations:

> Our goal is no less than a full knowledge-base of the biological diversity on our planet, by which we mean: knowledge of all Earth's species, and how they resemble and differ from each other (i.e. all their characters from detailed morphology to as much genomic information as is feasible to collect); a predictive classification of all these species, based on their interrelationships as inferred from all these characters; knowledge of all the places at which each of these species has been found with as much ecological data as are available from specimens in the world's collections (e.g. host data, microhabitat data, phenology, etc.); and cyberinfrastructure to enable the identification of newly found specimens (including automated identification systems based images and genomic information), the efficient description of species, and open access to data, information and knowledge of all species by anyone, amateur or professional, anywhere, any time.

Success in this mission depends on three groups. First, taxonomists who must be resolute in the mission. Second, natural history museums which must be committed to species exploration, growing collections representative of their diversity *and* engaging taxonomists to put collections to their highest uses. And third, consumers of taxonomic information—those who seek identifications of, information about and a phylogenetic context for species—who must support taxonomy from the outside as a fundamental, independent science. Not one of these groups is doing its part very well today. Taxonomists are overly compliant, going along with fads and funding trends, and failing to explain, advocate for and defend their own science. Museums foolishly minimize species exploration and taxonomic research to pursue a university-style, eclectic research agenda that fails to properly develop, curate, and take full advantage of collections. By forfeiting an identity based on collections and taxonomy that formerly differentiated and elevated them above competitors for recognition and funding, museums have, to an alarming degree, become also-rans. Finally, consumers of taxonomy are so eager to identify species, so narrowly focused on immediate needs, that they are unwittingly undermining the reliability and amount of information available.

In a disservice to everyone, priorities are being dictated by consumers rather than producers of taxonomic knowledge. How else can you explain 'minimal' approaches to taxonomy (e.g., Sharkey et al., 2021) when, as explained by Broad (this volume), deep knowledge could be had? This slows the growth of knowledge, undermines what conservation might achieve, reduces prospects for biomimicry and limits what we ultimately learn about species and evolutionary history (Wheeler, 2024). The biodiversity crisis means that the time for gathering specimens and evidence critical to taxonomy's mission grows short. The long-term costs of a molecular-based taxonomy, marked by limited knowledge and missed opportunities, will far exceed any short-term benefits. We should expect more from taxonomy.

For the best outcome, we must insist on the independence of systematic biology; commit to an unprecedented period of species exploration, discovery, description, and classification; demand that natural history museums reembrace their leadership role for systematics; educate colleagues and the public about the importance of fundamental taxonomy; inspire a new generation of taxon experts; and refuse to further dilute the identity of taxonomy. Sciences like ecology and genetics produce applied knowledge, too, yet are supported as fundamental sciences. Only taxonomy, with an important mission of its own, is singled out for servitude. It is clear that unless taxonomists defend the identity of their science, no one else will.

## TRADITIONS

Some traditions persist, weathering fads, and shifting paradigms. There is no sell-by date on facts, corroborated hypotheses, or excellence. There would be little reason to engage in science if it did not occasionally get things 'right,' creating knowledge, information, and practices that have enduring value. In a science as old as taxonomy, there has been time to get more than a few things right. Ignoring them because they were discovered in the distant past, are out of fashion, or didn't require the latest technologies makes no sense. Some of taxonomy's traditions worthy of respect and continuation are listed below.

*Natural History Collections.* An over-emphasis on DNA threatens to slow or derail the growth and development of collections at a time of accelerated extinction. We should continue to collect and preserve specimens until such time as collections are a complete reflection of earth's species and their characters. Collections of specimens are not relics to be shuffled off to remote warehouses. They are the single greatest assemblage of knowledge, and potential knowledge, of species and their attributes. They are dynamic, information-rich research resources that should be constantly studied by taxonomists, a permanent record of biodiversity otherwise lost and a priceless insurance policy against ignorance. No amount of genomic data can replace their role in science and, at present, they contain only a small fraction of the estimated number of species (e.g., IUCN, 2022; Wiens, 2023).

*Comparative, Descriptive, and Revisionary Studies.* Taxonomic revisions and monographs are how and when many taxonomic hypotheses are formulated and tested. They are taxonomy's gold standard for simple and enduring reasons. The simultaneous comparison of all species belonging to a higher taxon is the most efficient way to test the status of homologues, species, and clades. Advances in information science and digitization have paved the way for 'e-monographs' with up-to-the-minute completeness and accuracy.

*Phylogenetic Classifications.* Phylogenetic classifications are an optimal information storage system (Farris, 1979), the basis for predictions about attributes of poorly known species and clades, and a general reference system for biology (Hennig, 1966; Nelson and Platnick, 1981).

*Codes of Nomenclature and the Linnaean Hierarchy.* The introduction of Linnaean binomials was pure genius. The combination of genus name and specific epithet means that descriptive, memorable words may be used for many taxa without confusion. In addition, ranks are preadapted to express hypotheses about clades nested within clades. The Linnaean system is a nearly perfect balance of stability and flexibility, giving voice to taxonomic ideas (Humphries, 2001).

*Synthesis of Evidence.* We should be expanding, not narrowing, the spectrum of evidence for species, characters, and relationships, welcoming all data that are reliable and informative. Morphology, fossils, ontogeny, and sequence data each come with strengths and limitations. Gathering and synthesizing all relevant evidence is a hallmark of taxonomy (Simpson, 1961) that deserves to be maintained.

*Species Identification.* Taxonomists have always made species identifiable as a by-product of their research. What receives less attention from advocates of a molecular taxonomy are whether species are based on corroborated hypotheses and how much information about species exists. DNA barcodes are convenient but tell us little about species themselves.

As in astronomy, chemistry, and physics, the core aims of taxonomy have changed little in centuries—only the theories and methods used to pursue them. Like Aristotle, we still want to make sense of the pattern of similarities and differences among species. Like Linnaeus, we want to discover and characterize all species. Like Cuvier and Owen, we seek to understand the relationships among disparate morphological characters. Like Darwin, we desire to reconstruct the tree of life. And, like Hennig, we aim for classifications that reflect phylogenetic relationships. As millions of species are threatened with extinction, this is the worst imaginable time to reduce taxonomy to an identification service, limit its attention to one source of evidence, or slow the growth of natural history collections.

## BEYOND NEED AND GREED: THE FUTURE OF TAXONOMY

The trend toward molecular-based taxonomy is motivated by the immediate need for identifications, the lure of grant money and a desire to be accepted by colleagues who, unfortunately, neither understand nor respect taxonomy. It is obvious to taxonomists that we should continue to advance molecular studies, but few molecular geneticists speak up as descriptive taxonomy and natural history collections are increasingly marginalized.

To speed up species 'description,' some propose gathering minimal information beyond a DNA barcode (Sharkey et al., 2021). Others, like Byrne (2023), see collections as relics of 'the extractive, colonial science practices of the 20th century,' recommending that we only salvage dead animals as museum accessions in the future. She suggests that we 'emphasize conducting research on existing collections,' and make museums 'repositories of new specimen information, not bodies.' This fashionably 'woke' position on collections—linking collecting to the evils of colonialism—is as misguided as it is anti-scientific. It would condemn us to profound, permanent ignorance of species and evolutionary history; undermine the best hopes for effective conservation planning (e.g., Humphries, 2006); and stifle advances in biomimicry.

Most species are small-sized and their morphology is only studied in detail by careful, microscopic examination of well-preserved specimens. In the case of millions of unknown insect species, bug splatter scraped off automobile windscreens simply will not do. Neither science nor humanity is well-served by caring so little about soon-to-be-extinct species that we fail to honor them with recognition of what makes them unique, a name and at least some number of intentionally collected, well-preserved specimens. Allowing millions of species to go extinct, unknown, undescribed, unnamed, and unrepresented in museums, shows a cold indifference to the decimation of biodiversity and the long-term costs of ignorance, not compassion. In fairness, Byrne no doubt had vertebrate animals in mind, not insects, and we can agree that vulnerable species with small populations should not be collected beyond what is necessary and that animals should be handled humanely. Nonetheless, if her advice were taken literally, it would undermine science, diminish knowledge required for more effective conservation and allow us to avoid coming face to face with the species that we are driving to extinction.

Advances in theory and technology combine with the biodiversity crisis to indicate that it is time to reimagine taxonomy. Casting off constraints of the past, we can seize fantastic opportunities that will never be repeated. Taxonomy can, at last, realize its potential to inventory, describe, and classify every species on earth. It can assemble collections, data, and observations before millions of species have disappeared, preserving evidence of their distinctness, distribution in the biosphere and position in evolutionary history. Simply sequencing DNA and aiming to make species identifiable is thinking tragically small.

A molecular-based taxonomy tells us very little of what we should wish to know. It may make it possible to identify species and estimate their degree of relatedness, but to what end? If millions of species go extinct undescribed, we will be left with crippling ignorance of earth's biosphere and the evolutionary history of life. Addressing the urgent need of ecologists and conservationists to identify species deserves to be a high priority, but it need not detract from the broader mission of taxonomy. Without an emphasis on characters, we have token taxonomy. Science and humankind deserve more. The traditional goals of discovering, comparing, and describing characters; of tracing their step-by-step evolutionary transformations; of giving each species a unique name; and of synthesizing knowledge in a phylogenetic classification must be maintained during this era of mass extinction. We should accept nothing less.

Runaway rates of extinction dictate that we dream and act big, not small. Recognizing that species we do not collect, and characters we do not describe, may never be known, limiting knowledge to molecular data is one of the worst decisions we could make. Being the last generation with the opportunity to explore millions of species we have responsibilities to science and posterity. To miss this chance to discover and document the outcomes of evolutionary history, establish baseline knowledge of the biosphere, make evolutionary novelties available to biomimicry, inform conservation efforts and enrich our lives with knowledge of species and evolution would be a mistake of unconscionable proportions; to do so for short-term competitiveness for grants or peer pressure is even worse.

Great challenges call forth great leaders. It is difficult to imagine a greater challenge than a campaign to discover and make known every kind of organism on a planet undergoing rapid biodiversity decline, so where are the leaders? Individuals, organizations, and institutions who recognize what is at stake must find the courage to stand up to political correctness and meet the moment. One might assume that exploring, describing, and classifying millions of species would be taxonomy's greatest challenge, but these goals are clearly attainable. What may prove insurmountable are sociological forces: peer pressure to conform to popular fads, like molecular data; ignorance regarding the theoretical basis of taxonomy; irrational preferences for the new and technologically flashy; a distrust of non-experimental science; insatiable greed; and group-thinking.

## SYSTEMATICS REIMAGINED

Consumers of taxonomic information should be worried about how it is created and the quality of the science behind it—but most are concerned instead with quick, cheap and easy identifications. So, it is up to systematists to save systematics by reimagining their science for the challenges and possibilities at hand, with a combination of time-honored practices, innovative actions, and common sense.

*Clarity of Aims.* Misconceptions about taxonomy contribute to a lack of respect and support for its work. It is critical that we clearly, and unapologetically, communicate the identity, mission, and methods of taxonomy.

*Inspirational Vision.* The vision of knowing the millions of kinds of organisms on earth, and their evolutionary relationships, is as inspirational as cutting-edge physics or astronomy. Inexplicably, we have allowed taxonomy to be characterized as little more than an identification service. Sharing taxonomy's grand vision, and discoveries as they are made, we can engage the public in the excitement of species exploration and the goal of conserving as many, and as diverse, species as we can.

*Impactful Knowledge.* The perception that taxonomy exists to provide identifications must be refuted while, with more than a little irony, taxonomists make species identifiable for the benefit of science, conservation, and society. It is only with taxonomic knowledge that we can maximize species surviving, use knowledge of their attributes to solve problems, and fully appreciate and enjoy the natural world.

*Courage of Purpose.* Forces are lined up against taxonomy. While many taxonomists brilliantly integrate molecular with traditional evidence, peer pressure, and funding trends nonetheless are pushing the discipline toward a primarily, if not exclusively, molecular-based approach. Taxonomists must defend the study and synthesis of all relevant evidence.

*Planetary-Scale Inventory.* The biodiversity crisis means that we have one opportunity to explore and preserve evidence of millions of species. A planetary-scale inventory demands coordination among taxonomists, nation-states and institutions, an aggressive timetable, and an emphasis on the growth, development, and use of collections. Molecular data is a useful tool in the cause. Divorced from morphological (and other) evidence, however, it is like rescuing the card catalogue rather than the books in a burning library.

*e-Monography.* Technology allows us to reimagine revisions and monographs, making their production, revision, and translation into user-friendly resources, such as Web pages, distribution maps, checklists, and varied means of species identification, routine, and automatic. What is needed is a cyberinfrastructure designed by and for taxonomists to do taxonomy faster and better; one that also opens access to up-to-the-minute information by anyone who can use it.

*Language of Biodiversity.* Linnaean names and ranks are even more useful today than when first proposed. With tweaks, the codes of nomenclature continue to facilitate communication and stability through scientific names and classifications (Humphries, 2001). They should be refined, not supplanted by a less well-conceived 'phylo' code.

*General Reference System.* The one thing that all species share in common is evolutionary history making a phylogenetic classification both informative and predictive about the distribution of characters. A comprehensive phylogenetic classification of living things remains the best general reference system for biology, but Hennig's revolution is not yet finished (Hennig, 1966; Williams, Schmitt and Wheeler, 2016).

*Educating the Next Generation: Restoring, Modernizing, and Democratizing Curricula.* Both universities and major museums have a role to play in restoring robust taxonomic curricula that address theoretical, procedural and taxon-specific content. At the same time that formal degree programs are rebuilt, an opportunity exists to create self-guided, online curricula to elevate the knowledge and competence of amateurs who have a role to play in confronting the biodiversity crisis.

*Leadership and Natural History Museums.* No category of institution has a greater responsibility for taxonomic leadership than natural history museums. Being responsible for the growth, development, care, curation, *and use* of collections, they have a special role in both inventory work and revisionary taxonomy. The good news for museums is that this represents an opportunity to differentiate themselves from hundreds of universities and institutes with which they compete for funding, and to become flagship institutions in a scientific armada responding to the biodiversity crisis.

*Advocacy.* Taxonomists, professional organizations, and institutions engaged in taxonomy all share the responsibility to defend the integrity and mission of systematics *and* to be vocal advocates for it. Taxonomists must be reliable partners in addressing the biodiversity crisis while remaining committed to their own mission. No less importantly, taxonomists must appeal to the public for support of their mission. Sharing the excitement of discovery and a planetary-scale inventory is a fantastic opportunity to raise awareness and appreciation of taxonomy's mission. That specimens and morphology are more easily relatable to the average person than DNA sequences is a strength that should be exploited.

*Technology in Service to Knowledge.* It is reasonable to set out to describe and classify all species because of advances in technology, but it is important that the tail not wag the dog. Technologies, from digital imaging to DNA sequencing and artificial intelligence, represent means to create knowledge, not ends unto themselves.

*Biomimicry.* As characters of species are described, a vast number of ideas for biomimetic solutions (e.g., Benyus, 1997) will emerge. While phylogenetic classifications point to species related to biomimetic models that may share similar attributes, we must partner with information scientists to open access as well to unrelated species with similar properties (Wheeler, 2024).

## CONCLUSIONS: GREAT EXPECTATIONS

The obvious should not have to be said. But when it comes to systematics, apparently it does and Engel et al. (2021) have said it:

> Best practices in taxonomy are critical for addressing the current biodiversity crisis. Innovation is good, but it should improve on an existing situation rather than exacerbate difficulties or worse, revert to 18th- or 19th-century one-character (barcode or single gene) taxonomic approaches. Taxonomy needs to be rebranded to the scientific community as a modern, active and important discipline, and taxonomists need to come together to demand sufficient funding, training and employment for taxonomic researchers.

Accelerated extinction rates mean that there is no time to waste exploring species, biosphere, and evolutionary history. To get taxonomy back on course, we must defend its independence, clarify its vision and goals, and invest in its needs. The call for a new taxonomy (Wheeler, 2008) was a literary thumb in the eye of the New Systematics (Huxley, 1940; Mayr, 1942) that sowed seeds of the decline in support and prestige for taxonomy. Not meant to be taken literally, it was an argument for a fresh approach to goals that have motivated taxonomists for centuries. One constant in taxonomy's long history has been change, evidenced by increasing estimates of the number of species, the adoption of new technologies, advances in theory, and refinement of methods. At the same time, taxonomy's core goals remained constant: to know species, characters, and history, and to organize knowledge and make sense of diversity in a 'natural' classification. Such aims are timeless and as relevant today as ever. In what may be the greatest threat to taxonomy yet, its applications are confused with its mission, reinforced by an urgent need to identify species. Add to this a disproportionate preference for molecular techniques and the pressure to abandon its identity and conform to popular trends is tremendous.

Taxonomy must return to its traditional identity and established practice of pursuing all relevant and available evidence. Whether individuals pursue one or more sources of evidence in a particular project, it all is eventually synthesized in cladistic analyses and classifications. Taxonomy must be single-minded and undeterrable from a mission unlike that of any other science. And it must resist compliance with fads, being held hostage by funding trends or becoming subservient to the needs of other sciences. Because its 'raw materials'—species and characters—are rapidly disappearing, the extent to which it achieves its ultimate goal, to know earth's species and their history, depends on the extent to which it reasserts its independence. Of course, taxonomy cannot succeed in isolation. It needs the support of colleagues, the public and institutional leaders. And it has a solemn responsibility to do all that it can to address the biodiversity crisis, including making species identifiable and information about them accessible. The burden is on taxonomists to explain how their science is both rigorous and non-experimental, to enumerate the benefits of taxonomy to science and society, and to balance the challenges of pursuing its own mission while being a reliable partner to other disciplines.

Even if the wildest dreams of molecular taxonomists were realized—with entire genomes sequenced for every kind and variation of living thing—it would fall short of the aims and traditions of taxonomy. Genomes cannot tell us all that taxonomists want to know about complex characters and emergent properties. There is much to be learned from the comparative study and description of characters. Even if we knew enough about genes and ontogenetic pathways to predict what organisms look like from sequence data (ignoring, for the moment, epigenetic, and other confounding factors), we would be deprived of the intellectual joy and inspiration found in observing characters and elucidating the story of their transformational evolution. Recognizing copies of *On the Origin of Species* based on the sequence of letters and words in its text is not the same as reading Darwin's book (1859). We owe it to ourselves to aim higher than merely telling species apart: to observe, compare, experience, and understand the amazing panoply of species and characters that exist as results of billions of years of evolutionary transformations; not settle for simply detecting that species exist and telling them apart by DNA barcodes, or other similarly *minimally* informative methods. To do so, we must revive taxonomy and its traditions now, while time remains to substantially accomplish its agenda.

We can achieve what Linnaeus could only dream of: a global inventory of species and clades, complete with descriptions of the characters that make each unique. Or we can continue down the current path, focus only on immediate needs and one data source, and risk permanent ignorance of life and its history. The costs to science and humankind of failing to pursue taxonomy now, for its own sake and to its highest standards, are unthinkable, resulting in irreparable gaps in what we know and understand. Without such knowledge, we will certainly conserve less diversity among surviving species than we could have (Humphries, 2006; Vane-Wright, Humphries and Williams, 1991; Nixon and Wheeler, 1992); gather fewer biomimetic clues than we should have; impose ignorance on future generations; and impoverish our intellectual lives by failing to preserve evidence of the history of evolution. Sacrificing knowledge to address immediate needs, limiting taxonomy to molecular data, or making systematics subservient to general biology are grievous mistakes that we would soon and long regret.

As we enter earth's sixth mass extinction, we should do so with great expectations of taxonomy. We should ask for much more than the ability to identify species. We should demand to learn as much about species, characters, evolution and the biosphere as we can; vastly more than molecular data can teach us. We should expect nothing less than an inventory of *all* species, accompanied by detailed descriptions and specimens preserved in museum collections; comprehensive phylogenetic analyses and classifications; conservation priorities, informed by knowledge of relationships and distributions; knowledge of evolutionary innovations to inspire biomimicry innovators; and a deeper connection between humans and the natural world, borne out of credible knowledge of species, biosphere and evolutionary history. And, yes, making species identifiable along the way. The biodiversity crisis means that we have one last opportunity to explore, document, and understand earth's species and their history. It is time to think bigger than molecular taxonomy and to ask much more from systematics.

## NOTES

1 Taxonomy and systematic biology (or systematics) are terms used here as synonyms for one and the same science, in its full range of activities. Bisecting the science into systematics *sensu stricto*, for parts connected to evolutionary concerns, and taxonomy *sensu stricto*, restricted to practices of describing, naming, and classifying, has weakened the science as a whole and given license to fund the former and ignore the latter. Done to its highest standards, taxonomy cannot be divided. Descriptions provide evidence regarding species status and relationships, and hypotheses regarding homology, characters, and cladistic relationships inform names and classifications—that, in turn, function as an effective means of communicating cladistic knowledge.

2 This article is primarily directed at plant and animal species. Prokaryotes are a special case for several reasons. Until confusion regarding the relationship between microbial genetic variation and species status is resolved, it is difficult to make sense of estimates of numbers of microbial 'species' as high as one trillion (Locey and Lennon, 2016). *Escherichia coli*, with an estimated 16,000 genes, is an extreme, but illustrative, example of genetic variation within what is considered to be one species (Lukjancenko et al., 2010). Only twenty percent of the genome is shared by every strain of *E. coli*. In contrast, the genomes of humans and chimps are largely similar (Suntsova and Buzdin, 2020). The relation of variation to species status is not simple and may involve rates of evolution, time since divergence, and lateral gene transfer, among other factors. Unable to culture most microbes, it is challenging to gather information about them beyond what metagenomic environmental samples can tell us. Further, there is evidence to suggest that while microbes may undergo high levels of extinction, they appear to be less vulnerable to mass extinction episodes than plant or animal species (Louca et al., 2018). Molecular, field, and laboratory studies of microbes deserve strong support as the significance of their diversity, species boundaries, and evolutionary history are sorted out. This should not, however, impede progress in the study of most eukaryotic species that represent a qualitatively different challenge.

## REFERENCES

Benyus, J.M. (1997) *Biomimicry: Innovation Inspired by Nature*. Morrow, New York.

Byrne, A.Q. (2023) Reimagining the future of natural history museums with compassionate collection. *PLoS Biology*, 21: e3002102, https://doi.org/10.1371/journal.pbio.3002101

Darwin, C. (1859) *On the origin of species* [...]. John Murray, London.

Engel, M.S., Ceríaco, L.M.P., Daniel, G.M., Dellapé, P.M., Löbl, I., Marinov, M., Reis, R.E., Young, M.T., Dubois, A., Agarwal, I., Lehmann, A.P., Alvarado, M., Alvarez, N., Andreone, F., Araujo-Vieira, K., Ascher, J.S., Baêta, D., Baldo, D., Bandeira, S.A., Barden, P., Barrasso, D.A., Bendifallah, L., Bockmann, F.A., Böhme, W., Borkent, A., Brandão, C.R.F., Busack, S.D., Bybee, S.M., Channing, A., Chatzimanolis, S., Christenhusz, M.J.M., Crisci, J.V., D'Elía, G., Costa, L.M. da, Davis, S.R., Lucena, C.A.S. de, Deuve, T., Fernandes Elizalde, S., Faivovich, J., Farooq, H., Ferguson, A.W., Gippoliti, S., Gonçalves, F.M.P., Gonzalez, V.H., Greenbaum, E., Hinojosa-Díaz, I.A., Ineich, I., Jiang, J., Kahono, S., Kury, A.B., Lucinda, P.H.F., Lynch, J.D., Malécot, V., Marques, M.P., Marris, J.W.M., Mckellar, R.C., Mendes, L.F., Nihei, S.S., Nishikawa, K., Ohler, A., Orrico, V.G.D., Ota, H., Paiva, J., Diogo Parrinha, D., Pauwels, O.S.G., Pereyra, M.O., Pestana, L.B., Pinheiro, P.D.P., Prendini, L., Prokop, J., Rasmussen, C., Rödel, M.-O., Rodrigues, M.T., Rodríguez, S.M., Salatnaya, H., Sampaio, Í., Alba Sánchez-García, A., Shebl, M.A., Bruna S., Santos, B.S., Solórzano-Kraemer, M.M., Sousa, A.C.A., Stoev, P., Teta, P., Trape, J.-F., Van-Dúnem Dos Santos, C., Vasudevan, K., Vink, C.J., Vogel, G., Wagner, P., Wappler, T., Ware, J.L., Wedmann, S. and Kusamba, C.Z. (2021) The taxonomic impediment: A shortage of taxonomists, not the lack of technical approaches. *Zoological Journal of the Linnean Society*, 22: 381–387, https://doi.org/10.1093/zoolinnean/zlab072

Farris, J.S. (1979) The information content of the phylogenetic system. *Systematic Zoology*, 28: 483–519, https://doi.org/10.2307/2412562

Gaffney, E.S. (1979) An introduction to the logic of phylogeny reconstruction, pp. 79–111. In *Phylogenetic Analysis and Paleontology* (eds J. Cracraft and N. Eldredge). Columbia University Press, New York.

Hebert, P.D., Cywinska, A., Ball, S.L. and deWaard J.R. (2003) Biological identifications through DNA barcodes. *Proceedings of the Royal Society B: Biological Sciences* 270 (1512): 313–321. doi: 10.1098/rspb.2002.2218

Hennig, W. (1966) *Phylogenetic Systematics*. University of Illinois Press, Urbana.

Humphries, C.J. (2001) Systematics and taxonomy in modern biology, pp. 253–254. In *Linnaeus: The Compleat Naturalist* (ed. W. Blunt). Princeton University Press, Princeton.

Humphries, C.J. (2006) Measuring diversity, pp. 141–161. In *Taxonomy and Conservation* (eds E. Leadlay and S.L. Jury). Cambridge University Press, Cambridge.

Huxley, J. ed. (1940) *The New Systematics*. Oxford University Press, Oxford.

IUCN (2022) Number of described species, In *Our World in Data*. Retrieved April 10, 2024. https://ourworldindata.org/grapher/number-of-described-species

Johnson, K. (2007) Natural history as stamp collecting: A brief history. *Archives of Natural History*, 34: 244–258. https://doi.org/10.3366/anh.2007.34.2.244

Locey, K.J. and Lennon, J.T. (2016) Scaling laws predict global microbial diversity. *Proceedings of the National Academy of Sciences of the United States of America*, 113(21): 5970–5975. https://doi.org/10.1073/pnas.1521291113

Louca, S., Shih, P.M., Pennell, M.W., Fischer, W.W., Parfrey, L.W. and Doebeli, M. (2018) Bacterial diversification through geological time. *Nature Ecology and Evolution*, 2: 1458–1467. https://doi.org/10.1038/s41559-018-0625-0

Lukjancenko, O., Wassenaar, T.M. and Ussery, D.W. (2010) Comparison of 61 sequenced *Escherichia coli* genomes. *Microbial Ecology*, 60: 708–720. https://doi.org/10.1007/s00248-010-9717-3

Mayr, E. (1942) *Systematics and the Origin of Species*. Harvard University Press, Cambridge.
Meier, R., Shiyang, K. Vaidya, G. and Ng, P.K.L. (2006) DNA barcoding and taxonomy in Diptera: A tale of high intraspecific variability and low identification success. *Systematic Biology*, 55: 715–728. https://doi.org/10.1080/10635150600969864
Nelson, G. and Platnick, N.I. (1981) *Systematics and Biogeography: Cladistics and Vicariance*. Columbia University Press, New York.
Ness, E. (2008) SPIDA-Web. *Anthropocene*, 29 July. https://www.anthropocenemagazine.org/conservation/2008/07/spida-web/
Nixon, K.C. and Wheeler, Q.D. (1992) Measures of phylogenetic diversity, pp. 216–234. In *Extinction and Phylogeny* (eds M.J. Novacek and Q.D. Wheeler). Columbia University Press, New York.
Popper, K. (1959) *The Logic of Scientific Discovery*. Hutchinson and Co., London.
Russell, K.N., Do, M.T., Hugg, J.C. and Platnick, N.I. (2007) Introducing SPIDA-Web: Wavelets, Neural Networks and Internet Accessibility in an Image-Based Automated Identification System, pp. 131–152. In *Automated Taxon Identification in Systematics: Theory, Approaches and Applications* (ed N. MacLeod). CRC Press, Boca Raton.
Scotland, R., Hughes, C., Bailey, D. and Wortley, A. (2003) The *Big Machine* and the much-maligned taxonomist. *Systematics and Biodiversity*, 1: 139–143. https://doi.org/10.1017/S1477200003001178
Sharkey, M.J., Janzen, D.H., Hallwachs, W., Chapman, E.G., Smith, M.A., Dapkey, T., Brown, A., Ratnasingham, S., Naik, S., Manjunath, R., Perez, K., Milton, M., Hebert, P., Shaw, S.R., Kittel, R.N., Solis, M.A., Metz, M.A., Goldstein, P.Z., Brown, J.W., Quicke, D.L.J., Achterberg, C. van, Brown, B.V. and Burns, J.M. (2021) Minimalist revision and description of 403 new species in 11 subfamilies of Costa Rican braconid parasitoid wasps, including host records for 219 species. *ZooKeys*, 1013: 1–665. https://doi.org/10.3897/zookeys.1013.55600
Simpson, G.G. (1961) *Principles of Animal Taxonomy*. Columbia University Press, New York.
Suntsova, M.V. and Buzdin, A.A. (2020) Differences between human and chimpanzee genomes and their implications in gene expression, protein functions and biochemical properties of the two species. *BMC Genomics*, 21 (Suppl. 7): 535. https://doi.org/10.1186/s12864-020-06962-8
Vane-Wright, R.I., Humphries, C.J. and Williams, P.H. (1991) What to protect?—systematics and the agony of choice. *Biological Conservation*, 55: 235–254. https://doi.org/10.1016/0006-3207(91)90030-D
Wheeler, Q. (2024) *Species, Science and Society: The Role of Systematic Biology*. Routledge, Abingdon.
Wheeler, Q., Assis, L. and Rieppel, O. (2013) Heed the father of cladistics. *Nature*, 496: 295–296. https://doi.org/10.1038/496295a
Wheeler, Q.D., ed. (2008) *The New Taxonomy*. CRC Press, Boca Raton.
Wheeler, Q.D., Knapp, S., Stevenson, D.W., et al. (2012) Mapping the biosphere: Exploring species to understand the origin, organization and sustainability of biodiversity. *Systematics and Biodiversity*, 10: 1–20. https://doi.org/10.1080/14772000.2012.665095
Wiens, J.J. (2023) How many species are there on earth? Progress and problems. *PLoS Biology*, 21(11): e3002388. https://doi.org/10.1371/journal.pbio.3002388
Williams, D.M. and Ebach, M.C. (2020) *Cladistics: A Guide to Biological Classification*. 3rd ed. Cambridge University Press, Cambridge.
Williams, D.M., Schmitt, M. and Wheeler, Q. (2016) *The Future of Phylogenetic Systematics: The Legacy of Willi Hennig*. Cambridge University Press, Cambridge.

# Index

Note: **Bold** and *italic* page numbers refer to **tables** and *figures*. Page numbers followed by 'n' refer to notes.

## G



## H



## I

## J



## K



## L



## M



## N

## Q



## R



## S



## T